INTRODUCTION TO
Peptides and Proteins

INTRODUCTION TO

Peptides and Proteins

Ülo Langel
Benjamin F. Cravatt
Astrid Gräslund
Gunnar von Heijne
Tiit Land
Sherry Niessen
Matjaž Zorko

CRC Press
Taylor & Francis Group
Boca Raton London New York

CRC Press is an imprint of the
Taylor & Francis Group, an **informa** business

CRC Press
Taylor & Francis Group
6000 Broken Sound Parkway NW, Suite 300
Boca Raton, FL 33487-2742

First issued in paperback 2019

© 2010 by Taylor and Francis Group, LLC
CRC Press is an imprint of Taylor & Francis Group, an Informa business

No claim to original U.S. Government works

ISBN-13: 978-1-4200-6412-4 (hbk)
ISBN-13: 978-0-367-38487-6 (pbk)

Library of Congress Cataloging-in-Publication Data

Introduction to peptides and proteins / authors, Ulo Langel ... [et al.].
    p. ; cm.
Includes bibliographical references and index.
ISBN 978-1-4200-6412-4 (hardcover : alk. paper)
1. Proteins--Textbooks. 2. Peptides--Textbooks. I. Langel, Ülo. II. Title.
[DNLM: 1. Peptides. 2. Molecular Biology. 3. Proteins. QU 68 I819 2010]

QP551.I68 2010
572'.65--dc22

2009028736

Visit the Taylor & Francis Web site at
http://www.taylorandfrancis.com

and the CRC Press Web site at
http://www.crcpress.com

Contents

Introduction to Part I

Matjaž Zorko

From Building Blocks to 3-D Structure

Protein Biosynthesis and Posttranslational Modifications

Folding of Proteins

Introduction to Part III
Tiit Land

Introduction to Part IV
Ülo Langel

Introduction to Part V
Ülo Langel and Astrid Gräslund

Preface

Interest in the field of peptides and proteins has increased enormously in recent years. With the achievements in genomics as a background, the scientific community has now, for the first time, the opportunity to establish the role of peptides and proteins in health and disease at the molecular level.

There are several excellent textbooks on proteins, and a few on peptides. However, we feel that there is a need for an up-to-date textbook covering both peptides and proteins in a concise, introductory manner. We hope that we have written this book in the spirit of Tom Creighton's famous book *Proteins—Structures and Molecular Properties*. This book, known by many as *Creighton*, has inspired generations of young students and investigators, and we have taken it as a guide on how to cover modern bioscience based on molecular thinking. Here, we present our view on modern peptide and protein chemistry.

We have focused particularly on the rapidly developing fields of peptide synthesis, folding, protein sorting, protein degradation, methods in peptide and protein research, bioinformatics, proteomics, and clinical aspects of peptides and proteins. Among multiple topics, we describe some representative classes of peptides and proteins such as enzymes, cell-surface receptors, other membrane proteins, antibodies, fibrous proteins, and some bioactive peptide classes. We regard these as particularly important including clinical aspects of proteins and peptides, such as misfolding-based diseases (prion diseases), miscleavage-based diseases (Alzheimer's disease), and SNP-dependent diseases, together with the role of proteins in cancer development. We discuss the use of proteins and peptides as drugs and solid-phase synthesis for their production. Finally, we emphasize peptides as important functional biomolecules and research tools.

We are especially grateful to Professor Tamas Bartfai for encouraging us to write this textbook in its present form, for insightful ideas, and constant encouragement.

We are further grateful to several experts in the peptide and protein field who have volunteered their thoughts, comments, and corrections: Samir EL Andaloussi, Mattias Hällbrink, John Howl, Victor Hruby, Jaak Järv, Peter Järver, Bernard Lebleu, Toivo Maimets, Wojtek Makalowski, Roger Pain, Margus Pooga, John Robinson, Mart Saarma, Raivo Uibo, Anders Undén, and Mark Wheatley. We are also grateful to Imre Mäger and Indrek Saar for technical assistance.

Ü. Langel, B. Cravatt, A. Gräslund, G. von Heijne,
T. Land, S. Niessen, and M. Zorko

Further Reading

1. Creighton, T. E. 1993. Proteins. Structures and Molecular Properties, 2nd edition. New York: W. H. Freeman and Company.
2. Whitford, D. 2005. Proteins. Structure and Function. The Atrium, Southern Gate, Chichester: John Wiley & Sons Ltd.
3. Meyers, R. A., Ed. 2007. Proteins. From Analytics to Structural Genomics, vol. 1 & 2. Weinheim: Wiley-VCH Verlag Gmbh & Co. KGaA.
4. Bränden, C. and Tooze, J. 1999. Introduction to Protein Structure, second edition. New York: Garland Publishing, Inc.
5. Gutte, B., Ed. 1995. Peptides. Synthesis, Structures, and Applications. San Diego, New York, Boston, London, Sydney, Tokyo, Toronto: Academic Press, Inc.
6. Bründén, C. and Tooze, J. 1999. Introduction to protein structure. 2nd edition. Garland Publishing, Inc., New York, NY, USA.
7. Campbell, I. D. and Dwek, R. A. 1984. Biological spectroscopy. The Benjamin Cummings Publishing Co., Inc., Menlo Park, CA, USA.
8. Serdyuk, I. N., Zaccai, N. R. and Zaccai, J. 2007. Methods in molecular biophysics. Cambridge University Press, Cambridge, UK.
9. Greenfield, N. J. 2006. Using circular dichroism spectra to estimate protein secondary structure. Nature Protocols 1, 2876–2890.
10. Zimmer, M. 2005. Glowing genes. Prometheus Books, Amherst, NY, USA.
11. Ballou, B., Lagerholm, C., Ernst, L., Bruchez, M. and Waggoner, A. 2004. Noninvasive imaging of quantum dots in mice. Bioconjugate Chem. 15, 79–86.
12. Wüthrich, K. 1986. NMR of proteins and nucleic acids. John Wiley & Sons, Inc., USA.
13. Cavanagh, J., Fairbrother, W. J., Palmer III, A. G. and Skelton, N. J. 2007. Protein NMR spectroscopy: principles and practice. Elsevier/Academic Press, San Diego, USA.
14. Ferentz, A. E. and Wagner, G. 2000. NMR spectroscopy: a multifaceted approach to macromolecular structure. Quart. Rev. Biophys. 33, 29–65.
15. Palmer III, A. G. 2001. NMR probes of molecular dynamics: overview and comparison with other techniques. Annu. Rev. Biophys. Biomol. Struct. 30, 129–155.
16. Lindorff-Larsen, K., Rogen, P., Paci, E., Vendruscolo, M. and Dobson, C. 2005. Protein folding and the organization of the protein topology universe. Trends Biochem. Sci. 30, 13–19.
17. Nishimura, C., Dyson, H. J., and Wright, P. E. 2005. Identification of native and non-native structure in kinetic folding intermediates of apomyoglobin. J. Mol. Biol. 355, 139–156.
18. Yates III, J. R. 2004. Mass spectral analysis in proteomics. Annu. Rev. Biophys. Biomol. Struct. 33, 297–316.
19. Steen, H. and Mann, M. 2004. The ABC's (and XYZ's) of peptide sequencing. Nature Reviews, Molecular Cell Biology 5, 699–721.
20. Murphy, R. 2002. Peptide aggregation in neurodegenerative disease. Annu. Rev. Biomed. Eng. 4, 155–174.
21. Selkoe, D. J. 2003. Folding proteins in fatal ways. Nature 426, 900–904.
22. Cohen, F. E. and Kelly, J. W. Therapeutic approaches to protein-misfolding diseases. Nature 426, 905–909.
23. Temussi, P. A., Masino, L., and Pastore, A. From Alzheimer to Huntingdon: why is a structural understanding so difficult? The EMBO Journal, 22, 355–362.

24. Nelson, R. and Eisenberg, D. 2006. Recent atomic models of amyloid fibril structure. Curr. Opin. Struct. Biol. 16, 260–265.
25. Aguzzi, A., Heikenwalder, M., and Polymenidou, M. 2007. Insights into prion strains and neurotoxicity. Nature Reviews, Molecular Cell Biology 8, 552–561.
26. Wadsworth, J. and Collinge, J. 2007. Update on human prion disease. Biochim. Biophys. Acta 1772, 598–609.
27. Sipe, J. P. 2005. Amyloid proteins. The beta sheet conformation and disease. Wiley-VCH Verlag GmbH & Co. KGaA, Weinheim, Germany.
28. Goedert, M. and Spillantini, M. G. 2006. A century of Alzheimer's disease. Science 777–784.
29. Haas, C. and Selkoe, D. J. 2007. Soluble protein oligomers in neurodegeneration: lessons from the Alzheimer's amyloid β-peptide. Nature Reviews, Molecular Cell Biology 8, 101–112.
30. Masters, C. L. and Beyreuther, K. 2006. Alzheminer's centennial legacy: prospects for rational therapeutic intervention targeting the Aβ amyloid pathway. Brain 129, 2823–2839.
31. Vichinsky, E. 2002. New therapies in sickle cell disease. Lancet 360, 629–631.
32. Cianciulli, P. 2008. Treatment of iron overload in thalessemia. Pediatr. Endocrinol. Rev. 6 Suppl 1: 208–213.
33. Gadsby, D.C., Vergani, P. and Csanády, L. 2006. The ABC protein turned chloride channel whose failure causes cystic fibrosis. Nature 440, 477–483.
34. Enquist. K., Fransson, M., Boekel, C., Bengtsson, I., Geiger, K., Lang, L., Pettersson, A., Johansson, S., von Heijne, G. and Nilsson, I. 2009. Membrane-integration characteristics of two ABC transporters, CFTR and P-glycoprotein. J. Mol. Biol. 387, 1153–1164.

Biography

Ülo Langel is a professor and chairman of the Department of Neurochemistry, Stockholm University, Sweden. Prof. Langel graduated from Tartu University, Tartu, Estonia, as bioorganic chemist in 1974; he has received a Ph.D. degree twice: in 1980 from Tartu University (bioorganic chemistry) and in 1993 from Tartu University/ Stockholm University (biochemistry/neurochemistry). His professional experience includes a career at Tartu (from junior research fellow to associate professor, visiting professor, and adjunct professor from 1974 to the present); the Scripps Research Institute, La Jolla, California (associate professor and adjunct professor 2000 to the present); and Stockholm University (from research fellow to associate professor, professor, and chairman, from 1987 to the present). He is an honorary professor at Ljubljana University, Slovenia.

Prof. Langel has been selected as a fellow member of the International Neuropeptide Society (1995) and is a member of International Society for Neurochemistry, European Peptide Society, Swedish Biochemical Society, and Estonian Biochemical Society. He has been awarded the Order of the White Star, Fourth Class, by the Estonian Republic. He has been invited as lecturer at numerous international conferences, and is a coauthor of more than 270 scientific articles and 12 approved patents or patent applications.

Dr. Cravatt is a professor in the Skaggs Institute for Chemical Biology and Department of Chemical Physiology at the Scripps Research Institute. His research group is interested in mapping biochemical pathways in human disease using advanced proteomic and metabolomic technologies. Dr. Cravatt obtained his undergraduate education at Stanford University, receiving a B.S. degree in the biological sciences and a B.A. in history. He then trained with Drs. Dale Boger and Richard Lerner and received a Ph.D. in macromolecular and cellular structure and chemistry from the Scripps Research Institute (TSRI) in 1996. Professor Cravatt joined the faculty at TSRI in 1997 and is currently a professor in the Skaggs Institute for Chemical Biology and chair of the Department of Chemical Physiology. He has received multiple awards and honors, including the Eli Lilly Award in Biological Chemistry and Cope Scholar Award from the American Chemical Society, and the Irving Sigal Young Investigator Award from the Protein Society.

Astrid Gräslund is professor and chairman in the Department of Biochemistry and Biophysics, Stockholm University, Sweden. Prof. Gräslund graduated from the Royal Institute of Technology, Stockholm, with a master's degree in applied physics in 1967, and she received her Ph.D. degree in biophysics at Stockholm University in 1974.

Her professional career includes positions as associate professor in biophysics at Stockholm University, as professor of medical biophysics at Umeå University (1988–1993), and professor of biophysics at Stockholm University since 1993.

Prof. Gräslund is elected member of the Royal Swedish Academy of Sciences, Class of Chemistry, since 1993; she is at present secretary of the Nobel Committee for Chemistry and deputy member of the board of the Nobel Foundation. She has published more than 280 scientific articles in the fields of biophysics and biochemistry and has received national scientific prizes awarded by the Royal Swedish Academy of Sciences, the Swedish Chemical Society and Uppsala University.

Gunnar von Heijne is a professor at the Department of Biochemistry and Biophysics, Stockholm University, Sweden. He graduated in chemical engineering from the Royal Institute of Technology, Stockholm, in 1975 and received a Ph.D. in Theoretical Physics from the same institute in 1980. He did a postdoctorate at the University of Michigan, Ann Arbor, and was an associate professor at Karolinska Institutet, Stockholm, before he was recruited to Stockholm University in 1994.

Prof. von Heijne has worked mainly on problems related to protein sorting and membrane protein biogenesis and structure. The work includes both bioinformatics methods development (e.g., methods for prediction of signal peptides and other sorting signals as well as prediction of membrane protein topology) and experimental studies in *E. coli* and eukaryotic systems. He has been invited lecturer at numerous international conferences and is a coauthor of more than 280 scientific articles. He is a member of the Royal Swedish Academy of Sciences, the Royal Swedish Academy of Engineering Sciences, EMBO, and Academia Europaea.

Tiit Land is a professor at the Department of Natural Sciences, Tallinn University, Estonia. He graduated in bioorganic chemistry from the University of Tartu, Estonia, in 1989 and received a Ph.D. in neurochemistry and neurotoxicology from Stockholm University in 1994. He did his postdoctoral work at the National Institutes of Health, Bethesda, and has been working as an assistant professor and associate professor at Stockholm University.

Prof. Land has been working on issues related to cellular iron metabolism, role of iron in prion diseases, and signal transduction mechanisms in Alzheimer's disease. He is a coauthor of many scientific articles. He has been teaching biochemistry, recombinant DNA technology, and proteomics on undergraduate and graduate courses in Stockholm University, Tallinn University, and University of Tartu.

Sherry Niessen received her Ph.D. in Biology from the Scripps Research Institute (TSRI), La Jolla, California, in 2008. She did an M.Sc. in experimental medicine at McGill University, Montreal, Canada, in 2003. Currently, she is working at the Center for Physiological Proteomics at TSRI.

Matjaž Zorko is a professor at the Institute of Biochemistry, Medical Faculty, University of Ljubljana, Slovenia. He graduated in chemical engineering from the faculty of Natural Sciences and Technology, University of Ljubljana, Slovenia, in 1974 and received his Ph.D. in biochemistry from Medical Faculty, University of Ljubljana, Slovenia, in 1983. He is giving numerous lectures on biochemistry for undergraduate and postgraduate students of the University of Ljubljana and, from the year 2000, also at the University of Stockholm, Sweden, as a guest lecturer. He is a head of the research group studying the mechanism of receptors, G-proteins, and enzymes involved in signal transduction, and is a coauthor of more than 50 scientific articles, several book chapters, and one patent.

Introduction to Part I

Matjaž Zorko

The first part of our book covers the fate of peptides and proteins from assembly to the degradation. Starting from the amino acid building blocks and discussing the noncovalent interactions operating within the amino acid residues, we describe the peptide and protein structural organization, attempting to present the basis for the peptide and protein function that is elaborated in more detail in the following sections. In the chapter on protein biosynthesis, the introduction of noncoded amino acids into the protein sequence via ribosome is included, as this field is growing fast and can produce new proteins with yet unknown functions. Maturation of proteins by posttranslational modifications is covered in an overview of the most common modifications picked from the vast number of almost 100 known today. Protein folding is discussed in parallel with protein stability and protein sorting, showing how proteins are directed to their site of function. The section ends with a brief description of protein death and the recovery of amino acids to start a new protein and peptide life.

From Building Blocks
to 3-D Structure

1 Amino Acids

Matjaž Zorko

CONTENTS

The building blocks of peptides and proteins are amino acids. Most proteins consist of 20 standard amino acids that are coded by DNA; these amino acids are often called coded amino acids. They are joined into peptides and proteins by an amide bond called the peptide bond. The formation of this peptide bond is the condensation reaction in which water is released. What it is left of the amino acid as a part of the peptide or protein is called the residue. Besides standard amino acids, a large number of nonstandard amino acids are found in peptides and proteins, usually in limited amounts. The characteristics and properties of amino acids need to be understood in order to be able to comprehend the structure and behavior of polypeptides. Amino acids are not only structural elements of polypeptides, but they also have important precursor, transport, and metabolic roles. For example, histidine and tryptophan are precursors of the hormone histamine and the neurotransmitter serotonin, respectively. Carnitine can be acylated and, in this form, serves as a transporter of fatty acids across the mitochondrial membrane. Amino acids can be metabolized to release energy; some of them, called glucogenic amino acids, can also be converted into glucose to provide energy to the brain. Some of the amino acid derivatives are important in technology, medicine, and the food industry. Examples include the sodium salt of glutamic acid, sodium glutamate, as a food additive; aspartame, methylated dipeptide aspartyl-phenylalanine, as an artificial sweetener; and l-DOPA, l-dihydroxyphenylalanine, as a drug for the treatment of Parkinson's disease.

1.1 STANDARD AMINO ACIDS

There are 20 standard/coded amino acids (Table 1.1). Their names can be abbreviated using either the three-letter or the one-letter system, as shown in the table. The

6

TABLE 1.1
Standard Amino Acids—Abbreviations and Properties

Amino Acid	M_w V_r (Å³)	Occurrence in Proteins (%)[2]	pK$_1$	pK$_2$	pK$_R$	Hydropathy Index of R
Hydrophobic Aliphatic Radical						
Glycine	75	7.2	2.3	9.6		−0.4
Gly G	48					
Alanine	89	7.8	2.3	9.7		1.8
Ala A	67					
Proline	115	5.2	2.0	11.0		1.6
Pro P	90					
Valine	117	6.6	2.3	9.6		4.2
Val V	105					
Leucine	131	9.1	2.4	9.6		3.8
Leu L	124					
Isoleucine	131	5.3	2.4	9.7		4.5
Iso I	124					
Methionine	149	2.3	2.3	9.2		1.9
Met M	124					
Hydrophilic Uncharged Radical						
Serine	105	6.8	2.2	9.2		−0.8
Ser S	73					
Threonine	119	5.9	2.1	9.6		−0.7
Thr T	93					
Cysteine	121	1.9	2.0	10.3	8.2	2.5
Cys C	86					
Asparagine	132	4.3	2.0	8.8		−3.5
Asn N	96					
Glutamine	146	4.2	2.2	9.1		−3.5
Gln Q	114					
Hydrophilic Charged Radical						
Aspartate	133	5.3	1.9	9.6	3.7	−3.5
Asp D	91					
Lysine	146	5.9	2.2	9.0	10.5	−3.9
Lys K	135					
Glutamate	147	6.3	2.2	9.7	4.3	−3.5
Glu E	109					
Histidine	155	2.3	1.8	9.2	6.0	−3.2
His H	118					
Arginine	174	5.1	2.2	9.0	12.5	−4.5
Arg R	148					

(Continued)

TABLE 1.1 (Continued)
Standard Amino Acids—Abbreviations and Properties

Amino Acid	$M_w V_r$ (Å³)	Occurrence in Proteins (%)	pK₁	pK₂	pK_R	Hydropathy Index of R
		Aromatic Radical				
Phenylalanine	165	3.9	1.8	9.1		3.9
Phe F	135					
Tyrosine	181	3.2	2.2	9.1	10.1	3.2
Tyr Y	141					
Triptophane	204	1.4	2.4	9.4		1.4
Trp W	163					

one-letter abbreviations are used to record the amino acid sequences of proteins and larger peptides, while the three-letter codes are more common for most other purposes; the same convention will be employed in this book.

1.1.1 GENERAL PROPERTIES

All standard amino acids are α-amino acids with the general structure:

$$H_2N \blacktriangleright \overset{\displaystyle COOH}{\underset{\displaystyle R}{\overset{|}{\underset{|}{C^*}}}} \blacktriangleleft H \qquad (1.1)$$

They differ only in terms of the radical R, which is also called the amino acid side chain. According to their chemical nature, amino acids are usually divided into those with the hydrophobic aliphatic, the hydrophilic uncharged, the hydrophilic charged, and the aromatic radical.

In all standard amino acids except Gly, the α-carbon atom denoted in Equation 1.1 with an asterisk is asymmetric and in the L-form. Chirality is important in the three-dimensional structures of peptides and proteins, and particularly in the interactions with other molecules. It also results in optical activity—the ability to rotate linearly and circularly polarized light. The amount of rotation (i.e., the angle of rotation) depends on the concentration of the chiral molecule and other factors, including the temperature and the wavelength of the light employed. This makes it possible to determine high concentrations of free amino acids in solutions by polarimetric methods and is the basis of circular dichroism, a method for analyzing the secondary structure of peptides and proteins (see Chapter 13). Free in solution, pure L-amino acids are subjected to a very slow racemization into an equimolar mixture of both L- and D-enantiomers. The process is dependent on the pH, temperature, and structure of the radical. The racemization of L-Asp, for instance, would take around 3500 years in a neutral pH and at 25°C, but it is approximately 100 times faster at 100°C.

The racemization of amino acids can be applied to determine the age of archaeological and paleontological material of organic origin, and is used in forensic science to estimate the age of a cadaver. This method is called amino acid dating.

Free in solution, amino acids are always charged because the groups –COOH, –NH$_2$, and some other groups in radicals can release or bind the proton. The charge depends on the pH, as shown for Asp in the following equation:

$$(1.2)$$

To a lesser degree, charge is also dependent on the presence of ions in solution. The values of pK are determined by titration with acids and bases, and are given in Table 1.1. The presented values were obtained by titrating pure amino acids. The values of pK for the same groups in amino acids that are incorporated into the protein structure can vary considerably due to the inductive and other effects of groups in the proximity. This is particularly pronounced in the radical of His, with the assessed pK values in proteins ranging from 4 to 10. Different dissociation forms of amino acid are in equilibrium, as shown in Equation 1.2, and all are present at all values of pH, but at each pH, particular forms are predominant. The value of pH at which most of the amino acid is in the form that lacks net charge—form B for Asp (Equation 1.2)—is called the isoelectric point (pI). It is obtained as a mean value of the neighboring pK values; in the case of the Asp pI, it is $\frac{1}{2}(pK_1 + pK_R) = 2.95$. The value of pI depends on the composition of the solution in which it is determined; therefore, the presence of any additional ions or other substances must always be specified. The charge and its dependence on pH are pertinent to the separation and identification of amino acids by electrophoresis and isoelectrofocusing, methods that are based on the mobility of charged molecules in an electric field (see Chapter 9).

The size of the amino acid residues can be assessed from the residue masses and is strictly calculated by using the van der Waals volumes of the relevant atoms (Table 1.1). The van der Waals volume of the average residue is 114 Å^3, with Trp being the largest (163 Å^3) and Gly the smallest (48 Å^3). When the volume of the residue is measured by the amount of water displaced—that is, the increase in the volume of the water is determined after adding a known amount of residue to the solution—20% to 40% larger values are obtained for the residue volumes. This reflects the inability of the water molecules to fully surround the residue because it cannot penetrate into the crevices between the functional groups. Size becomes important in relation to the replacement of one amino acid by another in proteins, for instance, by site-directed mutagenesis. In proteins, residues are densely packed and in tight van der Waals contacts. The replacement of a residue by a bulkier one will result in destabilization of the protein structure because a larger residue will not fit in, and the structure of the protein will be distorted. The replacement of a larger residue by a smaller one is usually also unfavorable because it produces cavities that decrease the number of protein-structure-stabilizing interactions. This can be compensated by introducing solvent molecules—usually water—into the cavity, depending on the nature of the solvent and the nature of the groups that come into contact with the solvent.

Most amino acids are amphipathic structures, that is, a combination of polar or hydrophylic, and nonpolar or hydrophobic chemical groups. The differences in the hydrophobicity of amino acid side chains have been assessed by different methods, including a measurement of the partition coefficient of the model for each side chain distributed between the vapor and the water; a comparison of the solubility of amino acids and side-chain model compounds in water, ethanol, or dioxan; a measurement of the partition coefficient between water and octanol; and others. As observed, the obtained values for the hydrophobicity measured by various methods differ considerably. In general, amino acids Val, Ile, and Leu, containing aliphatic side chains, are the most hydrophobic, and those that form strong hydrogen bonds with water are the most hydrophilic, including Arg, Asp, Glu, and Lys, which have charged side chains at the physiological pH. When amino acids are incorporated into the peptide or protein structure, a part of the hydrophilicity is lost, but the –CO- and –NH- groups that participate in the peptide bond are polar, and thus retain some of their hydrophilic character. The hydrophobic/hydrophilic properties of amino acids as parts of the protein structure are probably best represented by the hydropathy index (Table 1.1). A hydropathy scale has been introduced by Kyte and Doolittle on the basis of a large amount of data from the literature, and is based on a computer program that determines the relative occupancy in the hydrophobic or hydrophilic environment within the proteins for each amino acid. A good correlation was demonstrated by comparing the obtained values and the position of a particular amino acid in known protein structures determined by crystallography.

Hydrophobicity is a very important factor in protein stability: the hydrophobic collapse of the nonpolar residues forming the nonpolar protein core is believed to play a fundamental role in the spontaneous folding of proteins. Hydrophobic residues with a large positive hydropathy index are found in the protein interior, while polar residues with a large negative hydropathy index occur on the protein surface exposed to the water. In membrane proteins, the surfaces in contact with the inner membrane lipid layer are also composed of nonpolar residues.

1.1.2 STANDARD AMINO ACIDS BY THE NATURE OF THE RADICAL

The hydrophobic amino acids with an aliphatic side chain are Gly, Ala, Val, Leu, Ile, Met, and Pro (Figure 1.1). Gly is the smallest amino acid, with just one hydrogen atom as a radical, which does not provide the molecule with a pronounced hydrophobic nature. Gly is also not chiral, and is therefore sometimes classified as a special amino acid not belonging to any particular group. Ala, Val, Leu, and Ile all have a short hydrocarbon side chain, which is branched, except in the case of Ala. The side chain of Ile has an extra asymmetric carbon atom. Met is a sulfur-containing amino acid that has a hydrocarbon chain and a methyl group attached to the sulfur. This makes Met a

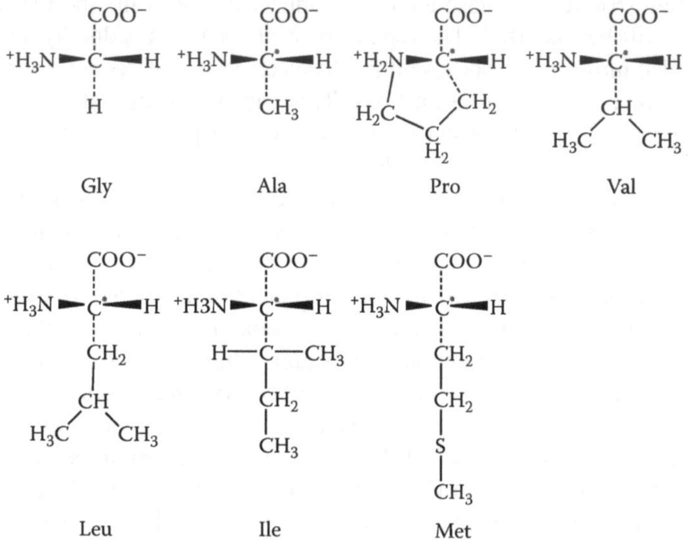

FIGURE 1.1 Structure of the hydrophobic amino acids with an aliphatic side chain.

suitable methyl-group donor in a number of methylation reactions in the cell. The sulfur atom can be oxidized by air into sulfoxide and sulfone. Although the Met side chain has a dipole moment, it also has a nonpolar and hydrophobic character. In proteins of prokaryotes, Met is the first amino acid in the sequence and also during the assembly of proteins in eukaryotes, but is later stripped away. Pro has an aliphatic but cyclic side chain; this results in special conformational properties relating to its location in proteins. The Pro ring is not totally planar; Cγ is out of the plane by about 0.05 nm.

The hydrophilic uncharged amino acids are Ser, Thr, Asn, Gln, and Cys (Figure 1.2). The polarity of the hydroxyl groups in Ser and Thr, and of the amide and carbonyl groups in Asn and Gln, ensures strong and multiple hydrogen bonding with water. Cys contains a thiol group that is related to the hydroxyl group but is chemically different—the hydrogen bonding is very weak, and the group is essentially hydrophobic, as is illustrated by the positive value of the hydropathy index (Table 1.1). However, it is able to ionize, and this is the reason why Cys is classified in this group. Ser is an important precursor in the biosynthesis of purines; pyrimidines; several amino acids, including Gly, Cys, and Trp; and other metabolites, including sphingolipids and folate. Ser is included in the active site of many enzymes, particularly in hydrolases, where it plays an important catalytic role as a nucleophile reagent. It is also the site of the glycosylation and phosphorylation of proteins. Ser is important in the mechanism of many kinases, and its phosphorylation modulates the activity of many proteins and is the control point of many cell-signaling cascades. Thr also contains the hydroxyl group in its side chain, and this results in an additional side chain chiral center. Its role in posttranslational modifications, kinases, and signaling is similar to that of Ser. Asn and Gln are the amidated forms of Asp and Glu, respectively, with side chains that do not ionize and are not very reactive. However, they do form strong hydrogen bonds and are used in proteins for

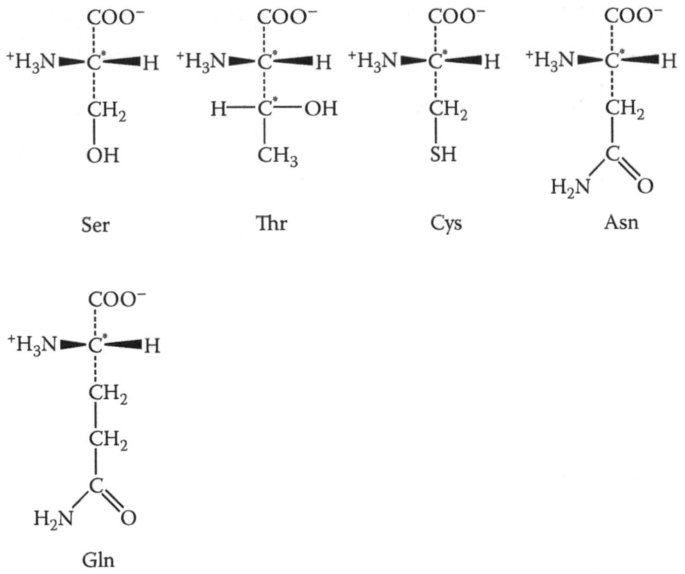

FIGURE 1.2 Structure of the hydrophilic amino acids with an uncharged side chain.

stabilizing the structure, for protein–protein interactions and for protein–DNA inter-actions (see Chapter 22, Figure 22.4). At very high and very low pH values and at high temperatures, Asn and Gln can be converted to Asp and Glu. Cys is part of the active center of the cysteine proteases, where, in general, it has the same role as Ser in hydrolases. The thiol group of Cys is relatively reactive, and can be alkylated by alkyl halides and added to double bonds. It also interacts with metal ions and forms complexes, as in zinc fingers (see Chapter 22). It is the target group in heavy-metal poisoning, particularly with the Hg^{2+} ion. A very important property of Cys is its ability to form disulphide bridges that play a key role in stabilizing the structure of many proteins.

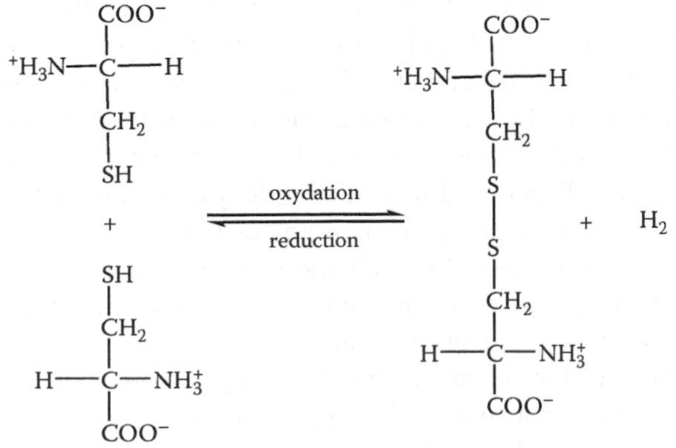

(1.3)

Disulphide bonds formed by Cys residues have a nonpolar surface area.

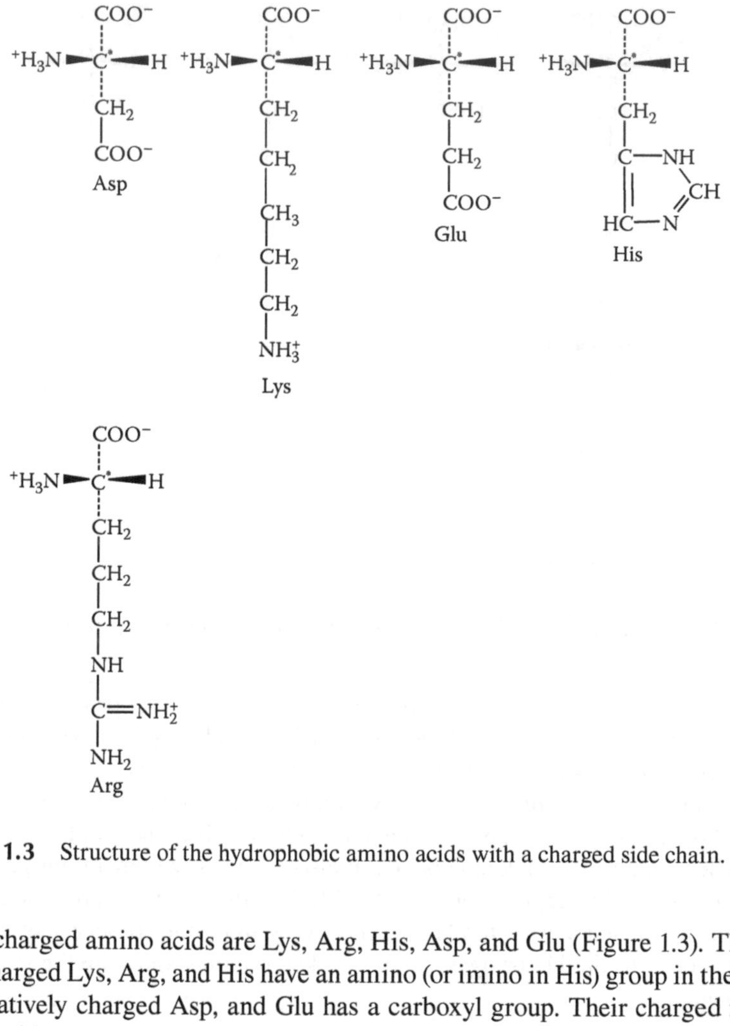

FIGURE 1.3 Structure of the hydrophobic amino acids with a charged side chain.

The charged amino acids are Lys, Arg, His, Asp, and Glu (Figure 1.3). The positively charged Lys, Arg, and His have an amino (or imino in His) group in the radical and negatively charged Asp, and Glu has a carboxyl group. Their charged radicals interact with water and other molecules by charge-to-charge interactions and hydrogen bonding. They are mostly found on the surface of the proteins. Lys is very basic and relatively reactive, particularly at higher pH values, where it is not charged. The important reaction of the Lys side-chain amino group is the formation of a covalent Schiff base with aldehydes, which is used to link the polypeptide chain with another one, for example, in collagen, and with other aldehyde groups containing molecules, for example, with pyridoxal phosphate in many enzymes. In proteins, it is the target for a number of posttranslational modifications, including methylation, acetylation, ubiquitination, hydroxylation, and glycosylation. ε-NNN-trimethylated Lys is a precursor of carnitine, which is involved in fatty-acid transport across the mitochondrial membrane. Lys is used for the treatment of the Herpes simplex infection because this virus needs Arg for its replication, and a supplement of Lys decreases Arg's availability. Arg has a planar guanido group in which the carbon atom is in the sp^2 hybridization state. The guanido group has the ability to form multiple hydrogen bonds and to attract negatively charged groups, both inside the protein structure and in the inter-action of proteins with other molecules. A notable example is the interaction of Arg in DNA-binding proteins with DNA base pairs (see Chapter 22, Figure 22.4). The guanido group is also the site of methylation, which can regulate several processes in the cell, including the binding of DNA and RNA to the DNA- and RNA-binding pro-teins, leading to a modified transcription and RNA processing, and signal transduc-tion by alternating protein–protein interactions. Arg also functions as a precursor of nitric oxide (NO), urea, an important regulator of blood pressure, by which nitrogen is excreted from the bodies of humans and many other organisms, and creatine, from which creatine phosphate, the energy-supplying molecule in muscles, is produced, as well as several other molecules. Because of its involvement in the production of NO, in the release of growth hormone, and in other processes, Arg supplements are marketed, though their beneficial effect is controversial. His has an imidazole ring in which two nitrogen atoms have different characters: one is slightly acidic and the other is basic, which can bind a proton. Because of the resonance structure of the ring, the location of the basic nitrogen is not fixed:

$$
\begin{array}{ccc}
\underset{\text{COO}^-}{} & & \underset{\text{COO}^-}{} \\
^+H_3N\!\!-\!\!C\!\!-\!\!H & & ^+H_3N\!\!-\!\!C\!\!-\!\!H \\
CH_2 & \rightleftharpoons & CH_2 \\
C\!-\!NH & & C\!-\!NH^+ \\
HC\!-\!NH^+ CH & & HC\!-\!NH\ CH
\end{array}
\tag{1.4}
$$

In enzymes, His is utilized in catalytic triads (see Chapter 21, Equation 21.3 and Figure 21.6) and to shuttle protons, for example, in carbonic anhydrase. In some proteins, it is used to form coordinate covalent bonds with metal ions; a well-known example is the zinc fingers in DNA-binding proteins (see Chapter 22). It is also a precursor of hormone histamine. His is the only amino acid with pK_R around the physiological pH (Table 1.1) and gives the proteins the properties of a buffer if it is present in considerable quantities, for example, in hemoglobin. Asp and Glu are both acidic amino acids that are present in the catalytic triad of hydrolases (see Chapter 21, Equation 21.3 and Figure 21.6). The shorter distance between the α-carbon atom and the carboxylic group in the radical in Asp than in Glu makes Asp the stronger acid (Table 1.1) and thus more efficient in the triad. The carboxyl groups of Asp and Glu have the same chemical properties as other organic acids, including acetic acid. These groups are used by proteins and peptides to bind metal ions, most notably Ca^{2+} and Zn^{2+}. Asp and Glu participate in the biosynthesis of several amino acids, including Ala, Met, Thr, Ile, and Lys. Glu has a key role in amino-acid transamina-tion, an important step in amino-acid degradation and synthesis. Together with the deamination of Glu, transamination is involved in nitrogen excretion via urea. Glu is also an important neurotransmitter in mammals and other species and a precursor of GABA (γ-amino butyric acid), the main inhibitory neurotransmitter in the mam-malian central nervous system. Glu can also be used as a food additive to enhance taste and flavor.

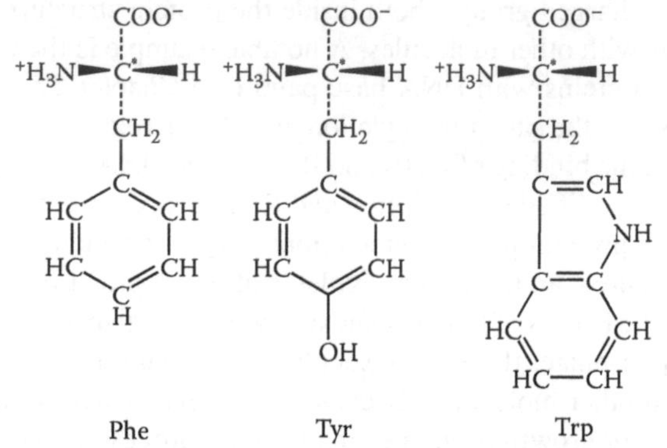

FIGURE 1.4   Structure of the amino acids with an aromatic side chain.

The aromatic amino acids are Phe, Trp, and Tyr (Figure 1.4), all of which are bulky and essentially hydrophobic, but the delocalized π-electrons in the ring systems can take part in weak electrostatic interactions (see Chapter 2). They all absorb ultraviolet light, most notably Trp. The molar extinction coefficients of Trp, Tyr, and Phe at the local maximum 280 nm and pH 7 are around 5500 cm^{-1}/M, 1200 cm^{-1}/M, and 200 cm^{-1}/M, respectively. This property is used to detect the proteins in solution. Trp, Tyr, and Phe are also the source of the intrinsic fluorescence in proteins. The quantum yield of the emitted light after excitation at 280 nm is 0.20 for Trp (emitted at 348 nm), 0.14 for Tyr (emitted at 303 nm), and 0.04 for Phe (emitted at 382 nm). The emission is very sensitive to the fluorophore surroundings, and this is employed to follow the changes in protein structure (see Chapter 11). The Tyr side chain is mainly hydrophobic, with the exception of the –OH group, which is capable of hydrogen bonding and dissociation at high pH (Table 1.1). This group is also the site of sulfonylation and phosphorylation, the latter being an important event in signal transduction via receptors with tyrosine kinase activity (see Chapter 23). Tyr is a precursor of the adrenal hormones dopamine, epinephrine, and norepinephrine, and the thyroid hormones triiodothyronine and thyroxin. Dopamine is also synthesized in various regions of the brain, where it functions as a neurotransmitter. Trp with an indole side chain is the largest amino acid. The nitrogen in the ring can participate in hydrogen bonds as a hydrogen donor; in other cases, the ring is hydrophobic. Trp is a precursor of the neurotransmitter serotonin and the neurohormone melatonin.

For the detection and identification of individual amino acids, mass spectrometry, chromatography, isoelectric focusing, optical and fluorescence spectroscopy (aromatic amino acids), and different combinations of these methods can be used. The amino group of amino acids reacts with ninhydrin, giving a colored product, and reacts with fluorescamine, giving a fluorescent product. A fluorescent derivative of the amino acids is also obtained in the reaction with dansyl chloride; however, this reagent is not specific just to the amino group but will also react with some other groups, including the hydroxyl group in phenols. These three classical reactions to detect amino acids are shown in Figure 1.5. Some of the reactions specific to the side chains of amino acids are

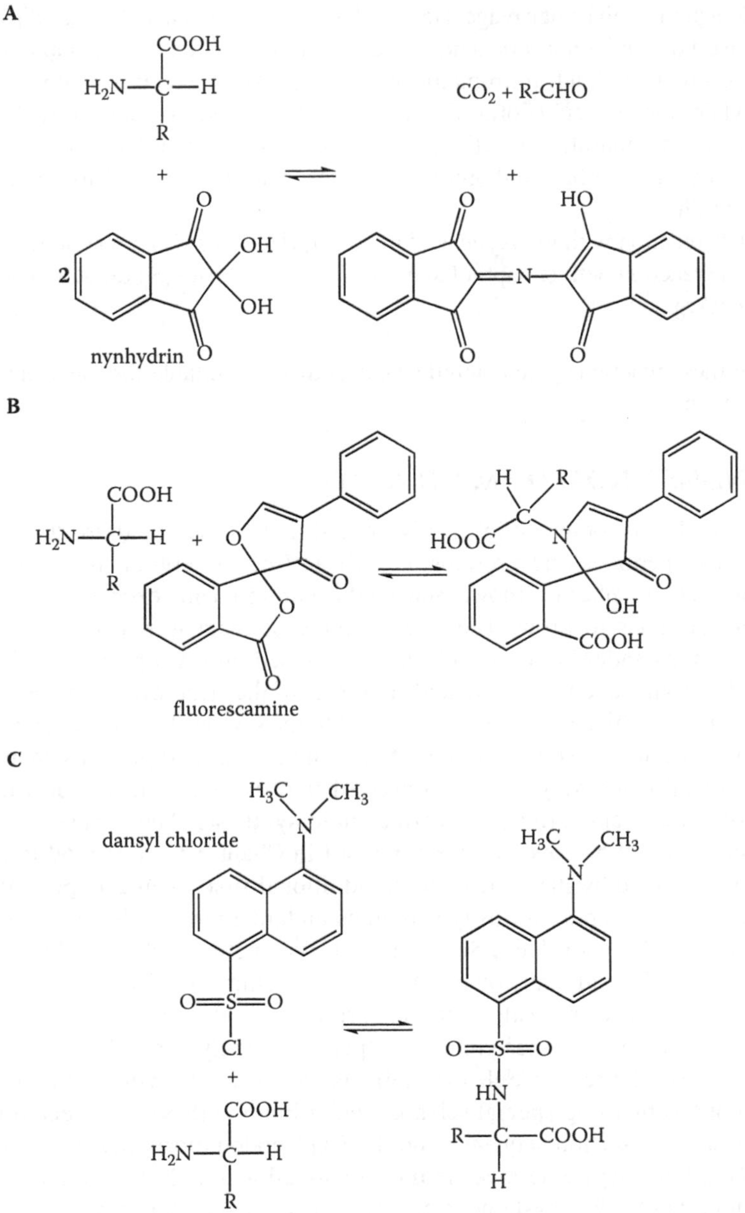

A

B

C

FIGURE 1.5 Reactions for the detection of amino acids by spectrophotometry and fluorescence: (A) ninhydrin reaction; (B) reaction with fluorescamine; (C) reaction with dansyl chloride.

- Reactions with Elman reagent (5.5′-dithio-bis-(2-nitrobenzoate); 2,2′-dipyridyl disulfide; and iodoacetic acid or iodoacetamide for the thiol group in Cys)
- Reactions with tetranitromethane; $HgNO_3$ (Millon reaction); phosphomolybdotungstic acid (Folin–Ciocalteu reaction); boiling concentrated nitric acid (xanthoproteic reaction); and α-nitrous-β-naphtol for the hydroxyl group in Tyr (the xanthoproteic reaction can also be used to detect Trp and Phe)
- The reaction with glyoxylic acid in H_2SO_4 (Hopkins–Cole reaction) for Trp
- The reaction with α-napthol and sodium hypochlorite (Sakaguchi reaction) for Arg

Most of these reactions give colored products that are suitable for spectrophotometric detections.

1.2   NONSTANDARD AMINO ACIDS

A vast number of other amino acids, in addition to the standard 20 acids, can be found in peptides and proteins as well as free in cellular and extracellular solutions. The number of known nonstandard amino acids exceeds 700, and this number continues to grow. The role of many of them is not well understood. However, nonstandard amino acids can be divided into two groups: amino acids that differ from standard amino acids in terms of the structure of the radical, and D-amino acids, which are structurally equal to the standard amino acids but are of different chirality. It seems that those found in peptides and proteins, synthesized via ribosomal machinery, are all converted into the nonstandard form from the standard amino acids after polypeptide chain synthesis. These posttranslational modifications will be covered in more detail in Chapter 5. Most modified amino acids are obtained by the attachment of additional functional groups to the radical. Important examples of this type of nonstandard amino acids are hydroxylated Pro and Lys; phosphorylated Ser, Thr, and Tyr; methylated Lys, His, and Arg; carboxylated Glu, and many others (for some structures, see Figure 1.6). The only known exception to this rule is the occurrence of selenocysteine in a few proteins, including the enzymes glutathione peroxidase, glycine reductase, and some hydrogenases (Figure 1.6). Selenocysteine is included in the ribosomal biosynthesis of proteins by using special selenocysteine-binding tRNA that recognizes the selenocysteine insertion sequence and the UGA codon (normally the stop codon) in mRNA. Selenocysteine is sometimes regarded as the 21st coded amino acid. The synthesis of selenocysteine is achieved by enzymatically catalyzed complex modifications of Ser that are attached to the tRNA. Mechanisms to convert Ser to selenocysteine and the required enzymes differ in terms of prokaryotes and eukaryotes.

Some peptides are synthesized in the cells nonribosomally by a sequence of enzyme-catalyzed reactions facilitating the incorporation of nonstandard amino acids and also non-amino-acid building blocks such as carboxy acids, heterocyclic rings, and fatty acids. Most of these peptides are synthesized by prokaryotes. The peptides in this category frequently include D-amino acids, and many of them are cyclic, for

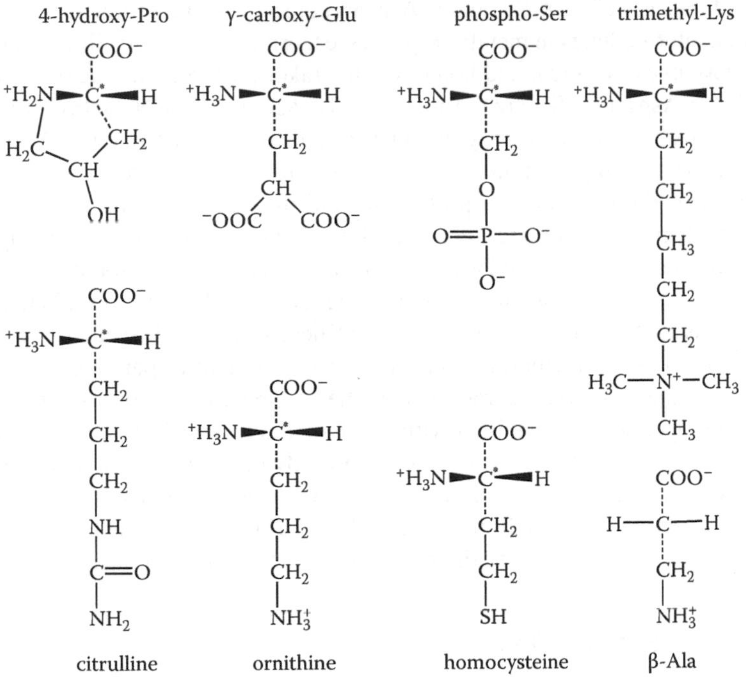

FIGURE 1.6 Some of the nonstandard amino acids.

instance, the peptide antibiotics valinomycin, gramicydin, and actinomycin D, and cyclosporin A. Microorganisms also synthesize other classes of antibiotics in which the conversion of L-amino acids into the D-form is a part of the posttranslational modifications, for example, in the so-called lantibiotics epidermin and cinnamycin. Higher organisms also synthesize D-amino-acid-containing peptides, but in these peptides the standard L-amino acids are in all known cases converted to the D-form after the synthesis is complete. Examples include muscle-contracting peptides such as achatin I from snails, opioid peptides such as demorphins, and antimicrobial peptides such as bombins from the skin of certain frogs and toads, as well as the peptide toxin from the platypus. D-amino acids are components of bacterial cell walls containing D-Ala and derivatives of D-Glu, while D-Tyr is a constituent of the wall of yeast spores. D-amino acids are also found in plants, for example, in some alkaloids such as N β-(D-Leu-D-Arg-D-Arg-D-Leu-D-Phe)-naltrexamine. Another nonstandard amino acid is β-Ala (Figure 1.6), common in the naturally occurring peptides carnosine (β-Ala-His) and anserine (β-Ala-methyl-His), and also in pantothenic acid (vitamin B_5), which is itself a component of coenzyme A. There seem to be two main functions of nonstandard amino acids in peptides and proteins. The first is to protect the polypeptide chain against decomposition by the proteolytic enzymes that are not able to recognize the unusual amino acids. The second function is to provide additional functional groups for special purposes. A number of nonstandard amino acids that have never been identified as peptide and protein constituents have been found in organisms. Known examples in humans include homocysteine, ornithine and citrulline, as well as D-Ser and D-Asp, and others. The first three (Figure 1.6) are important intermediates in metabolic processes, ornithine and citrulline are involved in the biosynthesis of urea, and homocysteine takes part in amino-acid metabolism. N-Methyl-D-Asp and D-Ser are the agonist and coagonist of the N-Methyl-D-aspartate receptors involved in excitatory glutamatergic synaptic transmission. D-Amino acids are of importance in peptidomimetic drugs, where their introduction into the structure enhances the resistance to proteolysis and increases the bioavailability of the drug (see Chapter 31). Peptide drugs based on D-amino acids that effectively inhibit the entry of the AIDS virus HIV-1 into cells are in common use today.

One obstacle to knowing more about the occurrence of D-amino acids in peptides and proteins is that they are not easily identified. Recently, Sweedler and coworkers have developed a method to analyze for them in natural peptides. The method is based on the selective degradation of the polypeptide chain with microsomal alanyl aminopeptidase, combined with mass spectrometry. The enzyme selectively degrades the peptides that lack D-amino acids. By comparing a sample before and after digestion, the D-amino-acid-containing peptides can be identified even when they are present in very small quantities. This approach should help to increase the number of detected D-amino acids in natural peptides and proteins.

1.3 THE PEPTIDE BOND

Polypeptides are made by binding amino acids into linear polymers via peptide bonds. A peptide bond is a covalent bond that is formed in a condensation reaction between the carboxyl group of one amino acids and the amino group of the other, releasing, as a result, a molecule of water. The peptide bond is a form of the amide bond:

$$
^+H_3N-\overset{\overset{\displaystyle H}{|}}{\underset{\underset{\displaystyle R_1}{|}}{C}}-COO^- + {}^+H_3N-\overset{\overset{\displaystyle H}{|}}{\underset{\underset{\displaystyle R_2}{|}}{C}}-COO^- \rightleftharpoons {}^+H_3N-\overset{\overset{\displaystyle H}{|}}{\underset{\underset{\displaystyle R_1}{|}}{C}}-\overset{\overset{\displaystyle O}{\|}}{C}-\overset{}{\underset{\underset{\displaystyle H}{|}}{N}}-\overset{\overset{\displaystyle H}{|}}{\underset{\underset{\displaystyle R_2}{|}}{C}}-COO^- + H_2O \qquad (1.5)
$$

The free-energy change for this reaction is positive ($\Delta G \approx 10 \ kJmol^{-1}$). The equilibrium of this reaction is shifted toward the reactants, meaning that the peptide bond has a tendency to be hydrolyzed. However, the activation energy of the hydrolysis is very high ($80-130 \ kJmol^{-1}$, depending on the participating amino acids and other factors), and the decomposition of the peptide bond is thus very slow. The half-life of the peptide-bond hydrolysis at pH 7.0 and 25°C is measured in years, and the bond has properties of a partial double bond due to the delocalization of an electron pair from the nitrogen to the carbonyl oxygen:

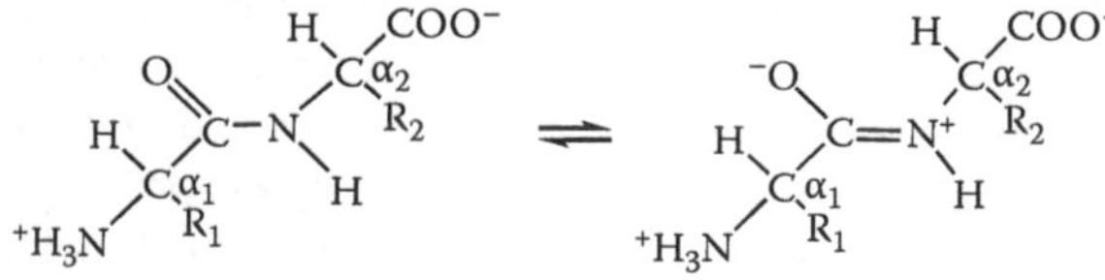

$$
\qquad (1.6)
$$

The two structures shown in Equation 1.6 are in resonance, and an approximately 40% double-bond character is assumed. The partial double-bond character has several consequences. First, the bond is planar, and rotation around the C–N axis is impossible. This forces six atoms—the C and N participating in the bond and all the four atoms connected to them (O, H, Cα_1, and Cα_2; see Equation 1.5)—to be located on the same plane. Second, the partial double bond can be in two configurations—trans and cis:

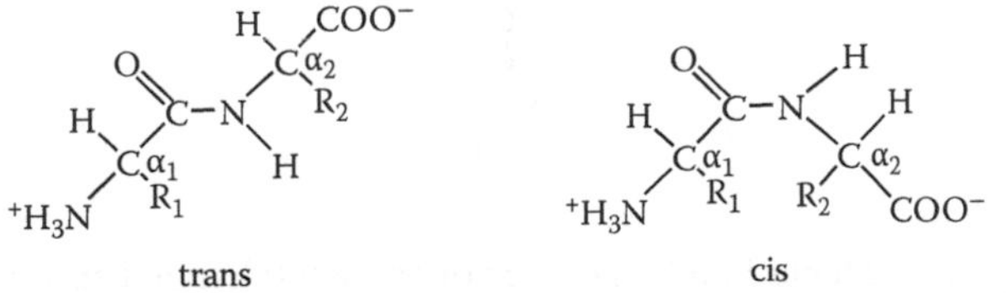

$$(1.7)$$

trans   cis

The preferred configuration is trans, and the ratio of cis to trans was found to be about 1:1000 in peptides and proteins, except when the Pro is on the N side of the peptide bond, in which case the ratio of cis to trans is 1:20. The trans configuration is favored because, in this configuration, the overlap and repulsion of the radicals is minimized. In the case of Pro, in which the peptide-bond-forming nitrogen atom is a part of the ring structure, the overlap of radicals in the cis configuration is considerably decreased, resulting in about 10% of these peptide bonds in proteins being cis. In Pro, the lack of a hydrogen atom attached to the nitrogen of the peptide bond that could help in the resonance stabilization of this bond also corroborates the increased occurrence of the cis configuration. However, the ratio of cis to trans is also dependent on other factors, and is considerably lower in more flexible and less structured small peptides, where up to 30% of proline-following peptide bonds can be in the cis configuration. Another consequence of the partial double-bond character of the peptide bond with a C–N spacing of 0.132 nm is that this bond is shorter than the single C–N bond (a C–N distance of 0.145 nm) and longer than the double C–N bond (a C–N distance of 0.125 nm). The distances and angles between the atoms in the peptide bond are shown in Figure 1.7. An additional important feature of the peptide

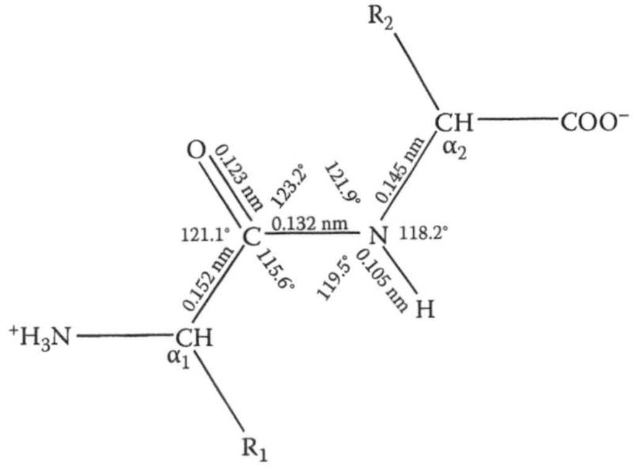

FIGURE 1.7 Distances and angles between the atoms in the peptide bond.

bond is the polarity. Because of the delocalization of a nitrogen electron pair and the differences in the electronegativity of the participating atoms, the peptide bond shows a dipole moment of around 3.5 Debye units:

$$C\alpha_1-C \overset{O}{\underset{\substack{\delta^+ \\ H}}{\parallel}}{}^{\delta^-} N-C\alpha_2 \tag{1.8}$$

Finally, the partial double-bond character of the peptide bond results in the absorbance of light in the UV region at wavelengths between 190 and 230 nm. The properties of the peptide bond are important for the structure of peptides and proteins, as shown, for example, in Chapter 3.

FURTHER READING

1. Barrett, G. C. 1985. Chemistry and Biochemistry of Amino Acids. New York: Chapman and Hall.
2. Amino Acids, Peptides and Proteins. Specialist Periodical Reports of the Royal Society of Chemistry, Volumes 1–30.
3. Wolosker, H., Dumin, E., Balan, L., and Foltyn, V. N. 2008. D-Amino acids in the brain: D-serine in neurotransmission and neurodegeneration (Minireview), FEBS J. 275, 3514–26.

2 Noncovalent Interactions

Matjaž Zorko

CONTENTS

The polypeptide backbone is formed by covalent peptide bonds, as presented in Chapter 1. All the higher structures of peptides and proteins arise from the interactions of amino-acid residues protruding from this backbone. These interactions are as follows: the electrostatic interaction, the van der Waals interaction, the hydrogen bond, the hydrophobic interaction, and the –S-S– bridges between Cys residues. All these interactions—with the exception of the covalent –S-S–bridges—are weak and strongly affected by their environment. However, the interactions are numerous and do not only stabilize the physiologically functional native structure of the proteins and peptides but also provide the basis for their dynamic behavior.

2.1 THE ELECTROSTATIC INTERACTION

Electromagnetic interactions provide the fundamental basis for all the different bonded and nonbonded interactions between atoms and molecules. We will discuss here the interactions between charged atoms, functional groups, and molecules. According to Coulomb's law, the force between two interacting charges A and B can be described by the equation

$$F = \frac{q_1 \cdot q_2}{D \cdot d^2} \tag{2.1}$$

where q_1 and q_2 are charges, d is the distance between the charges, and D is the relative dielectric constant of the medium in which the charges are situated. The values of the dielectric constants of some common solvents are listed in Table 2.1. In crystals the value of D is considered to approach 1—the value in vacuum—and

TABLE 2.1
Relative Dielectric Constants of Some Common Solvents

Solvent	Dielectric Constant	Solvent	Dielectric Constant
Vacuum	1	Dichloromethane	9.1
CCl$_4$	2.23	Dimethylformamide	36.7
Benzene	2.27	Water	80
Ethylacetate	6.02	Formamide	109

the electrostatic interactions, also called ionic bonds, in ionic crystals are strong, the energy of the bond being over 100 kJ mol^{-1}. In a protein interior, the value of D is 2–4, and the interaction is also relatively strong; however, because of solvation, charged groups do not tend to enter the protein interior, and such interactions are rare. Although strong, the internal electrostatic interactions actually destabilize the protein structure because the solvated state of the charged groups in an unfolded protein is more thermodynamically favorable than the nonsolvated state in the interior of the folded protein. In an aqueous environment with a high dielectric constant of around 80, the electrostatic interaction becomes a weak interaction. This is due to the shielding effect of water dipoles enclosing charged ions and functional groups (Figure 2.1). Most of the electrostatic interactions between charged amino-acid residues—Asp, Glu, Lys, Arg, and His at the physiological pH (see the corresponding values of pK in Chapter 1, Table 1.1)—occur mainly on the protein or peptide surface in close contact with water, and are, as a result, weak. For instance, a positively charged Lys or Arg residue can interact with negatively charged side chains of Asp or Glu. In proteins this interaction is referred to as a salt bridge. As observed in the solved three-dimensional structures of proteins, salt bridges are relatively rare

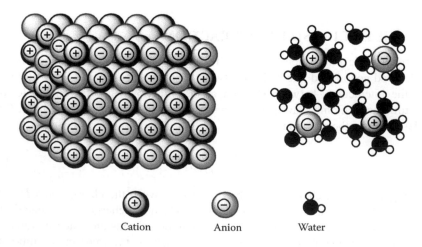

Cation   Anion   Water

FIGURE 2.1 Electrostatic interactions in a crystal and in a water solution.

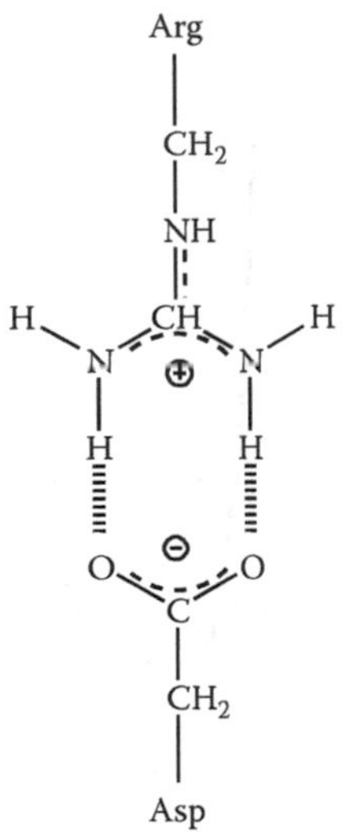

FIGURE 2.2   Interaction of Arg with a carboxylic group.

in such material, and they normally occur on the surface. An exception is when an internal salt bridge is involved in the catalytic mechanism of an enzyme such as in the Asp–His–Ser triad of serine proteases (see Chapter 21.) An additional problem of salt bridges in proteins is the delocalization of charge. Coulomb's law is invariable for point charges; however, in amino-acid residues the charges can be distributed across the broader surface of the residue, and the interaction is thus weakened. One such example is the residue of Arg, where the positive charge is shared by three amino groups. When Arg is interacting with the carboxyl group of Asp or Glu, where the negative charge is also delocalized and shared by both carbonyl oxygens in the ionized group, the electrostatic interaction is much weaker than in the corresponding hydrogen bonds that are also formed between these two groups (see Figure 2.2). Salt bridges on the protein surface encounter another problem: they are situated on the border between the less dielectric protein interior and the more dielectric solvent surroundings. Two charges on the spherical protein surface have the shortest mutual distance via the protein interior but, as was observed, they interact across the surface via a high dielectric solvent, which makes the interaction much less strong. Bulky, nonpolar groups in the vicinity of the charge and with limited solvent accessibility can further reduce the interaction energy.

2.2   THE VAN DER WAALS INTERACTION

Two opposite forces operate between two atoms approaching each other: the repulsive and the attractive van der Waals forces. The magnitude of each of these two forces depends differently on the interatomic distance. With a decreasing interatomic distance, the repulsive force increases much faster than the attractive one, as shown in Figure 2.3. The resulting curve is the sum of both forces and shows two regions of net interaction: the attractive and the repulsive. It is hard to give a precise

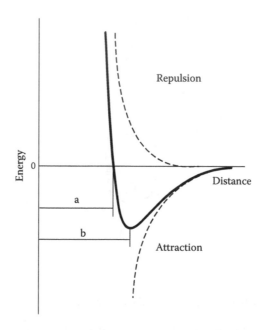

FIGURE 2.3 Schematic presentation of the van der Waals interaction—repulsive and attractive forces as the function of distance. Label a is the sum of the van der Waals radii; b is the distance between the centers of the two atoms in the optimal van der Waals interaction.

mathematical presentation of the resulting curve since any generalization will lead to an approximation. The net potential energy of the interaction is usually given as the Lennard–Jones potential:

$$E = \frac{C_n}{d^n} - \frac{C_6}{d^6}$$ (2.2)

where C_n and C_6 are constants, and d is the distance between the centers of the participating atoms. The first term represents repulsion, and the second term attraction. Since this is an approximation, n can be chosen arbitrarily to some extent, and a value of 12 is usually taken for n, for mathematical reasons. As a consequence, the Lennard–Jones potential is frequently referred to as the 12–6 potential.

The interatomic distance where the attractive and repulsive forces are equal to one another is called the sum of the van der Waals radii. If measured for the diatomic molecule of an element, usually in crystals, the van der Waals radius for a particular atom can be obtained. For the atoms of the main bioelements H, O, C, N, P, and S, it ranges from 0.14 nm for a hydrogen atom to 0.2 nm for a sulfur atom. A value of 0.19 nm is commonly obtained for carbon. These values are approximate and dependent on the system in which they are measured. Another distance to be considered is the optimal van der Waals contact distance, defined by the minimum of the curve in Figure 2.1. The van der Waals radius defines the size of an atom as modeled by a hard-shell sphere. Squeezing two atoms below the sum of the van der Waals radii results in a large energy cost as the electron orbitals start to overlap. The Pauli

Exclusion Principle states that no two electrons can share the same state, so half the electrons of the system would have to go into orbitals with an energy higher than the valence state. The repulsive force prevents matter from collapsing and defines the size and shape of all molecules, including amino acids, peptides, and proteins. The van der Waals volumes of the coded amino acid residues are given in Chapter 1, Table 1.1.

Attractive van der Waals interactions are usually divided into three classes according to the degree of polarization of the participating molecules or their parts: (1) between permanent dipoles, (2) between a permanent dipole and an induced dipole, and (3) the dispersion interaction. Molecules or parts of molecules having a nonequal partition of charges due to the different electronegativities of the atoms are called dipoles. The separation of charges in a molecule determines its dipole moment μ, which is a vector defined by the magnitude of the separated excess charge Z and the distance d by which it is separated:

$$\vec{\mu} = Z \cdot \vec{d} \tag{2.3}$$

The dipole points from the negatively charged atom toward the positively charged one. Molecules composed of atoms with widely different electronegativities will tend to form permanently charged ions that are held together by the electrostatic interactions already discussed. When the difference in the electronegativity is moderate, the electrons in the bonds are unevenly distributed so that the centers of the negative and the positive charges do not coincide and the molecule is a permanent dipole. The permanent dipoles that are of particular relevance to polypeptide chemistry are the peptide bond, the polar groups in amino acid radicals, and water. Dipole–dipole interactions between freely rotating polar molecules tend to cancel out because the attraction between poles of different charges is compensated by the repulsion of poles with the same charge. However, the electric fields of the dipoles will not, in reality, allow a totally free rotation, and the energy of the interaction is proportional to the inverse sixth power of the distance between the dipoles, as shown for the attractive portion of Equation 2.2. When polar molecules or groups are not free to rotate, as in the amino acid residues of peptides or proteins, the energy of the dipole–dipole interaction is a complicated function that includes the dipole moments of both dipoles as well as the distance and the angle between them. In a simple case, when the dipoles are aligned as below,

$$\begin{array}{cccc} q_1 & -q_1 & q_2 & -q_2 \end{array} \tag{2.4}$$

the potential energy of the interaction in vacuum is

$$E = -\frac{\mu_1 \cdot \mu_2}{2\pi D_0 \cdot d^3} \tag{2.5}$$

where μ_1 is the dipole moment of the dipole $q_1/-q_1$, μ_2 is the dipole moment of the dipole $q_2/-q_2$, D_0 is the dielectric constant in a vacuum, and d is the mean distance between the dipoles. Similarly, the interaction between the dipole $q_1/-q_1$ and the point charge q_2, an approximation of the dipole–ion interaction, in the aligned positions

$$\overset{q_1}{\ominus}\!\!-\!\!\overset{-q_1}{\ominus}\cdots\cdots\overset{q_2}{\oplus}\cdots \tag{2.6}$$

is

$$E = -\frac{\mu_1 \cdot q_2}{4\pi D_0 \cdot d^2} \tag{2.7}$$

Another type of van der Waals interaction occurs between permanent dipoles and neutral molecules in which a dipole is induced. The electric field of the ion, the charged group or the dipole, will affect the distribution of electrons in the vicinal neutral molecule, and, as a result, it will become transiently polarized. The magnitude of the induced dipole moment depends on the strength of the electric field and the polarizability of the molecule, which is a function of the structure and volume of the molecule. In general, larger molecules composed of bigger atoms with electrons that are less tightly controlled by the nuclear charges will be more affected by the external electric field than smaller ones, and will, therefore, be more strongly polarized. In principle, the induced dipole will interact with the permanent dipole or point charge in accordance with Equations 2.4–2.7, but the value of the induced dipole moment should be estimated separately, and this is usually not easy to do. It is interesting that the alignment of the induced dipole in relation to the permanent dipole is always opposite because it is defined by the orientation of the permanent dipole. The van der Waals interactions are not restricted to the polar molecules, and because of the fluctuation of the electrons in the atoms and molecules, nonpolar molecules can also transiently acquire a dipole moment, and, in turn, they can induce a dipole moment in neighboring nonpolar molecules. The interactions between induced dipoles are called dispersion interactions, or London interactions.

The strength of all forms of van der Waals interactions is in accordance with the term C_n/d^6 in Equation 2.2, but C_n is composed of different parameters and has a different magnitude in each particular type of interaction. At 25°C, the energies of the interaction of dipoles in close proximity—which are practically in contact—are comparable or smaller than the thermal kinetic energy, which is equal to 3/2RT = 3.7 kJ mol^{-1}. Although very weak and of short term, they operate within all molecules, groups, and atoms in contact and represent an important contribution to the stability of proteins and peptides.

Aromatic rings such as benzene molecules are electroneutral and are not dipoles. However, the charges in the ring are not distributed evenly. The negative charges of π-electrons are concentrated on both sides of the ring, protruding perpendicularly out of the ring plane, while positive charges are situated near the ring edge (see Figure 2.4). Molecules with this type of charge distribution are known as quadrupoles. The interaction of two aromatic structures where they lie on each other, that is, "face to face," is usual for heterocyclic structures such as Trp, particularly when several rings are stacked one above the other in a hydrophobic environment and

A   B   C

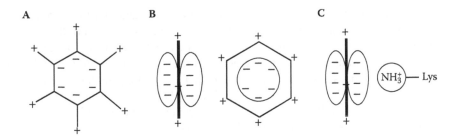

FIGURE 2.4 Interactions between aromatic rings, and the ion π interaction between an aromatic ring and an ion.

additionally stabilized by hydrophobic interactions and the resonance of adjacent π-electron rings. Studies of the crystal structure of benzene have demonstrated a "face-to-edge" packing where the edge of the ring, composed of partially positive H-atoms, interacts with the partially negative π-electrons in the aromatic structure. Phe and Tyr residues often participate in this type of interaction. The negative charge of π-electrons also attracts positively charged ions and functional groups. The residues of Phe, Tyr, and Trp, interacting with the positively charged amino group in the residues of Lys and Arg, have been identified in peptides and proteins. As determined by an analysis of these interactions in DNA-binding proteins, Arg–Phe and Arg–Tyr are the two most frequent pairs involved in cation–π interactions.

Interestingly, in DNA-binding proteins the average energy of the interaction of these pairs is around 20 kJ mol^{-1}, while in other proteins, it is approximately 40% lower. Cation–π interactions are of importance for the stability of peptides and proteins but can also be the basis of the interaction of proteins with other molecules. The neurotransmitter acetylcholine binds to the acetylcholine receptor predominantly via the interaction of its positively charged trimethylamino group with the Trp ring in the binding site of the receptor. The interaction between aromatic quadrupoles and the interaction of the quadrupole with the charged group are shown in Figure 2.4.

2.3   THE HYDROGEN BOND

The hydrogen bond can be regarded as a specific dipole–dipole interaction in which a hydrogen atom covalently attached to an electronegative atom, for example, oxygen, is approached by another electronegative atom with a lone pair of electrons, for example, carbonyl oxygen, as in the example below:

$$\underset{\text{O}}{\overset{\delta^-}{}}-\underset{\text{H}}{\overset{\delta^+}{}}\cdots\cdots\underset{\text{O}}{\overset{\delta^-}{}}=\underset{\text{C}}{\overset{\delta^+}{}} \qquad (2.8)$$

Being more electronegative than hydrogen, oxygen draws the electrons in the O–H bond toward itself. The hydrogen atom is then left with a net positive charge, and the oxygen is, as a result, negative. The bond between the hydrogen and the carbonyl oxygen (Equation 2.8) is of a double nature because, in addition to the attractive electrostatic interaction between partial charges, a quasimolecular orbital between the hydrogen and the carbonyl oxygen is also formed. In the presented case, the oxygen in the −OH group is a hydrogen donor, and the carbonyl oxygen is a hydrogen acceptor. In proteins and peptides, the groups that can participate in hydrogen-bond formation are the carbonyl group in the peptide bond as a hydrogen acceptor and the −NH group in the peptide bond as a hydrogen donor, as well as the following groups in amino-acid residues: the −OH of Ser, Thr, and Tyr; the amido and carbonyl groups of Gln and Asn; the imino group in the rings of His and Trp; the carboxyl group of Asp and Glu, the amino group of Lys; and the guanido group of Arg. Groups containing hydrogen can play both the role of acceptor and of donor, whereas others can function only as hydrogen acceptors. The energy and length of the bond depend on a combination of the hydrogen acceptor and the hydrogen donor and also on the configuration of the bond. When a donor, the hydrogen and acceptor atoms are lying on the same line, and the hydrogen bond is the strongest, and it becomes weaker—though not very much—when this configuration is distorted because of the increased repulsion between the acceptor and donor atoms; the distortions can form an angle of up to 40°. The energy of the bond is between 10 and 40 kJ mol^{-1}, and the length between the donor and acceptor atoms is typically around 0.28 ± 0.03 nm. An important hydrogen-donor and hydrogen-acceptor molecule is water, which can form numerous hydrogen bonds either with other water molecules or with functional groups of the polar amino-acid residues. The ability of a molecule to form hydrogen bonds with water defines its solubility in water, and the inability to form hydrogen bonds with water is the basis of hydrophobic interactions. Hydrogen bonds are of great importance for the stabilization of the peptide and protein structures; however, they are weak enough to be broken during the conformational changes of protein molecules, thus enabling the necessary protein and peptide dynamics. Hydrogen bonding of the backbone carbonyl oxygen to the backbone amino groups leads to the formation of different secondary structures such as alpha helices and beta sheets. In addition, hydrogen bonds of the groups in amino-acid radicals contribute substantially to the stabilization of all the higher levels of the protein and peptide structures, from secondary to quaternary (see Chapter 3). Interactions can also take place between groups carrying a formal charge and hydrogen-bonding atoms; this is an especially strong variant of the hydrogen bond.

2.4 THE HYDROPHOBIC INTERACTION

In water solutions, nonpolar molecules tend to aggregate in order to minimize the amount of surface exposed to water. This process is known to be entropically, rather than enthalpically, driven. The energy needed for the ordering of nonpolar molecules in the hydrophobic interaction that keeps nonpolar molecules together is outweighed mainly by the decreased ordering of the surrounding water molecules (see Figure 2.5A). The results of a more detailed analysis of the transfer of a nonpolar cyclohexane molecule from gas, liquid, and solid phases to an aqueous environment in terms of free-energy change, and its enthalpic and entropic

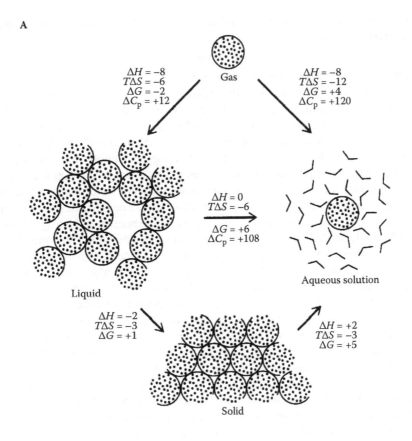

FIGURE 2.5 ΔG, ΔH, and $T\Delta S$ changes in hydrophobic interactions: (A) thermodynamic parameters for the transfer of a nonpolar molecule (cyclohexane) in water at 20°C. Units for ΔG, ΔH, and $T\Delta S$ are kcal mol^{-1}; units for ΔC_p are kcal K^{-1} mol^{-1}. (B) hydrophobic interaction is entropy driven. (Part A Reproduced from Creighton, T. E. Proteins Structure and Molecular Properties, 2nd ed., W. H. Freeman, New York, 1993. With permission.)

energy contributions are summarized in Figure 2.5B. The fundamental equation of thermodynamics

$$\Delta G = \Delta H - T\Delta S \tag{2.9}$$

is applied, in which the enthalpy change ΔH reflects the binding energy, and the entropy factor $T\Delta S$ is a measure of the ordering of the system during the transfer. It is evident that the transfer of cyclohexane to water from any phase is a thermodynamically unfavorable process characterized by a positive value of ΔG because of the relatively high negative value of the entropy factor $T\Delta S$. It also shows that the hydrophobic interaction does not arise from the binding of nonpolar molecules to each other but rather from preventing water molecules from achieving optimal hydrogen binding. In pure water at 25°C, each water molecule tends to form an average of 3.4 hydrogen bonds with the neighboring water molecules. The introduction of a nonpolar substance in a water solution decreases the average number of hydrogen bonds in water by steric hindrances, and thus the system is destabilized. The optimal stability of the system is achieved by keeping nonpolar molecules together in as small a volume as possible. Two important consequences of this process are that the hydrophobic interaction depends on the relative polarity of both the solute and the solvent, and that the magnitude of the interaction is proportional to the amount of polar solvent displaced by the nonpolar solute, or, more specifically, to the surface area of the nonpolar solute exposed to water. The hydrophobic interaction has a complex temperature dependency due to the opposite effect of the increased temperature on the enthalpic factor in Equation 2.9, which increases, and the entropic factor, which approaches zero. The resulting ΔG for the transfer of the polar solute into the water increases with temperature and reaches its maximum at around 140°C.

Different ways exist to quantify the energy contribution of the hydrophobic effect. In addition to a calculation of the free energy of transfer, discussed earlier, the most common and simplest procedure is to determine the partition of a molecule between water and a water-insoluble organic solvent. In that case the organic solvent serves as a model for the hydrophobic environment. The solute, for example, an amino acid, is added to the mixture of nonmixable solvents, for example, water and n-octanol, and the concentration in each phase of the solvents is measured. The result is presented as the partition coefficient K_p:

$$K_p = \frac{C_{\text{solvent}}}{C_{\text{water}}} \tag{2.10}$$

where C_{solvent} and C_{water} are the equilibrium concentrations of the tested substance in the organic solvent and in water, respectively. Instead of concentrations, the solubilities of the compound in a nonpolar solvent and in water can be used. The result is frequently given in logarithmic form π ($\pi = \log_{10}K_p$).

Another measure of hydrophobicity is the distribution of the tested substance in vapor and water, where the distribution constant K_D is determined in accordance with the equation

$$K_D = \frac{C_{\text{vapor}}}{C_{\text{water}}} \tag{2.11}$$

and can also be given in logarithmic form π ($\pi = \log_{10}K_D$). Different model systems that can be applied can lead to some confusion and to different scales of the obtained values. To avoid this, the system and type of measurements must be precisely defined. The values of π for the nonpolar groups are practical because they can be independent of the interacting group and are additive. For a methylene group, the value of π is 0.5, and this corresponds to a ΔG of 2.8 kJ mol^{-1} for each additional methylene group in a molecule that participates in a hydrophobic interaction. For amino acids the hydropathy index is usually used because it is considered to illustrate the behavior of amino acid residues in the peptide and protein structures more realistically (the explanation and values of the hydropathy indexes for the coded amino acids are given in Chapter 1, Table 1.1).

Hydrophobic interaction is of great importance for the stabilization of the peptide and protein structures and seems to be the most important driving force for protein folding (see Chapter 6). This is illustrated by the well-known fact that transferring a protein from a water environment into a nonpolar solvent will result in the disruption of hydrophobic interactions in the protein core, leading to protein denaturation.

2.5   MULTIPLE INTERACTIONS AND COOPERATIVITY

The special characteristics of the noncovalent interactions in peptides and proteins are that

- Many interactions are operating in the same molecule.
- The interactions are interdependent (or at least are not independent).
- The concept of "effective concentration" can be applied.
- Interactions show cooperativity.

Although weak interactions, when considered separately, are of low strength, they are numerous, and some of them are always present. Weak interactions should be regarded as a tight network that has a large impact on the stabilization of a three-dimensional structure of peptides and proteins. Additionally, the weak nature of the interactions enables the fast and effective transition in a three-dimensional structure, which is extremely important for protein and peptide functions. Since the interacting functional groups are not free in the solution but are connected to the polypeptide backbone, the probability of an interaction is affected as a consequence.

In proteins and peptides, many interactions occur in close proximity, and are, therefore, not totally independent. Consider the two interacting groups, A and B, first, as free in the solution

$$A + B \underset{}{\overset{K_{\text{inter}}}{\rightleftharpoons}} A\text{'''''}B \tag{2.12}$$

and, second, as bound to the backbone:

$$A\text{———}B \underset{}{\overset{K_{\text{intra}}}{\rightleftharpoons}} A\text{'''''}B \tag{2.13}$$

It is useful to define the effective concentration to describe the effect of the backbone on the efficacy of the interaction. The effective concentration, denoted here as $^{(AB)}C_{eff}$, of the groups A and B, when attached to the backbone, is

$$^{(AB)}C_{eff} = K_{intra}/K_{inter} \qquad (2.14)$$

An example is the reaction of carboxylic groups to form an anhydride. The reaction of free molecules of acetic acid would be

$$H_3C-\overset{\overset{\displaystyle O}{\|}}{C}-OH + HO-\overset{\overset{\displaystyle O}{\|}}{C}-CH_3 \overset{H_2O}{\rightleftharpoons} H_3C-\overset{\overset{\displaystyle O}{\|}}{C}-O-\overset{\overset{\displaystyle O}{\|}}{C}-CH_3 \qquad (2.15)$$

$$K_{inter} = 3 \times 10^{-12} \text{ mol L}^{-1}$$

and the reaction between two attached carboxylic groups, for example, in succinylic acid would be

$$(2.16)$$

$$K_{intra} = 8 \times 10^{-7} \text{ mol L}^{-1}$$

The effective concentration of the carboxylic group is 3×10^5 mol L^{-1}. Although these examples include a covalent bond formation, they are also relevant for weak interactions, showing how the equilibrium of the reaction is shifted because of the proximity and proper orientation of the reacting groups. However, because of the dynamic fluctuations in the conformation of the succinylic acid, the orientation of the carboxylic groups is not ideal. This can be improved if the carboxylic groups are fixed at the proper distance and in the proper orientation for the reaction, as in the following example:

$$(2.17)$$

In this case, the effective concentration is 5×10^9 mol L^{-1}. The value of the equilibrium constant of the interaction is increased by a factor of over 1 billion in comparison to the interaction of free acetic acid. In all the examples shown, the effective concentration had a value larger than 1, meaning that the intramolecular reaction proceeds with an equilibrium that is shifted toward the products when compared with the equilibrium of the intermolecular reaction. However, this is not necessarily so: if two potentially interacting groups are attached to the backbone in such a way that the interaction is hindered or even prevented, the effective concentration will have a value of less than 1, and the intermolecular reaction between these two groups is favored. As follows from Equation 2.14, the value of the equilibrium constant for the binding of two interconnected groups K_{intra} is obtained by multiplying the equilibrium constant for the binding of free groups K_{inter} with the corresponding effective concentration $^{(AB)}C_{eff}$. When several weak interactions occur in the same macromolecule, which is the usual situation in peptides, proteins, and nucleic acids, they are cooperative—the strength of each interaction depends on several other interactions due to the changes in the values of the effective concentrations of the interacting groups, mainly in close proximity. In terms of the effective concentrations, the sequence of interactions 1 to 5 presented in Equation 2.18

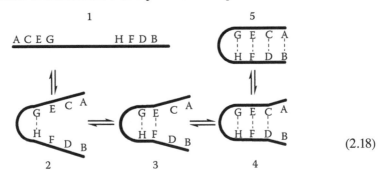

$$(2.18)$$

will have the overall equilibrium constant expressed as:

$$K_{overall} = K_{GH} \cdot {}^{(GH)}C_{eff} \cdot K_{EF} \cdot {}^{(EF)}C_{eff} \cdot K_{CD} \cdot {}^{(CD)}C_{eff} \cdot K_{AB} \cdot {}^{(AB)}C_{eff}$$

Note that the sequence of reactions 1 to 5 is only one of several possible sequences in this system, but the final outcome (Equation 2.19) is valid for all possible sequences. If the value of $K_{overall}$ is larger than 1, state 5 will be more stable than state 1. The first few interactions are normally not favorable because of the decrease of the entropy. Only with a sufficient number of weak interactions does the enthalpy start to prevail, and the efficient concentration starts to increase with each additional interaction, eventually exceeding the value of 1. In this case, each interaction will affect and increase the effective concentrations of the groups involved in the subsequent interactions. This is called positive cooperativity, in contrast to negative cooperativity, which is characterized by a decrease in the effective concentrations. It should be remembered that, in proteins and other macromolecules in organisms, the number of interacting groups is very large and that cooperativity plays an important role in stability. In cooperativity, not all the interactions are equally affected by the others. In general, the most stable interaction is between the groups that are held most rigidly by the other interactions. However, the stability of each interaction depends on the stability of all the other interactions.

FURTHER READING

1. Fersht, A. 2002. Structure and Mechanism in Protein Sciences. New York: W. H. Freeman.
2. Atkins, P. W. 1994. Physical Chemistry. Oxford: Oxford University Press.
3. Panigrahi, S. K. and Desiraju, G. R. 2007. Strong and weak hydrogen bonds in the protein–ligand interface. Proteins 67, 128–41.
4. Baldwin, R. L., 2005. Weak interactions in protein folding: hydrophobic free energy, van der Waals interactions, peptide hydrogen bonds, and peptide solvation. In Protein Folding Handbook (Buchner, J. and Kiefhaber, T., Eds.). New York: John Wiley & Sons, pp. 127–162.
5. Gromiha, M. M., Santhosh, C., and Ahmad, S., 2004. Structural analysis of cation-π interactions in DNA binding proteins. Int. J. Biol. Macromol. 34, 203–211.

3 Structural Organization of Proteins

Matjaž Zorko

CONTENTS

In 1951, while lecturing at Stanford University, the Danish biochemist Kaj Ulrik Linderstrøm-Lang was the first to propose the four-level, three-dimensional organization of protein structures. These structures can be described as follows:

1. The primary structure: the covalent chemical structure of the polypeptide chain or chains in a given protein, that is, the number and sequence of the amino acid residues linked together by the peptide bonds.
2. The secondary structure: any such folding that is brought about by linking the carbonyl and the imides groups of the backbone together by means of hydrogen bonds. This structure includes helices, sheets, turns, loops, and so on.
3. The tertiary structure: the organization of the secondary structures linked by "looser segments" of the polypeptide chain stabilized primarily by the side-chain interactions. The disulfide bonds are included in this level.
4. The quaternary structure: the aggregation of the separate polypeptide chains into a functional protein.

The criteria for these classifications have been modified over the years. In fact, today, most authors consider the disulfide bonds as well as any posttranslational modifications to be a part of the covalent structure of the polypeptide and so include them as part of the primary structure. Additional levels of the structure have also been introduced, and all these rules can be summarized as follows:

1. The primary structure: the linear amino acid sequence of the polypeptide chain, including posttranslational modifications and disulfide bonds.
2. The secondary structure: the local structure of the linear segments of the polypeptide backbone atoms without regard to the conformation of the side chains. This level includes two sublevels: the super-secondary structure and the domains. The super-secondary structure means the association of the local secondary structural elements through side-chain interactions. The elements of the super-secondary structure are also called the motifs. A domain is usually a larger part of the protein sequence that can function and exist independently of the rest of the protein chain.
3. The tertiary structure: the three-dimensional arrangement of all the atoms in a polypeptide chain. In the single-chain proteins, this structure is functional.
4. The quaternary structure: the arrangement of the separate polypeptide chains (subunits) into the multi-subunit functional protein.

3.1   THE PRIMARY STRUCTURE

The primary structure of a protein is the exact sequence of the amino acids joined by the peptide bond in a given polypeptide. In multichain proteins, each chain is referred to separately. By convention the primary structure is reported using three-letter or one-letter amino-acid coding (see Chapter 1), starting from the aminoter-minal (N) end and finishing at the carboxyl-terminal (C) end. The primary structure also requires specifying the crosslinking cysteines involved in the protein's disulfide bonds as well as all the posttranslational modifications in the polypeptide chain (see Chapter 5). Frequently, the modifications include cleavage of the polypeptide chain; therefore, the sequence of the polypeptide does not always correspond to the sequence of its mature mRNA. The linear polypeptide chain folds in a particular arrangement, giving a defined three-dimensional structure. Proteins unfolded in vitro fold back to their original native state when the solution conditions are returned to those in which the folded protein exists. All the information for the native three-dimensional structure of the protein appears, therefore, to be contained within the primary structure. On this basis, a prediction of the higher structures of the proteins and their functions is possible (see Chapter 19). Proteins are self-folded by a process called self-assembly; however, in vivo the polypeptide folding is often assisted by molecular chaperones. The assembly of the primary structure of peptides and proteins, and the determination of the sequence are discussed in Chapter 4.

There is a practically infinite number of different possible primary structures. This is the basis for the great diversity of the three-dimensional structures and functions of proteins. However, it is apparent that only a fraction of the possible primary structures actually exist (or have ever existed). It is also very unlikely that two proteins with similar amino-acid sequences have evolved independently. Such similarities would therefore suggest that the two proteins must be related and share a common ancestor. These related proteins are termed homologous. Over evolutionary time spans, proteins mutate, and their primary structure becomes altered, generally by one amino acid at a time, although more drastic single modifications can also occur. Such alterations are caused by mutations in the genes that encode them. However, it is not the case that only point mutations may occur. A protein sequence may lose some of its amino acids (deletion mutation) or have amino acids inserted (insertion mutation). If two primary sequences are more than approximately 20% identical, then they are assumed to be homologous. Generally, a particular type of protein has the same, or a very similar, sequence within one species of an organism. However, there are cases of polymorphism, where several different functional sequences exist for a given type of protein within the population. A detailed approach to the sequence comparison and analysis is given in Chapter 18.

3.2   THE SECONDARY STRUCTURE

As initially proposed by Pauling et al. in 1951, the secondary structure is assigned on the basis of hydrogen-bonding patterns. There are three common secondary structures in proteins, namely, helices, β-structures (β-sheets), and turns or bends. A part of the protein structure that cannot be classified as one of the standard three classes is usually grouped into a category called "other" or "random coil." This last designation is unfortunate, as no portion of the protein's three-dimensional structure is truly random, and it is usually not in the form of a coil. Another, though not much better, expression for the random coil is the unordered secondary structure, which is an attempt to imply that this structure is not random but also that it does not exhibit the elements of the common types of secondary structure, such as helices, β-structures, and turns.

As discussed in Chapter 1, Section 1.3, the peptide bond is usually planar due to its partial double character, and, as a result, it is rigid. Rotation is limited to the bonds on both sides of the C_α atom and is defined by two torsion or dihedral angles Φ (phi) and Ψ (psi):

(3.1)

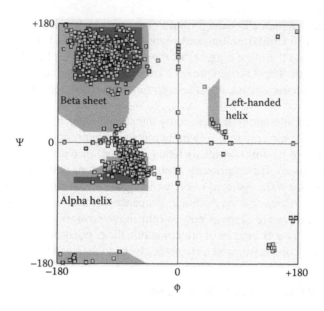

FIGURE 3.1 Ramachandran plot. The white area is a restricted area, the light gray area shows possible angle formations that include Gly, while the dark gray areas are for formations that exclude Gly. Squares are the calculated values for Φ and Ψ coordinates from the human DNA clamp PCNA, PDB ID 1AXC, using the VMD computer program with the Ramaplot extension. Freely accessible and released to public domain at http://en.wikipedia.org/wiki/Ramachandran_Map.

When the planes of two neighboring peptide bonds are covering each other, the values of Φ and Ψ are 0 degrees, but in the fully extended conformation, as shown in Equation 3.1, both are equal to 180°. The values of Φ and Ψ are constrained geometrically due to the steric interactions between the atoms in the neighboring amino acid radicals. By using computer models of small polypeptides, G. N. Ramachandran analyzed the allowed values of Φ and Ψ in order to extract the stable protein conformations. For each conformation, the structure was examined for close contacts between atoms, which were treated as hard spheres, with their dimensions corresponding to their van der Waals radii. The results were plotted on a two-dimensional map on the Φ–Ψ plane (Figure 3.1). The white areas in the diagram in Figure 3.1 correspond to the sterically disallowed conformations where atoms in the polypeptide come closer than the sum of their van der Waals radii. The dark regions show the allowed conformations where there are no steric constraints; they mainly correspond to the α-helical and β-structure conformations. The light-gray areas show the allowed regions if slightly shorter van der Waals radii are used in the calculation and Gly is included. This brings out an additional region that corresponds to the left-handed α-helix. Although it appears likely that larger side chains would result in more restrictions and consequently a smaller allowable region in the Ramachandran plot, the disallowed regions generally involve steric hindrance only between the methylene groups at the β-position in the side chain. Gly has a hydrogen atom with a very small van der Waals radius instead of a methyl group at the β-position, and it is, therefore, the least restricted. Because of this, the allowed values of Φ and Ψ for Gly are marginally limited, and the conformations of Gly-containing polypeptides, particularly the poly-Gly chains, can take large areas in the Φ–Ψ plane in all four quadrants of the Ramachandran plot. Gly also frequently occurs in the turn regions of proteins, where any other residue would be sterically hindered. In contrast, due to the ring structure, Pro can take only a very limited number of possible combinations of Φ and Ψ.

Note that the generation of the Ramachandran plot was based purely on theoretical calculations, and it can be considered as a prediction. After only the first few three-dimensional protein structures were resolved, the theoretical work of Ramachandran was confirmed, and today this is regarded as the cornerstone of protein-structure research. This is illustrated in Figure 3.1, which shows a Ramachandran plot generated from the protein PCNA, a human DNA clamp protein. It should also be remembered that the Ramachandran plot considers the polypeptide backbone and does not refer to the orientation of the amino-acid side chains, which largely determines the biological properties of the polypeptides related to the arrangement of the side chains in a three-dimensional space called χ (chi) space. Although they point out of the backbone, amino-acid side chains are frequently not free to rotate, and this is particularly true for the branched side chains. Their positions in chi-space can be defined by the torsion angles χ_1, χ_2, etc., denoting the torsion angles around the bonds C_α-C_β, C_β-C_γ, etc (see Equation 3.1). These torsion angles can be determined for all amino-acid side chains except Gly, Ala (symmetry problem), and Pro (ring structure), although this is quite difficult in some cases, for example, in Ser, Thr, and Cys. The values of the torsion angle χ_1 for all the residues except Gly, Ala, and Pro, as well as for χ_2 and χ_3 for larger amino acids, are available in databases.

3.2.1 THE HELICES

Helices are a repetitive secondary structure in which the relationship of one peptide unit to the next is the same for all α-carbons. This means that the dihedral angles Φ and Ψ are the same for each residue along the helix. The helix can be defined by three parameters: n, the number of residues per helical turn; d, the rise per helical residue; and p, the rise per full turn (the last of these is also called the pitch—see Figure 3.2). By convention, a positive value of n denotes a right-handed helix, while a negative value corresponds to a left-handed helix. Historically, helices were also designated by the number of residues per helical turn and the number of atoms in one hydrogen-bonded ring. This nomenclature persists only for the 3.10 helix (three residues per helical turn and ten atoms in one hydrogen-bonded ring) since it can lead to ambiguities (e.g., 3.6(13) and 3.7(13) are both α-helices). Helices are the most abundant form of secondary structure, accounting for approximately 30%–40% of the residues in globular proteins. There are basically three types of helices that have been observed to exist in proteins: the α-helix, the 3.10 helix, and the π-helix. The γ-helix, defined as the 3.6(14) helix, was also predicted by Pauling and Corey, but it has never been observed.

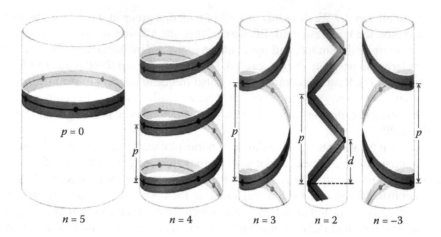

FIGURE 3.2 Characteristic parameters for different types of helices: n, the number of residues per helical turn; d, the rise per helical residue; and p, the rise per full turn. Positive value of n denotes a right-handed helix. (Illustration: Irving Geis/Geis Archives Trust, copyright Howard Hughes Medical Institute. Reproduced with permission.)

3.2.1.1 The α-Helix

The right-handed α-helix (Figure 3.3) is the most common helical conformation found in globular proteins, accounting for more than 30% of all residues. The dihedral angles Φ and Ψ for the geometrically pure α-helix are −57.8° and −47.0°; however, the average values in the real α-helices calculated from the large number of resolved protein structures seems to be $-64 \pm 7°$ and $-41 \pm 7°$, respectively. However, no matter how big the distortions from the ideal values are, it appears that the sum $\Phi + \Psi$ tends to be close to −105°, and the rotation angle per residue (Ω) is in accordance with Equation 3.2:

$$3\cos\Omega = 1 - 4\cos^2\left(\frac{\Phi + \Psi}{2}\right)$$ (3.2)

The fact that the angles Φ and Ψ of the α-helix lie in the center of the allowed minimum-energy region of the Ramachandran plot is an important contribution to the stability of the helix. The ideal α-helix would be characterized by 3.6 amino acid residues per turn (n), 0.15 nm of rise per helical residue (d), and a rise of 0.54 nm per full turn (p). The average length of an α-helix is between 10 and 11 residues or three turns. The α-helix is stabilized by hydrogen bonds connecting each backbone carbonyl oxygen of the residue i with the backbone imino group residue $i + 4$, that is, four residues away in the direction of the C-terminus. In the ideal α-helix, the hydrogen bonds are in near-perfect alignment, parallel to the helix axis, and in the real helices, they are only slightly distorted. The dipole moments of the peptide bonds (see Chapter 1) are also aligned nearly parallel to the helix axis, rendering the α-helix together with the polar hydrogen bonds to the net dipole with a partially negative charge of 0.5 to 0.7 units at the C-terminus, and a partially positive charge of the

Side view   **Top view**

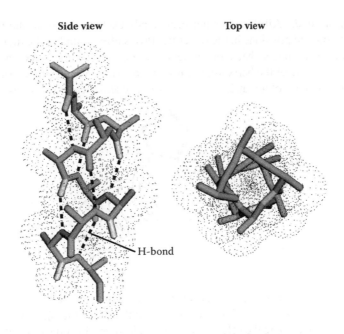

H-bond

FIGURE 3.3 The polypeptide chain of 10 Ala folded in the right-handed α-helix. Hydrogen atoms are omitted for clarity. Hydrogen bonds are indicated as the dashed lines. Dots indicate the space-filling spheres corresponding to the van der Waals radii of the atoms. (Generated by the Pymol computer program using coordinates available in PDB.)

same size at the N-terminus. These partial charges are frequently compensated for by positively charged basic amino acids near the C-terminus of the helix, and negatively charged acidic amino acids at the N-terminus. In the α-helix, the radius of the helix (0.27 nm) allows for favorable van der Waals interactions across the helical axis. This makes the helix very compact and contributes to its stability.

Representations of the α-helix using ball-and-stick or wireframe models can lead to a false impression of the hole inside the helix structure into which water can enter. Use of space-filling models reveals the compactness of the helix core and the close Van der Waals proximity of the backbone atoms (Figure 3.3). The side chains point out of the helical core into the solution, and they are slightly inclined toward the N-terminus. Such staggering minimizes steric interference. However, bulky amino-acid radicals in close proximity might still be in a collision, and not all conformations defined by the dihedral angles χ_1, χ_2, etc., are allowed. This and some other reasons lead to different propensities of amino acids for forming the α-helix.

The amino acids from the group called MALEK (from their one-letter codes), including Met, Ala, Leu, Glu, and Lys, are very frequently found in the α-helix. On the other hand, Tyr and Ser, and especially Gly and Pro (in general), do not tend to be in the α-helix. To keep the small and conformationally very flexible Gly in the constrained α-helix is ectopically unfavorable. However, just the opposite is true for Pro. Its flexibility is limited by the ring structure, resulting in a rather unfavorable value of the dihedral angle Φ (around −70°), and it has no amide hydrogen to participate in the hydrogen bonds. All this does not necessarily exclude Pro from the α-helix but assigns it to special positions in the structure. Pro is often found in the first turn of the α-helices, as its helix-like backbone dihedral angles can get the helix started. When Pro is in the middle of the helix, dihedral-angle restrictions and ring constraints lead to a kink in the helix of about 25° from linearity, schematically shown as follows:

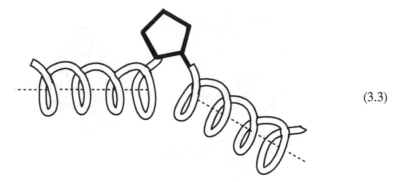

$$(3.3)$$

If, however, the proceeding residue is Gly, there is no such effect. Pro-induced kinked helices are almost exclusively long helices (more than four turns), which might be a requirement to overcome the effect of the kink. However, in some cases it was found that the kink in the helix also persists when Pro is replaced with some other amino acid. This helix curvature is another common deviation found in globular proteins. The majority of α-helices contain some curvature of the helical axis with a radius of around 6 nm. The curvature is very often observed in the amphipathic α-helices and is centered on the hydrophobic side of the helix, resulting in slightly different Φ and Ψ angles for the hydrophobic and hydrophilic faces. The bending of the helix arises from the additional hydrogen bonds formed between the backbone carbonyl oxygen and the water in the solution, which cannot possibly occur on the helix side facing the protein interior. The curvature is not expected to be energetically expensive, and has been estimated to involve less than 10 kJ mol^{-1} in a five-turn helix.

Many helices found in globular proteins are amphipathic, mostly those that are located parallel to the surface of the globular proteins, making them prime candidates for serving the function of a barrier between inside and out. As such, they should necessarily be amphipathic with the hydrophobic face to satisfy the protein core and the hydrophilic face to interact favorably with the solvent. A simple way to analyze the occurrence of an amphiphatic helix is to plot the peptide sequence as a helical wheel where the angle between each amino-acid residue is 100° in three-dimensional space, giving a value of around 75° in a two-dimensional projection (Figure 3.4). Some naturally occurring peptides also have a sequence that can be organized into the amphipathic α-helix (see Chapter 27). Peptides containing segments of potentially amphipathic structures are largely unstructured when in solution but will spontaneously adopt the helical conformation if provided with an amphipathic environment. These peptides have been shown to undergo highly cooperative monomer–oligomer equilibrium transitions for micromolar concentrations in an aqueous solution. At low concentrations, they are typical random conformers, which become helical as the concentration is increased and oligomers are formed.

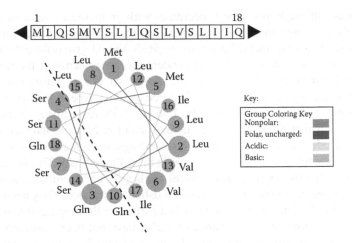

FIGURE 3.4 Helical wheel presentation of an amphipatic peptide. It was generated by a Java Applet written by Edward K. O'Neil and Charles M. Grisham (University of Virginia in Charlottesville, Virginia). The applet is accessible at http://cti.itc.Virginia.EDU/~cmg/Demo/wheel/wheelApp.html, and can be used by the visitors to generate a helical wheel of a custom sequence. The figure is a snapshot from this home page and is originally in color. Different colors indicate differences in the hydrophilic/hydrophobic nature of the amino-acid residues. In this gray scale presentation, hydrophilic residues on the left side are separate from the hydrophobic residues on the right side by the dashed line.

The stability of the α-helix is very much influenced by its surroundings. Isolated α-helices appear not to be very stable in water solutions. The nucleation of the helix is the most difficult and entropically costly, while adding additional amino-acid residues to the helix is much less energetically demanding. The carbonyl oxygen of the residue already in the α-helical conformation is reasonably close to the amido group hydrogen of the approaching residue, and the effective concentration for hydrogen-bond formation is thus increased (see Chapter 2). The growth of the helix is supported by the positive cooperativity of the additional hydrogen-bonds formation, assisted by the favorable dipole-moment alignment and, frequently, also by the favorable interactions between the adjacent amino-acid side chains. The cooperativity of the isolated α-helix formation is reflected in the equilibrium constant K for the transition of the random coil (C) into the α-helix (H_n):

$$K = \frac{[H_n]}{[C]} = \sigma \cdot s^n \tag{3.4}$$

where n is the number of residues in the helix, σ is the nucleation factor with a value between 2×10^{-3} and 3.5×10^{-3}, and s is the ratio of the rate constants for adding and removing the residue, respectively. As calculated for some model systems, K increases with an increased value of n, but n must be very large to obtain a value of K above unity. These calculations also revealed the dynamics of the isolated α-helix folding and unfolding, showing that this transition can occur around a million times per second—although this number decreases with an increased value of n. Most of the calculations were undertaken for model peptides composed of the same type of amino acids. It is clear that different amino acids should give different values for s, which in most cases should also be dependent on the sequence of the nearby-lying amino acids. However, not all these values are known. For an illustration, the values of s for Ala and Gly that are the least influenced by the polypeptide sequence because of their small size were estimated to be 1.56 and 0.15, respectively. It follows from this discussion that the α-helices are mainly stabilized by the interaction of their amino-acid side chains with the helix surroundings. In proteins the stabilization is provided by the tertiary structure discussed in the following text. This is supported by observations that the short natural peptides are largely unstructured in solution even when composed of amino acids that favor an α-helix formation. They need additional stabilization to fold in the helix, as described earlier for the amphipathic peptides.

Extended regions of the left-handed α-helix have not been observed in natural proteins composed almost exclusively of L-amino acids. However, a survey of 7284 proteins from the Protein Data bank (PDB) revealed 31 mostly very short (four residues) left-handed helices, of which more than 80% were shown to be important for protein stability and function. These helices were found in several enzymes where they participated in the active-site architecture and catalytic processes, and in other proteins where they contributed to ligand binding. Besides this, occasionally individual residues at the end of a three- or four-residue β-turn can adopt a conformation corresponding to the left-handed helix. These residues are usually Gly but can also be Asn or Asp, where the side chain forms a hydrogen bond with the main chain and, as a result, stabilizes this otherwise unfavorable conformation. In the synthetic all-D-peptides and proteins, the left-handed α-helix is normally found, as predicted by the studies of Ramachandran and others. In naturally occurring peptides containing both L- and D-amino acids, for example, in the antimicrobial peptide tolaasin, an extended left-handed amphipathic α-helix was found.

3.2.1.2   The Other Helices

Since the α-helix is so abundant, all other types of helical structures seem to be of marginal importance. Two types of helices will be discussed here: the 3.10 helix and the π-helix.

The 3.10 helix (sometimes referred as the 3_{10} helix) occurs in proteins with a frequency of approximately 3%. Nearly 96% of these contain four residues or less. Hydrogen bonds within a 3.10 helix also display a repeating pattern in which the backbone carbonyl of the residue i hydrogen bonds to the backbone imino group of the residue $i + 3$. In the ideal 3.10 helix c, the values of the angles Φ and Ψ are −74.0° and −4.0°, but real helices show average values of −71.0° and −18.0°, respectively. The result of this distortion is a larger radius (0.20 nm versus 0.19 nm) and a larger number of residues per helical turn (3.2 versus 3.0). Alpha helices sometimes begin or end with a single turn of a 3.10 helix. There are several reasons why the helices different from α-helix are not observed more often: (i) the values of Φ and Ψ lie at the edge of an allowed minimum-energy region in the Ramachandran plot; (ii) the dipoles of the hydrogen-bonding backbone atoms and peptide bonds are not aligned;

(iii) van der Waals repulsion can occur between the backbone atoms (because being thinner than the α-helix); and (iv) the amino-acid side chains might exhibit steric interference.

The π-helix is an extremely rare secondary structural element in proteins. It is larger than the α-helix because of the hydrogen bonds between the residues i and $i + 5$. As in the case of the 3.10 helix, one turn of the π-helix is sometimes found at the ends of regular α-helices. The Φ and Ψ values of −57.1° and −69.7°, respectively, lie almost in a forbidden minimum-energy region of the Ramachandran plot. The larger radius of the helix does not allow any van der Waals contact between the atoms of the polypeptide backbone across the helical axis, forming an axial hole that is too small for water molecules to fill. All this accounts for the low stability and thus the infrequent occurrence of this helix.

3.2.2 THE β-STRUCTURE

In contrast to the helices, polypeptide chains in the β-structure are aligned in parallel ribbons of the almost fully extended backbones, which are laterally connected by the hydrogen bonds between the peptide-bonded carbonyl group from one chain to the peptide-bonded imino group from the adjacent chain, in this way forming a folded surface called the β-pleated sheet. The folding of the pleated sheet is perpendicular to the direction of the backbones and is a consequence of the tetrahedral configuration of the carbon atom. Each polypeptide chain in the β-pleated sheet is named a β-strand and usually consists of 5 to 10 amino-acid residues. The orientation of the amino-acid side chains in each β-strand is alternating: if the given side chain is pointing above the plane of the β-pleated sheet, the neighboring two side chains in the strand point below the plane, and vice versa. The orientation of the β-strands from the N-terminus to the C-terminus give rise to three types of β-structure: antiparallel, parallel, and mixed (Figure 3.5). In all these types, the backbone in the β-strands is not fully extended, as revealed by the distance between the neighboring α-carbon atoms and the values of Φ and Ψ being approximately 0.30 nm, −135°, and 135°, respectively (in the fully extended conformation, these values are 0.38 nm, −180°, and 180°, respectively). The Φ and Ψ values vary a little in the different types of β-structure, but in all cases they are energetically optimal and situated in the middle of the allowed area in the Ramachandran plot (see Figure 3.1). In an antiparallel β-structure, the Φ and Ψ values are −139° and +135°, respectively. The antiparallel β-sheets are thought to be intrinsically more stable than the parallel sheets because of the more optimal linear orientation of the interstrand hydrogen bonds and because the peptide-bond dipoles of the nearest neighbors within a strand cancel each other out, whereas in the parallel sheet, the components of the dipoles parallel to the strands align and may interact unfavorably.

In the slightly less stable parallel β-structure, the atoms participating in hydrogen bonds are not aligned and are not perpendicular to the individual strands. Because of this a component of approximately one-third of the peptide dipole is oriented parallel to the β-strand, which sums up to an overall dipole with an effective charge of approximately +0.06 unit elemental charge at the N-terminus and a −0.06 charge at the C-terminus. The Φ and Ψ values in the parallel β-structure are

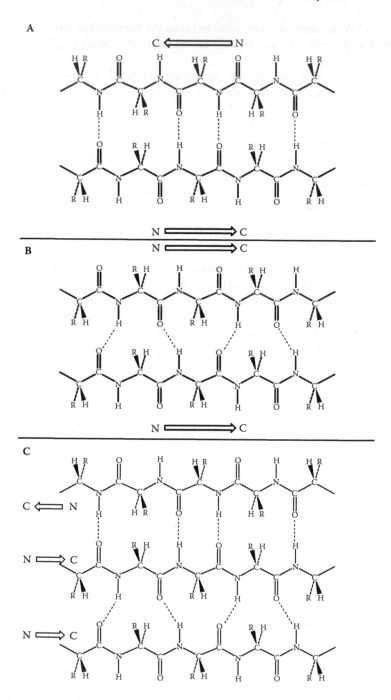

FIGURE 3.5   Schematic presentation of β-structure: (A) antiparallel; (B) parallel; and (C) mixed type.

−120° and +115°, respectively. Parallel β-sheets are less common, while antiparallel sheets are very often found. The frequency of the mixed β-structure, composed of antiparallel and parallel strands, lies between the pure types. The number of strands making up the β-structure is very variable, from 2 to over 15. Parallel pleated sheets of less than five strands are rare. This might be the result of its somewhat lower stability. The tendency of amino acids to build a β-structure is not as strong as in the case of the α-helix. The bulky amino acids, Trp, Tyr, and Phe, and the branched amino acids, Thr, Val, and Ile, that show some restrictions for building the α-helix, are readily found in the β-structure as well as the Pro, which is common in the edge strands. It has been suggested that this position of the Pro is related to the prevention of the aggregation of proteins via β-strand association, as in an amyloid formation.

The originally proposed, classical β-sheets are planar, but most sheets observed in globular proteins are twisted (0° to 30° per residue). In addition, antiparallel β-sheets are more often twisted than parallel sheets, and two-stranded β-structures show the largest twists. At least two separate explanations have been offered as reasons for the observed twist. One, based on statistical reasoning, suggests that the greater amount of allowed Φ and Ψ space to the right of the classical β-structure region favors a right-handed twist. Another possible explanation lies in the distortion of the tetrahedral nitrogen, seen in the x-ray crystal structures, leading to a right-handed twist. Some of the larger twisted β-sheets are observed in the interior of the proteins. The mutual interactions of carbonyl and imino groups via hydrogen bonds decrease the ability of these polar groups to interact with water, favoring their shift into the hydrophobic protein core. Another notable distortion of the β-structure is the β-bulge, found mostly in the antiparallel β-sheets (only about 5% of bulges occur in parallel strands). The β-bulge arises from the hydrogen bonding of the imino groups of the two consecutive residues from one strand with the single carbonyl oxygen of a residue from the other strand. As it seems β-bulges stem from the insertion of an amino acid into the sequence as a consequence of the mutation. The bulge allows the accommodation of the additional residue without any disruption to the secondary structure. In immunoglobulins the bulges are conserved and function by helping in the dimerization of the immunoglobulin domains. In some enzymes, for example, in superoxide dismutase, they support the formation and function of the active site. In some cases a β-bulge was found near the β-hairpin loops joining two adjacent antiparallel β-strands.

3.2.3 THE TURNS

Turns are the third of the three classical secondary structures with approximately one-third of all residues in globular proteins being contained in turns that serve to change the direction of the polypeptide chain, usually by connecting adjacent antiparallel β-strands. The diameter of the average globular protein domain is around 2.5 nm. Since this distance can be spanned by an extended polypeptide chain composed of only seven amino-acid residues, the turns seem to be a necessary element, providing a compact globular structure of the proteins. Turns are located primarily on the protein surface and, accordingly, contain polar and charged residues. Antibody recognition, phosphorylation, glycosylation, hydroxylation, and intron/exon splicing are found frequently at or adjacent to turns. However, it is not clear if this is due to specific recognition or the surface location of turn conformations. Turns were originally identified in model-building studies by Venkatachalam in 1968, who proposed three distinct conformations (designated I, II, and III) based on the Φ and Ψ values and their related mirror turns (I', II', and III'), which have the signs of the Φ and Ψ values reversed. In each of these turns a hydrogen bond was proposed and later found between the carbonyl oxygen of residue i and the amide nitrogen of $i + 3$. Additional turns disclosed later introduced some confusion to the turn nomenclature. Today, turns are usually classified on the basis of their hydrogen-bonding pattern into α-, β-, γ-, δ-, and π-turns. The common turns are shown in Figure 3.6. Theoretically, the smallest turn is the δ-turn, containing only two adjacent residues connected by a hydrogen bond. Because it occurs only rarely, this turn is not included in all classification schemes.

The most compact, commonly approved turn is a three-residues-containing γ-turn, with the hydrogen bond between the residues i and $i + 2$. It can exist in two forms: the classical form with Φ and Ψ values for the central residue of 75° and −65°, respectively, and the mirror image or inverse form with Φ and Ψ values of −75° and 65°, respectively (Figure 3.6). The first and the last residues in a γ-turn usually represent the terminal residues of the two consecutive antiparallel β-strands. The middle residue, which is constrained due to the compactness of the turn, is frequently, the most flexible amino acid conformationally—Gly. As well as existing in some proteins, the γ-turn was proposed to exist in some specialized peptides, for example, in the antifreeze peptides and glycopeptides. Four residue turns are known as the β-turns. The hydrogen bonding is between residues i and $i + 3$. At least eight normal and several inverse types of the β-turn exist, including all the types proposed by Venkatachalam. The structures of the type I, I', II, and II' β-turns are shown in Figure 3.6. The most frequently occurring β-turn is the type I. The inverse types are rare, but type I' was found in β-hairpins. There exists a position-dependent amino-acid preference for the residues in the β-turn conformations. Type I can tolerate all the residues in the positions i to $i + 3$, with the exception of Pro at the position $i + 2$. Pro is favored at the position $i + 1$ and Gly is favored at $i + 3$ in the type-I and type-II β-turns. The polar side chains of Asn, Asp, Ser, and Cys often populate position I, where they can hydrogen bond to the backbone NH of the residue $i + 2$. Ideally, the type-I' turns have Gly at the positions $i + 1$, and $i + 2$ and the type-II' turns have Gly at the position $i + 1$, as the presence of a C_β atom would cause a steric interference with the peptide carbonyl oxygen. The α-turn is larger, containing five residues with the hydrogen bond between the i and $i + 4$ residues. This resembles the α-helical pattern of the hydrogen bonding and hence the name. Although considered to be rare, Pavone and his coworkers found over 350 α-turns in a study that comprises 193 protein structures. They classified α-turns into nine different types on the basis of the dihedral angles of the inner three residues.

Besides the number of residues and the position of the hydrogen bond, another common feature of all types of α-turns is the distance of less than 0.7 nm between the C_α atoms of the i and $i + 4$ residue. Because the α-turns are usually exposed to the solvent, they are composed of mainly hydrophilic amino acids. Sometimes they

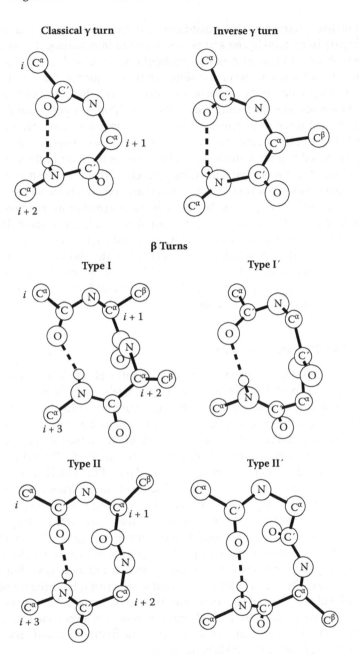

FIGURE 3.6 The structures of turns, showing are the classical and the inverse γ-turn as well as β-turns of the type I, I′, II, and II′. The hydrogen bonds between *i* and *i* + 2 amino-acid residues in the γ-turns, and the hydrogen bonds between *i* and *i* + 3 amino-acid residues in the β-turns, are indicated by a dashed line. (From Rose, G. D., et al. 1985. Turns in peptides and proteins. Adv. Protein Chem. 37, 1–109. With permission.)

form hook-like protrusions that extend outward from the protein and are thought to participate in protein–ligand and protein–protein interactions. The α-turns have also been observed in small peptides, particularly those with a circular structure, containing between seven and nine residues in their sequence, for example, in amatoxins. The largest turn is the π-turn of six residues, with the hydrogen bond between the i and $i + 4$ residues. In a more recent study in 2000, Dasgupta and Chakrabarti reported the existence of different π-turn types, some with more than one hydrogen bond. Some of the π-turns encompass multiple β- and α-turns. Although it was initially proposed that the π-turns should be found at the ends of the α-helices, the investigations of Dasgupta and Chakrabarti revealed that they are mainly included in the hairpin structures between β-strands and are sometimes located in the cavities of the enzyme-active sites, where they participate in ligand or ion binding and in the catalytic process. In the broader sense, we can also include the loops in the group of turns. However, they are longer than the turns, consisting of up to 30 residues. The best known is the Ω-loop, which is characterized by the short distance between its beginning and end, named after its shape, and found frequently in many proteins. It does not have a defined composition and size, but many Ω-loops contain a substantial number of internal hydrogen bonds.

3.2.4   POLY-PRO AND POLY-GLY

As already discussed, Pro shows unique conformational restrictions when incorporated into the polypeptide chain and is found relatively rarely in helices and sheets, but it frequently occurs in turns. In addition, a pure poly-Pro chain can adopt two specific helical conformations called the poly-Pro helix I and the poly-Pro helix II. The left-handed poly-Pro helix I has all peptide bonds in the cis configuration and is characterized by values of the dihedral angles Φ and Ψ of −75° and 160°, respectively, and 3.3 residues per turn. The right-handed poly-Pro helix II is more extended. It has only three residues per turn, all peptide bonds in the trans configuration and values of the Φ and Ψ angles of −75° and 150°, respectively. Although no internal hydrogen bonds have been observed in any of these two helices, they are stable because of the rotational restrictions. Interconversion between both poly-Pro helices is possible, but it is slow because of the high activation energy for the isomerization of the cis peptide bond to the trans configuration and vice versa. Similar to the poly-Pro helices, Gly can also form two regular structures, but because of the high conformational flexibility of Gly, these are stable only in the solid state or, as calculated by computer simulations, in a solvent-free system in a vacuum. One conformation, called State I, is extended and is similar to the β-structure, and another, called State II, is helical with three residues per turn.

3.2.5   THE UNORDERED SECONDARY STRUCTURE

In some proteins a large portion of the protein structure is not in the distinct secondary structure discussed, and this structure is referred to as an unordered secondary structure, or a random coil. However, in the defined environment, the unordered structure of each particular protein is always the same. Each atom is in the precisely defined position; otherwise, the protein is not functional. As in the cases of other secondary structures, the residues in the unordered structure are kept in their positions by numerous interactions, including hydrogen bonds between the carbonyl and imino groups of the peptide bond. The only difference is that no repeatable regularity can be observed in the bonding patterns, and the structure is thus called unordered. According to this, the unordered structure has sometimes been more properly referred to as the randomly ordered structure. However, because of the obvious contradiction in this expression, it has not been adopted in general use.

As more and more protein structures have been revealed in detail, computer programs have been developed to use this data for predicting the secondary structure from the protein sequences. An approach to the prediction of protein structure is discussed in Chapter 19.

3.3 THE TERTIARY STRUCTURE

If the secondary structure is regarded as a more local organization of the polypeptide chain, the tertiary structure is a global, precisely defined three-dimensional organization of the polypeptide chain, in which every atom takes an exact position in space, characterized by the atomic coordinates. In mono-polypeptide chain proteins the tertiary structure is functional, and it is called the native state. The tertiary structure is largely maintained by the interactions between amino-acid side chains—a short review of these interactions is given in Chapter 2. Two features determine the tertiary structure: the primary structure and the environment. The only information required for the correct protein structure is encoded in the DNA as a sequence of amino acids. This information is sufficient for the protein to take the defined conformation, which is dependent on the environment because a given sequence may adopt different tertiary structures in different environments. The environment must be regarded in the broad sense, and it comprises not only the solution composition but also all the available enzyme systems for the posttranslational modifications, the transport systems for the transfer of a protein between different cell compartments in which modifications are accomplished, and the chaperones that help in acquiring the protein's native conformation, to list only the most important players involved. When forming the tertiary structure, the amino-acid residues are very tightly packed in the protein. The fraction of the total space occupied by the atoms according to their van der Waals radii is called the packing density and, in proteins, this is comparable to that found in crystals (0.74 for proteins, and 0.70 to 0.78 for crystals). The tertiary structure is assembled from the substructural elements. The smaller substructures are called motifs, while the larger substructures are called domains.

3.3.1 THE SUPERSECONDARY STRUCTURE—MOTIFS

The motif is a frequently encountered combination of the secondary structure found in different proteins. It is composed of a limited number of secondary-structure elements. The motifs are not very strictly defined, and a unique classification system for the motifs does not exist. Because of the relatively large number of different

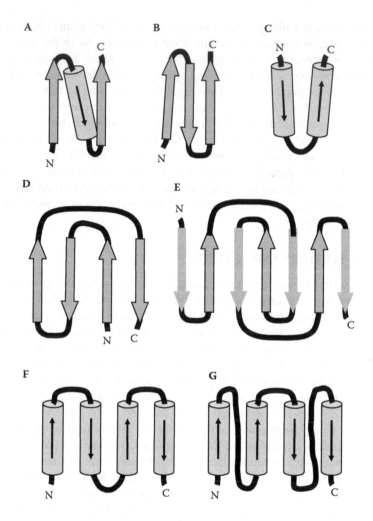

FIGURE 3.7 Common motifs of the supersecondary structure: (A) a β-α-β motif; (B) a β-hairpin motif; (C) a helix-turn-helix motif (αα motif); (D) a Greek key motif; (E) the immunoglobulin fold; (F and G) two types of a four-helix bundle. α-helices are represented by cylinders and β-strands by arrows.

motifs, only some examples can be given here (Figure 3.7). Motifs can be of all-α, all-β, or of mixed type. Examples of the all-α types are helix-turn-helix, helix-loop-helix, and the four-helix bundle. The helix-turn-helix is composed of two antiparallel α-helices connected by a turn. It is a common motif found in many DNA-binding proteins since it fits nicely into the DNA major groove. In the helix-loop-helix motif, two α-helices are joined by a loop. In some proteins, a loop consisting of about 12 amino acids can accommodate a metal ion. A special case is the basic helix-loop-helix motif found in transcription-factors. Here, one helix is composed of basic residues, allowing for the DNA binding. A flexible loop is important in dimerization transcription-factor domains. The most abundant of the all-α motifs is the four-helix bundle, composed of four tightly joint α-helices packed in a coiled-coil arrangement with the hydrophobic core in the center.

The orientations of the helices are dependent on the protein, but the nearby helices are usually antiparallel. Bundles of many more than four α-helices are known. Helices in a coiled-coil arrangement interact via the heptad repeat of seven amino acids in which the first and the fourth residues are hydrophobic, and the others are hydrophilic, as in Leu-zipper (see Chapter 22). The simplest all-β motif seems to be the β-hairpin, consisting of two antiparallel β-strands connected by a turn. Other common all-β motifs are the Greek key, the immunoglobulin fold, and the β-barrel. The standard Greek key is composed of four antiparallel β-strands linked by three turns. Strands 1 and 2 are inside the structure, while strands 3 and 4 form the edges. There are different Greek key motifs possible, consisting of up to 13 strands. The immunoglobulin fold is an expanded Greek key motif composed of seven antiparallel β-strands and is characteristic of the variable and hypervariable immunoglobulin domain (see Chapter 25). Two four-stranded Greek key motifs build an eight-stranded β-barrel.

Barrels of many more antiparallel β-strands are known, for example, in porins, where the barrels consist of up to 18 β-strands. Sometimes this large motif is regarded as a domain. A typical mixed motif is the β-α-β motif composed of two parallel β-strands joined by an α-helix. Two conformations of this motif are possible, but the right-handed one is the most common (Figure 3.7A). Another known mixed motif is the α/β barrel or the TIM barrel. TIM stands for triosephosphate isomerase, where this motif was first observed. Basically, it consists of eight parallel β-strands connected by eight α-helices, but several variants exist. Strands consisting of hydrophobic residues are organized in the core of the barrel and are supported from outside by the amphipathic helices. A somewhat smaller variation of this arrangement is the Rossman fold, consisting of three parallel β-strands connected by two α-helices, which form a nucleotide-binding structure found in many proteins.

It is important to realize that by no means do all the helices and strands in proteins belong to supersecondary structures. For example, proteins of the globin family consist of eight α-helices in contact, but the helices do not pack against other helices, which are adjacent in the sequence, with the exception of the final two, which form an antiparallel helix-turn-helix motif.

3.3.2 THE DOMAINS

The term domain was coined by Wetlaufer in 1973. He defined domains as stable units of the protein structure that could fold autonomously. However, the definition of a domain is not strict and tends to overlap with the definition of motifs. Some authors see a domain as a functional unit that evolved, while others simply see it as compact structure units with more intensive and numerous interactions between the residues within it than with other parts of the protein. The size of a domain is variable, ranging from around 40 to around 700 residues, but most of the domains (around 90%) do not exceed 200 residues. Often, a domain is composed of a single stretch of polypeptide chains, and the contact between two domains is limited to only two or three short segments of the chain, which might also be quite flexible.

In multidomain proteins, domains are a part of the same polypeptide chain. Proteins' tertiary structures are usually classified by their domain structure into four main classes: all-α, all-β, $\alpha + \beta$, and α/β. All-α domains are usually small and built exclusively from α-helices in different arrangements. Examples of all-α proteins are myoglobin and hemoglobin, two members of the large protein family of globins. All-β domains have a core of an almost entirely β-structure, mostly in antiparallel arrangements. Topologically different all-β domains comprise β-sandwiches composed of two β-sheets packed in different ways (aligned, orthogonal, or mixed); β-barrels, β-propellers which consist of four to eight β-sheet "blades" (six in neuraminidase, for example); the β-helix, found typically in antifreeze proteins but also in some others, in which parallel β-strands are stacked one above the other in a helical arrangement with each strand oriented perpendicular to the long axis of the domain; and many other all-β domains. Mixed $\alpha + \beta$ and α/β domains consist of both α-helical and β-strand structures. In the $\alpha + \beta$ domains, the all-α and all-β motifs are mixed, but the β-α-β motif is lacking. This domain type is difficult to define precisely due to its overlapping with other three classes, and so it is sometimes not included in the classification systems. The α/β domains are characterized by combining the β-α-β motifs in layers and barrels.

Most eukaryotic proteins show a multidomain structure. Multidomain proteins, particularly those in which repeated domains exist, are often termed mosaic proteins. The domains seem to have evolved by a process of gene transformation, including gene duplication and fusion. Many domains could have existed in the past as independent proteins but have been subsequently altered and used as modules that have been combined in new multidomain proteins. The mechanisms of genetic shuffling were employed to move domains within and between species. The simplest multidomain organization is the tandem repetition of a single domain. In the large evolutionary analysis of the human genome in 2001, Li and collaborators identified proteins with up to nine different domains, and proteins with up to 130 identical domain repetitions. By a process of domain swapping, proteins can acquire additional properties and functions. No unique classification system of domains has been adopted by the scientific community yet. In a commonly used database called CATH (CATH is an acronym of the four levels of protein classification: class, architecture, topology, and homologous superfamily), over 900 protein domains are registered.

3.4   THE QUATERNARY STRUCTURE

Some proteins consist of a single polypeptide chain, such as myoglobin. The biologically functioning form of others is an aggregation of several folded polypeptide chains, termed monomers, protomers, or subunits into the oligomeric protein. For example, the quaternary structure of hemoglobin consists of four chains, two called α and two called β, each of which is similar to a myoglobin molecule. Other proteins exist that are an aggregation of one or more copies of different chains. Regarding the quaternary structure more broadly, it can also include viruses, which are composed of a large number of the same or different monomeric protein copies associated with the nucleic acid. The tobacco mosaic virus, for example, is composed of 2130 identical subunits of coat protein and a single genomic RNA molecule. Proteins are often organized symmetrically in the quaternary structure. This allows the formation of large complexes with only a few different subunits. In the quaternary structure, the subunits are in multiple contacts with each other via the same set of predominantly weak interactions that maintain the tertiary structure, and in some cases also via inter-subunit disulfide bridges. The residues in the contact interfaces are very tightly associated; the packing density is comparable to that of the interior of the subunits. The dissociation constants describing the breakup of the associated oligomeric structure into the individual subunits are very small, usually in the nanomolar to femtomolar range. Two types of interaction between identical monomers are possible: isologous and heterologous. The isologous association involves the same surfaces on both monomers, which associate to produce a dimer with a twofold rotation axis of symmetry. The two monomers are equivalent. The heterologous association involves two different sites that are complementary and largely nonoverlapping. In such a dimer, the two monomers are not equivalent, and each has a single unpaired binding site. The assembly of proteins into quaternary structures provides enhanced, multiple, or novel functional roles, often associated with the conformational change transmission. This aspect of the quaternary structure is described in several chapters, most notably in Chapters 21 and 23. Furthermore, proteins such proteasomes, chaperons, and large enzyme complexes exemplify the advantages of multimeric complexes that combine different functions. Finally, large assemblies play not only functional but also structural roles on the cellular level.

Peptides, often said to not contain any secondary structure, have also been found to be structured; it is believed that these structural elements of the peptides are responsible for their biological activity. However, in small peptides the structure is poorly stabilized and can be very flexible. Often, it is strongly stabilized only after binding to proteins (receptors, etc.) or when in an interaction with a lipid bilayer. Peptides do not usually exhibit the quaternary structure. A notable exception to this rule is insulin, consisting of two polypeptide chains that arise from the proteolytic cleavage of the proinsulin molecule (see Chapter 5).

3.5   THE DATABASES OF PROTEINS

The most prominent database for protein structures is the Protein Data Bank (PDB). It was founded in 1971 at The Brookhaven National Laboratory, United States, by Meyer and Hamilton, and is today managed by Rutgers University, New Jersey. The PDB is a repository for protein and nucleic acid structural data obtained mainly by x-ray diffraction or the nuclear magnetic resonance (NMR) method. Files in the PDB contain the coordinates of all the atoms in the structure, with the exception of the coordinates of the hydrogen atoms. By the end of 2008, about 55,000 structures were stored in the PDB, of which the vast majority were proteins. The released data can be freely accessed via the Internet using a number of mirror Internet sites. The data files can be downloaded and used for analyses and presentations by means of a suitable computer program such as Rasmol, Molscript, VMD, and many others. Attempts to introduce some order and establish a comprehensive classification of protein structures deposited in the PDB resulted in the formation of different protein classification databases. The best known are the SCOP, CATH, and FSSP databases, based on three different methods for protein classification: purely manual, combined manual and automated, and purely automated. The databases use different rules of classification, but in spite of this, about two-thirds of the proteins are arranged into common classes in all three databases. Much larger are the databases containing protein sequences, the most important being Swiss-Prot and the International Sequence Database Collaboration. More on all these and other similar bases can be found in Chapters 18 and 19.

3.6 DIVERGENT AND CONVERGENT EVOLUTION OF THE PROTEIN STRUCTURE

If the differences between two homologous proteins are examined, there is a general tendency for chemically similar amino-acid residues to be found at the same position. The substitution, for example, of one acidic residue (e.g., Glu for Asp) is likely to be of less consequence to the interactions with nearby residues than would the substitution of Glu for Val, a hydrophobic residue. Mutations to dissimilar residues are more likely to lead to the destabilization of the protein structure or even to the inability of the mutant polypeptide chain to ever fold. In such cases, the function of the protein is therefore impaired or disabled, which is likely to be a disadvantage for the organism. The result is that such mutations tend to be lost from the population. Consequently, homologous proteins have similar secondary, tertiary, and, if present, quaternary structures because the differences in the primary structure do not result in a drastic rearrangement of the folded conformation. However, some mutations are favorable, and they lead to the altered three-dimensional structure of the protein and to its more efficient functioning. This is a divergent evolution of protein structure. On the other hand, similar structural/functional designs appear to have evolved independently in some instances in proteins with a totally different primary structure. For example, the active sites of two families of functionally similar enzymes contain a specific arrangement of Ser, Asp, and His residues. These families are the trypsin family (mammalian serine proteases) and the subtilisins (bacterial serine proteases). Even though the two utilize a similar mechanism dependent on this structure, their primary structures exhibit no evidence of homology because the bacterial and mammalian lines diverged from a common ancestor very early in evolution, and this is seen as evidence for convergent evolution.

FURTHER READING

1. Cowell, S. M., Lee, Y. S., Cain, J. P., and Hruby, V. J. 2004. Exploring Ramachandran and Chi Space: conformationally constrained amino acids and peptides in the design of bioactive polypeptide ligands. Curr. Med. Chem. 11, 2785–2798.
2. Yu Zheng and Daiwen Yang 2005. STARS: statistics on inter-atomic distances and torsion angles in protein secondary structures. Bioinformatics 21, 2925–2926.
3. Bhargavi, G. R., Sheik, S. S., Velmurugan D., and Sekar K. 2003. Side-chain conformation angles of amino acids: effect of temperature factor cut-off. J. Struct. Biol. 143, 181–184.
4. Novotny, M. and Kleywegt, G. J. 2005. A survey of left-handed helices in protein structures. J. Mol. Biol. 347, 231–241.

5. Jourdan, F., et al. 2003. A left-handed alpha-helix containing both - and D-amino acids: the solution structure of the antimicrobial lipodepsipeptide tolaasin. Proteins 52, 534–543.
6. Golovin, A. and Henrick, K. 2008. MSD motif: exploring protein sites and motifs. BMC Bioinformatics 9, 312.
7. Pavone, V. et al. 1996. Discovering protein secondary structure: classification and description of isolated alpha-turns. Biopolymers 38, 705–721.

Protein Biosynthesis and Posttranslational Modifications

4 The Biosynthesis of Proteins

Matjaž Zorko

CONTENTS

4.1 ASSEMBLY OF THE PRIMARY STRUCTURE

The protein structure in DNA is coded by the genetic code, with the exception of some viruses that use RNA for this purpose. A very complex cell apparatus has evolved to translate the language of nucleotides in DNA into the language of amino acids in proteins. A multistep biosynthesis of the polypeptide chain starts with the transcription of DNA into RNA. The transcribed RNA, called the primary transcript, is then transformed in a series of molecular events into the mRNA, which is used as a template for the next step of the biosynthesis, termed translation. This step takes place in the ribosome and is assisted by a large number of accessory proteins, tRNAs, and energy-providing molecules. The three-step procedure of the protein biosynthesis is basically the same in all organisms. All the steps of the protein biosynthesis are strictly regulated in order to get the proper repertoire of functional proteins present in the cells and in the extracellular solutions. Recently, techniques have been developed to additionally include into the protein sequence some of the noncoded amino acids in order to design and produce proteins with novel functions. In addition to the proteins, the majority of physiologically important peptides are also synthesized via a transcription and translation process; however, some of the important peptides are assembled nonribosomally via a series of enzyme-catalyzed reactions.

4.1.1 THE GENETIC CODE

Each amino acid in DNA is encoded by a triplet of nucleotides called codons. DNA consists of nucleotides composed of phosphate; deoxyribose; and a nucleotide base adenine (A), cytosine (C), guanine (G), or thymine (T). In RNA, all the building blocks are the same; only the deoxyribose is replaced by ribose and thymine by uracil (U). Since the template for the polypeptide chain is actually mRNA, the genetic code is given in terms of RNA-consisting nucleotides. Although the total number of coded amino acids is 20, all 64 possible codons (combinations of three nucleotides out of four available give $4^3 = 64$ codons) are used—61 to code the amino acids, and three as STOP-codons. The genetic code presented in Table 4.1 is the standard genetic code or canonical code, which is regarded as being universal for all organisms. Variant codes do exist: an example is a code valid for the synthesis of proteins in mitochondria. As seen in Table 4.1, only two amino acids with a single codon exist, Met and Thr, while all the other amino acids are encoded by several codons. Variant codons for the same amino acid usually differ only in the last base in the triplet. This variation is called the degeneration of the genetic code. Exceptions to this rule also exist: Arg, Leu, and Ser are encoded by six codons, two of which differ in the first base. There is only one start codon, AUG, which at the same time also encodes Met, and three stop codons, UAG, UGA, and UAA. All the codons for a particular protein in DNA are organized in a gene.

Although the human genome is widely known (a large part of the work to disclose the human genome was done in the scope of the Human Genome Project), the number of genes is not. Estimates range from 20,000 to 30,000 protein-coding genes, stored in 46 chromosomes. In humans, the part of the DNA that codes for the proteins represents less than 1.5%, the rest is the gene-regulatory sequences, which are only partly known, the sequences coding transfer RNAs (tRNA) and ribosomal RNAs (rRNA), also called tRNA and rRNA genes, and about 97% of the genome,

TABLE 4.1
The Genetic Code

Amino Acid	Codons	Amino Acid	Codons
Ala	GCU, GCC, GCA, GCG	Leu	UUA, UUG, CUU, CUC, CUA, CUG
Arg	CGU, CGC, CGA, CGG, AGA, AGG	Lys	AAA, AAG
Asn	AAU, AAC	Met	AUG
Asp	GAU, GAC	Phe	UUU, UUC
Cys	UGU, UGC	Pro	CCU, CCC, CCA, CCG
Gln	CAA, CAG	Ser	UCU, UCC, UCA, UCG, AGU, AGC
Glu	GAA, GAG	Thr	ACU, ACC, ACA, ACG
Gly	GGU, GGC, GGA, GGG	Trp	UGG
His	CAU, CAC	Tyr	UAU, UAC
Ile	AUU, AUC, AUA	Val	GUU, GUC, GUA, GUG
START	AUG	STOP	UAG, UGA, UAA

which does not have a well-understood function. Contrary to earlier beliefs, the non-coded part of DNA is largely transcribed to RNAs with special functions that are, at present, understood only to a very limited extent.

4.1.2   TRANSCRIPTION OF DNA TO MRNA

The transcription of DNA to RNA is a process in which the sequence of a DNA segment called a transcription unit is used as a template to synthesize a complementary sequence of RNA. Each transcription unit is preceded by a regulatory region called a promoter. In most cases—the exception is the DNA in some viruses—the RNA is transcribed from a double-stranded DNA, but only one strand is used as a template. As a result, it is called the template strand (the other strand is called the coded strand due to the identity to the synthesized RNA with T in the place of U). The transcription unit also contains the regulatory sequences upstream and downstream of the protein-coded sequence. The transcripted RNA is synthesized in the $5' \to 3'$ direction from the corresponding ribonucleoside-5'-triphosphates as the substrates. The process is catalyzed by RNA polymerase and is usually divided into three main stages: initiation, elongation, and termination. Note that more detailed divisions of the transcription recognize some additional interstages. Prokaryotes have only one type of RNA-polymerase, which is used for the synthesis of all types of RNA: both protein-coded and non-protein-coded. In eukaryotes, three distinct types of the nuclear RNA-polymerase exist, designated by the Roman numerals I, II, and III, with the additional type IV present only in plants. Only RNA-polymerase II is used for the transcription of protein-coded genes. All these RNA-polymerases (and some others found in mitochondria and chloroplasts) use DNA as a template and are thus classified as DNA-dependent RNA-polymerases. Besides these, RNA-dependent RNA-polymerases exist, which are involved in the synthesis of small interference RNA (siRNA). RNA-polymerases are large multi-subunit enzymes. The prokaryotic enzyme of around 400,000 Da consists of six subunits: two α, β, β', ω, and transiently bound σ. The eukaryotic RNA-polymerase II is much larger (around 550,000 Da) and composed of 12 subunits. The initiation step of the transcription starts with the binding of the RNA-polymerase to the promoter region of the DNA sequence. In prokaryotes this region is usually recognized by a subunit σ. However, after the binding of the complete RNA-polymerase to the DNA and before the elongation phase starts, the subunit σ dissociates from the RNA-polymerase, which then switches from the so-called closed complex to the open complex. This conformational change results in the local dissociation of the two DNA strands in the range of approximately 13 nucleotides.

In eukaryotes the initiation starts with the binding of the TATA-binding protein (TBP) to its DNA recognition region in the promoter called the TATA-box. TBP binds the DNA inside the saddle-like curved β-sheet region so that fits to the minor groove of the DNA in the TATA-box. This initiates the formation of the transcription complex composed of the TATA-binding protein, RNA-polymerase II, and a number of transcription factors (see Section 4.2 and Chapter 22), the whole complex being attached to the promoter. The binding of one of the transcription factors (TFHII) induces a shift from the closed to the open complex and the unwinding of the short DNA region. In both prokaryotes and eukaryotes, ribonucleotides are then added to the open-template DNA strand according to the complementarity rule. For any further action, the RNA-polymerase requires, in addition to the template DNA, four ribonucleoside-5'-triphosphates as the substrates, Mg^{2+} and Zn^{2+}. The elongation step in prokaryotes is started by the dissociation of the subunit σ and in eukaryotes by the phosphorylation of the carboxy-terminal domain of the RNA-polymerase II as well as by the release of some of the transcription factors and the binding of the elongation factors. In the elongation step, the ribonucleotide-5'-phosphates are added sequentially and are joined together by the phosphodiester bonds formed via the catalytic action of the RNA-polymerase, which moves along the template DNA in the $3' \rightarrow 5'$ direction. This step proceeds until the transcription is terminated.

In prokaryotes the specific termination DNA sequences have been identified. One consists of a series of 4 to 10 A·T base pairs with the adenines on the template strand. The other codes are for the G·C paired palindrome that forms a hairpin structure in the transcribed RNA. Some, but not all, transcription processes in prokaryotes need a helicase protein ρ (rho) for the transcription termination. In eukaryotes the termination sites and the termination mechanisms are not well understood. In the transcription of RNA from DNA, many RNA-polymerases work successively in tandem in order to produce a large number of RNA copies. The rate of the transcription determined in *E. coli* is between 20 and 50 nucleotides per second. In contrast to DNA-polymerases, RNA-polymerases cannot repair any errors that occur during the transcription. It is estimated that the wrong base is incorporated once for approximately every 10,000 transcribed bases, meaning that the frequency of the errors is a thousand times higher than in DNA replication. These errors are not fatal since many copies of the RNA are usually produced, first, because the degenerated genetic code provides several related codes for most of the amino acids; second, because a substantial part of the RNA sequence is usually excised during the maturation; and third, because the replacement of one residue for another may have a limited impact on the protein structure and function, depending to a large extent on the site of the replacement and the nature of the replaced residue.

In prokaryotes, most of the transcribed RNA molecules can be immediately used in translation without any further modifications. In eukaryotes the primary transcript is also called the precursor mRNA or the heterogeneous nuclear RNA, and it is subjected to a multistep maturation process comprising the addition of the cap and poly(A) tail, splicing, and sometimes also the editing of the sequence. At the 5' end of the primary transcript, a 7-methylguanine nucleotide is enzymatically readily added during the transcription. This structure is called the 5'-cap and serves to protect the 5'-end against the 5'-exonucleases, but it is also important for the proper interaction of mRNA with the ribosome. An unusual 5'-5' connection between the cap and the first nucleotide in RNA is achieved by a three-phosphate bridge. At the 3'-end a poly(A) tail is added, consisting of 80 to several hundred adenine nucleotides. First, the primary transcript is cleaved approximately 20 nucleotides downstream of the highly conserved sequence AAUAAA, and then the adenine nucleotides are added by the stepwise catalytic action of the polyadenylate polymerase using ATP as the substrate. The poly(A) tail is not directly required for the translation process, but its length is important for the efficiency of the translation. In addition, the poly(A) tail

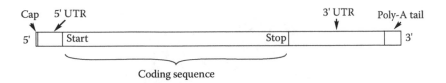

FIGURE 4.1   The structure of a typical human protein-coding mRNA. 5'UTR and 3'UTR are the untranslated regions on both sides of the coding region.

is essential as protection against mRNA degradation and for the transport of mRNA out of the nucleus.

During splicing, parts of the primary transcript sequence are removed, and the parts that remain can be recombined. The precursor mRNA consists of protein-coding regions called exons and protein-noncoding regions called introns. Introns can be classified into four groups and are removed by either of two main mechanisms: the autocatalytic action of the RNA itself (introns of groups I and II) or by means of a spliceosome, with the structure consisting of five specialized RNA and protein complexes (snRNPs). While prokaryotic mRNA is monocistronic (codes only for one protein), eukaryotic mRNA is polycistronic, meaning that it carries information that can be translated to several proteins. This is achieved by the alternative splicing of a precursor mRNA, a process in which the number of introns that are excised is varied and/or the remaining exons are combined in different ways to produce templates for a number of different proteins. The alternative splicing of the Drosophila gene DSCAM, for example, can result in over 38,000 splice variants. This alternative splicing is believed to be the main mechanism by which several hundred thousand different proteins can be synthesized from less than 30,000 genes in the human genome.

Finally, the mRNA sequence is sometimes modified, mainly by the enzymatically catalyzed alteration of bases, although insertion and deletion in the sequence have also been observed. An example of editing is the deamination of cytidine by cytidine deaminase, producing uridine. One known consequence of such editing is the change of the codon CAA into the stop codon UAA, which results in the production of a shortened protein, for example, in human apolipoprotein B. All these modifications yield a mature mRNA (Figure 4.1) that is transported out of the nucleus and is used for the translation.

4.1.3   TRANSLATION OF MRNA TO THE POLYPEPTIDE CHAIN

In translation, a mature mRNA molecule is used as a template for synthesizing a new protein. Besides mRNA, very many additional players are introduced into this process, including ribosomes; activated transfer RNAs (tRNAs); and initiation, elongation, and termination factors or releasing factors, GTP and Mg^{2+}. Ribosomes are the supramolecular machines in which the synthesis is carried out. They were first observed in the 1950s by G. E. Palade, a Nobel Prize winner in 1974. Ribosomes are composed of two subunits: one large and one small. Although not equal in size, prokaryotic and eukaryotic ribosome subunits are structurally very similar, composed of the ribosomal RNA (rRNA) and proteins. Prokaryotes have 70 S ribosomes

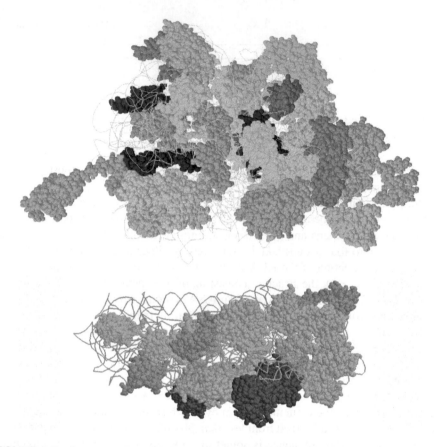

FIGURE 4.2 Structure of the 50 S (top figure: PDB id 2jl6) and 30 S ribosome subunits (bottom figure: PDB id 2vqf). The backbone of RNA is shown as a wireframe accommodating proteins shown in the space-fill presentation. Generated by Rasmol computer program.

with subunits of 50 S and 30 S. Here, S denotes a Svedberg unit, a unit of sedimentation coefficient that is related to the particle size (see Chapter 9). The larger subunit consists of two rRNAs (5 S and 23 S) and 36 proteins (33 different), while the smaller subunit has only one rRNA (16 S) and 21 proteins (see Figure 4.2). Eukaryotic ribosomes are larger (80 S), with subunits of 60 S and 40 S, but these numbers are variable and species dependent. The large subunit is composed of 5 S, 5.8 S, 28 S rRNA, and around 50 proteins. The small subunit has one rRNA (18 S) and more than 30 proteins. A functional ribosome is obtained by joining both subunits with the incorporated mRNA molecule. In the active cleft, situated in the ribosome interior, no proteins can be found near the site where the peptide bond between the amino acids is formed, indicating that the catalyses of the bond formation is performed by rRNA.

The function of tRNA is to bind the proper amino acid to its 3′-end and to recognize the corresponding codon on the mRNA with its anticodon arm (Figure 4.3). In this way, tRNA acts as an adapter, essential for the translation of the coded information brought by the mRNA into the protein sequence. tRNAs are short, less than

FIGURE 4.3 Structure of tRNAPhe with indicated anticodon region and CCA-arm where amino acid (Phe) is attached. Generated by Pymol computer program.

100 nucleotides long, polyribonucleotide chains, usually represented by a clover-leaf shape, although in reality they have an L-shaped form (Figure 4.3). The anticodon of each tRNA is composed of three unpaired bases that are complementary to the codon it reads. Although there are only 61 different codons specifying 20 amino acids, 497 genes encoding the cytoplasmic tRNA molecules were identified in the human genome. When the tRNA binds to its complementary codon in an mRNA strand, the ribosome ligates its amino-acid cargo to the new polypeptide chain, which is synthesized from the amino terminus to the carboxyl terminus. During protein synthesis, the ribosome travels along the mRNA in the 5'→3' direction. The translation is shown schematically in Figure 4.4.

The complete biosynthesis of a protein is divided into five stages: (i) activation of the amino acids, (ii) initiation, (iii) elongation, (iv) termination, and (v) posttranslational modification and folding. The first phase, the activation of the amino acids, provides the binding of the amino acids to the corresponding tRNA. This ensures the incorporation of the amino acid into the position of the polypeptide defined by the sequence of the corresponding codons in the mRNA. Attaching the proper amino acid to the proper tRNA is the critical point with regard to the fidelity of protein synthesis. The attachment is catalyzed by the enzyme aminoacyl-tRNA synthetase. This phase is called the activation of the amino acid because the binding is coupled to the hydrolysis of the ATP, and the amino acid is charged at the expense of the energy provided by the ATP. The $\Delta G^{\circ\prime}$ for the hydrolysis of the bond between the amino acid and the tRNA is -29 kJ mol^{-1}. In *E. coli* and in most other organisms, 20 aminoacyl-tRNA synthetases were identified, one for each of the 20 amino acids. At least 32 different tRNAs are required to recognize the codons for all 20 amino

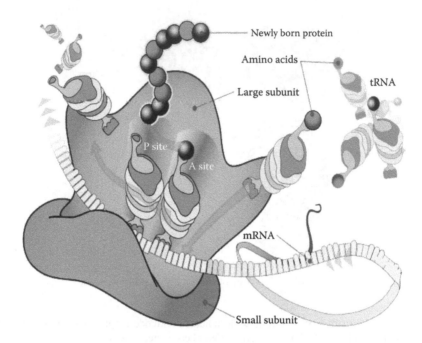

FIGURE 4.4 Schematic presentation of the translation. Freely accessible and released to public domain at http://en.wikipedia.org/wiki/Translation_(genetics).

acids; however, some cells do have more. To decrease the frequency of binding of the incorrect amino acid to the tRNA, some of the amino-tRNA synthetases have a double-function active site—one part is used to catalyze the binding of the amino acid to the tRNA, and the other is a proofreading device that can eliminate a wrongly bound amino acid.

The next step, initiation, is slightly different in prokaryotes and eukaryotes. In prokaryotes, first, two prokaryotic initiation factors, IF1 and IF3, bind to the small ribosome subunit; second, the mRNA is properly positioned in the small ribosomal subunit cleft by the binding of the Shine–Dalgano sequence in the mRNA to its complementary region in the 16 S rRNA. The start codon AUG is thus placed into the peptidyl site (P site) of the small ribosome subunit, while IF1 sits in the amino-acyl site (A site). In the next step, the first tRNA carrying N-formylMet is bound to the start codon in the P site. This tRNA is designated as fMet-tRNA^{fMet} and is the only tRNA that binds directly to the P site. Along with fMet-tRNA^{fMet}, the GTP-bound initiation factor 2 (IF2) is also associated to this complex. IF2 is a GTPase that subsequently releases GDP and, together with IF1 and IF3, dissociates from the small ribosomal subunit. The large ribosomal subunit then joins, and a functional ribosome is formed that holds the mRNA and fMet-tRNA^{fMet}. This also results in the formation of the third site in the ribosome called the exit site (E site). The complete structure is called the initiation complex. The eukaryotic initiation step differs from the prokaryotic one because of the presence of many more acces-sory proteins (at least nine eukaryotic initiation factors are known), the different form of mRNA start-codon recognition, and because the first tRNA is associated with Met instead fMet.

The elongation step is similar in prokaryotes and eukaryotes and comprises three substeps. In the first one, the incoming aminoacyl-RNA is first associated with the elongation factor Tu (EF-Tu), to which the GTP is bound. This complex is then bound to the A site in the ribosome, the GTP is hydrolyzed, and the EF-Tu-GDP complex is released. The EF-Tu-GTP needed to assist the binding of the next incoming aminoacyl-tRNA is regenerated by another elongation factor EF-Ts and the GTP. The second elongation substep involves the formation of the peptide bond between the carbonyl group of the first Met (fMet in prokaryotes) and the amino group in the second amino acid attached to the tRNA sitting in the A site of the ribosome. The amino group of the second amino acid actually displaces the Met or fMet from the first tRNA, and the A-site-bound tRNA is then associated with a dipeptide. The next substep of the elongation is the translocation of the ribosome for one codon in the $5' \rightarrow 3'$ direction. The start codon and the empty tRNA^{Met} are thus shifted to the E site; next, the tRNA with the attached dipeptide is shifted from the A site to the P site, and so the A site is free to accept the next tRNA in the next turn of the elongation step. All three phases of the elongation repeat until the ribosome reaches one of the stop codons (UAA, UAG, or UGA). When a stop codon is in the A site, it is recognized by the termination or release factor (RF). In prokaryotes, there are three RFs, and in eukaryotes there is only one. The binding of the RF to the A site promotes the hydrolysis of the bond between the completed polypeptide chain and the last tRNA. This is followed by the decomposition of the functional ribosome into two subunits with the release of the synthesized polypeptide, the last "empty" tRNA and the mRNA. The synthesized polypeptide chain is then ready for posttranslational modifications, folding, and sorting (see Chapters 5–7).

4.2 REGULATION OF PROTEIN BIOSYNTHESIS

Protein biosynthesis is regulated on several levels. Which proteins are present in the cells and in the extracellular solutions and in what amounts determine the phenotype and physiological activities of the organism. The main regulatory point in protein biosynthesis is the start of the transcription. Some genes are transcribed regularly at a constant rate for the production of proteins that are always required. These are called housekeeping genes. The others are only transcribed after some external signal, giving rise to a protein that will only occasionally be needed. These proteins are called inducible. For every gene there exists a regulatory region upstream of the transcription start point. A number of proteins and other molecules are involved in the process of the proper positioning and conditioning of the RNA polymerase to start the transcription.

The regulatory mechanisms differ considerably in prokaryotes and eukaryotes. In prokaryotes, many genes are clustered and, together with the regulatory region of the DNA that includes a promoter, they are called an operon (see Figure 4.5). All the genes in an operon are transcribed together. In principle, the regulation of the transcription of the operon's structural genes can be positive or negative. In the positive regulation, proteins called activators bind to the region near to the promoter and increase the activity of the RNA-polymerase bound to the promoter sequence.

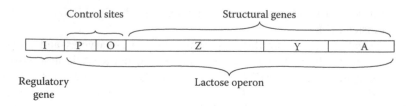

FIGURE 4.5   *Escherichia coli* lac operon. The structural genes Z, Y, and A code for the enzymes that control lactose metabolism: β-galactosidase, galactoside permease, and thiogalactoside transacetylase. These genes are expressed together and are controlled by the control sites P and O and by the regulatory gene I.

The activator can be in two conformations: one when bound to DNA, and the other when free in solution. Both conformations are in equilibrium. Signal molecules that actually start the transcription modify the binding of the activator to the DNA in two ways (Figure 4.6A, B). In one mechanism, the signal molecule binds only to the free activator but not to the DNA-bound one. The conformation of the free activator is thus stabilized and, because of the shift of the equilibrium, the activator is released from the DNA. This attenuates the activity of the RNA-polymerase, and the transcription is stopped.

In another mechanism, the signal molecule binds only to the DNA-bound activator. When the concentration of the signal molecule is decreased, it is released from the activator. This promotes the conformational change of the activator, and it dissociates from the DNA activator-binding region, in this way stopping the transcription. The presence of a signal molecule will thus stop the transcription in the first mechanism and promote the transcription in the second one. The negative regulation is mediated by the binding of the protein called the repressor to the repressor-binding region in the DNA. The presence of the repressor in this region prevents the binding of the RNA-polymerase to the promoter and thus prevents the transcription. Similar to activators, repressors are in two conformational states that are in equilibrium: in one state when they are bound to the DNA, and in the other when they are unbound. Again, two mechanisms can regulate the binding of the repressor to the DNA (Figure 4.6C, D). In one mechanism, the signal molecule binds only to the free repressor, stabilizes its free conformation, and thus, by preventing binding to the repressor region in the DNA, promotes the transcription. In the other mechanism, the signal molecule only binds to the DNA-bound conformation, and when it is present, the transcription is stopped.

A very well-known example of the transcription regulation in prokaryotes is the negative regulation of the lactose operon (lac operon) in *E. coli*. Its basic regulation is in terms of the binding of lactose to the lac repressor in the unbound conformation, which results in the transcription of three structural genes coding for the three enzymes required for the lactose internalization into the cell and its conversion into usable products. All these enzymes are inducible since they appear in the cell only when lactose is present as a nutrient. However, other factors can be involved in additional regulation of the lac operon. One of these is the presence of an alternative nutrient, glucose. If glucose is present in a high concentration, there is no need for the other

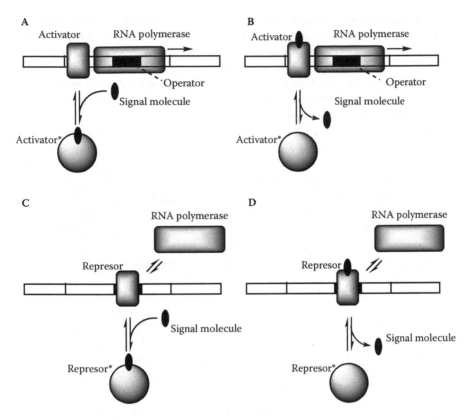

FIGURE 4.6 Positive and negative regulation of operon by activators (A, B) and repressors (C, D). See text for details.

sugars to be used in catabolism, and the transcription of the lac operon is suppressed. This is achieved via a positive regulation by the cAMP receptor protein (CRP). This protein is a transcription activator that binds DNA when it is interacting with cAMP (Figure 4.6 B). It interacts with the RNA-polymerase, enhancing its activity. During times of high glucose concentration, the production of cAMP in the cell is decreased, the complex CRP–cAMP decomposes, the CRP is released, and the transcription is very slow even in the presence of a high lactose concentration. Several other complex mechanisms for the fine-tuning of the operon transcription in prokaryotes are also known, including transcription attenuation, coordinated gene regulation (called the SOS response), regulation by small RNAs, and DNA recombination.

In eukaryotes the regulation of the gene transcription is very complex. To understand this complexity and the regulation of the transcription, it is necessary to look at the transcription initiation more closely. In contrast to the prokaryotic RNA polymerase, the eukaryotic RNA polymerase II has a very low affinity for the promoter and needs a number of activator proteins to be able to bind to DNA. This is in principle comparable to the positively regulated promoters in prokaryotes and is thought to arise from the great complexity of the eukaryotic genome. A larger number of transcription factors and other accessory proteins that bind to several regulatory genes in

DNA increases the specificity of the RNA polymerase II and decreases the possibility of random binding to an inappropriate site. The binding of the RNA polymerase II to the promoter site is assisted directly or indirectly by four types of proteins: (i) transcription activators, (ii) chromatin modification proteins, (iii) coactivators, and (iv) basal transcription factors. Transcription activators bind first to the region of the DNA called the enhancer, which can be quite far away from the TATA-box in the upstream direction. Assisted by the binding of different nonhistone proteins, including the high-mobility-group proteins (HMG), the bending of DNA occurs, forming a DNA loop that brings several parts of the DNA sequence into closer proximity. Into this loop a number of chromatin modification proteins bind. Eukaryotic DNA is associated with histones that keep DNA strands together and do not allow their separation.

The chromatin-modification proteins are enzymes that modify histones by acylation and methylation, targeting mainly Lys residues. Some of these enzymes also induce histone reorganization. The remodeled chromatin makes possible the binding of the transcription factors with a higher affinity. The next step is the binding of 20 or more transcription coactivator proteins to the DNA loop region, forming a complex called a mediator. Only then can the formation of the preinitiation complex (PIC) at and around the TATA-box start by binding the TBP to the TATA-box, followed by several basal transcription factors and RNA-polymerase II. Protein–protein interactions keep together the proteins bound in the promoter (TATA) region with proteins bound to the DNA elsewhere in the "transcription loop." The binding of each new protein to this large structure is usually cooperative. The preinitiation complex and the accessory proteins are shown schematically in Figure 4.7. The regulation of the transcription is possible at several steps in the formation of PIC. Transcription activators are subjected to regulation by the binding of the signal molecules. In the well-known example of the regulation of the galactose metabolism in yeast, it was found that three proteins, Gal3, Gal4, and Gal80, form a complex that, once galactose is bound to one of them (Gal3), can act as the transcription activator, promoting the expression of at least six genes that code for the enzymes required for galactose metabolism. This system also includes a repression mechanism that operates when glucose is available as a nutrient to switch off galactose metabolism. The complex of chromatin modification enzymes and mediator complex formation are also regulated by Gal4 when Gal4 is not bound to Gal80.

Transcription factors can bind regulatory molecules of cellular or extracellular origin, steroid hormones being a well-known example. The intracellular steroid receptors usually change the conformation and dimerize after the binding of the steroid hormone. Altered in this way, they function as the transcription factors, modifying the gene transcription by binding to the hormone response elements in DNA and by interacting with other transcription factors. Nonsteroid hormones can also affect gene expression, for example, via the action of cAMP. When the concentration of cAMP in the cell is high, for example, as a result of the presence of nonsteroid hormones, the protein kinase A is activated, and after the translocation into the nucleus, it phosphorylates CREB (the CRE-binding protein). Phosphorylated CREB acts as the transcription factor that can bind to the cAMP response element (CRE) in DNA, activating in this way the expression of the adjacent genes. Regulation of protein synthesis can also be achieved at other levels. Posttranscriptional regulation occurs

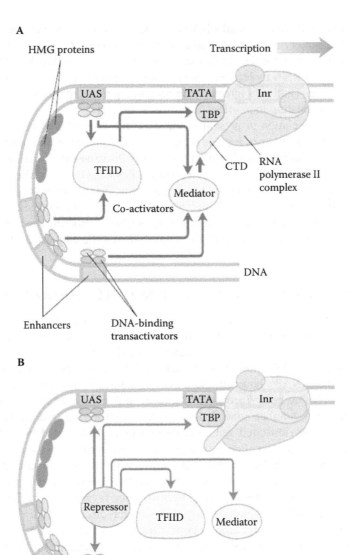

FIGURE 4.7 Regulation of transcription in eukaryotes is very complex. (A) The activators act via mediator protein. Numerous interactions are indicated by arrows. (B) Repressors can use different mechanisms and interact directly with DNA or with other proteins; arrows indicate possible interaction points. (Reproduced from Nelson, D. L. and Cox, M. M. 2008. Lehninger Principles of Biochemistry. W. H. Freeman, New York. With permission.)

with a number of different, small RNA molecules that bind to the mRNA and inhibit translation or direct mRNA to degradation. Some of these RNAs emerge only during development and are thus called small temporal RNAs (stRNAs). The regulatory role of small RNA molecules is very complex, and the complete picture is only slowly emerging. They also interact with proteins, for example, with the transcription factors, modulating their effect on protein synthesis.

Finally, the regulation of protein synthesis at the level of translation should be mentioned. One mechanism for this regulation is the phosphorylation of the translational initiation factors by kinases. This usually decreases the binding affinity of the initiation factors and thus their efficiency. Phosphorylation is also involved in the regulation of the activity of the binding proteins that interfere with the interaction between elongation factors. When phosphorylated by the kinases, they cannot prevent the interaction between elongation factors. Some proteins can prevent translation by direct binding to the mRNA, which inhibits the formation of the proper initiation complex.

4.3 RIBOSOMAL ASSEMBLY OF NONCODED AMINO ACIDS

In general, 20 standard or coded amino acids are involved in protein synthesis. There are, however, two exceptions to this. It has been shown that some of the archaea can incorporate seleno-Cys into the sequence of the protein and that this amino acid is coded in the DNA by the stop codon UGA, called opal. The codon is recognized by the tRNA^{Ser}; however, the tRNA-bound Ser is enzymatically converted to seleno-Cys before being incorporated into the polypeptide chain. In some methanogenic archaea, another noncoded amino-acid pyrrolysine, a derivative of Lys, is incorporated by ribosomal synthesis into proteins. It is coded by another stop codon UAG, called amber. For both seleno-Cys and pyrro-Lys (sometimes abbreviated Pyl), specific tRNAs are used, but while seleno-Cys is generated from Ser on tRNA^{Ser}, it seems that pyrro-Lys has its own tRNA and also its specific tRNA amino-acid synthetase. Seleno-Cys and pyrro-Lys are regarded as the 21st and 22nd coded amino acids. These findings, together with the discovery that ribosomes are not very restrictive and can accept a large variety of different amino acids, fueled the idea of developing a method for the ribosomal incorporation of noncoded amino acids into a specific point in the protein structure.

Two important benefits of the introduction of noncoded amino acids into protein were expected. The enlargement of the repertoire of amino acids would provide additional functional groups and makes possible potentially new functions of proteins. It can also provide a new approach for studying the mechanisms of protein action, structure–function studies, exploring protein–protein interaction, labeling and cross-labeling of the protein at defined points within the protein, folding and stability studies, and many others. The method commonly used today is called nonsense suppression. It comprises the introduction of the nonsense codon, usually the stop codon, into the site in the mRNA where a noncoded amino acid is planned to appear, and the construction of the tRNA molecule that is able to recognize the nonsense codon as well as to carry the required amino acid at its 3′ end. The modifications in the mRNA and tRNA can be achieved by a molecular biological approach in suitable

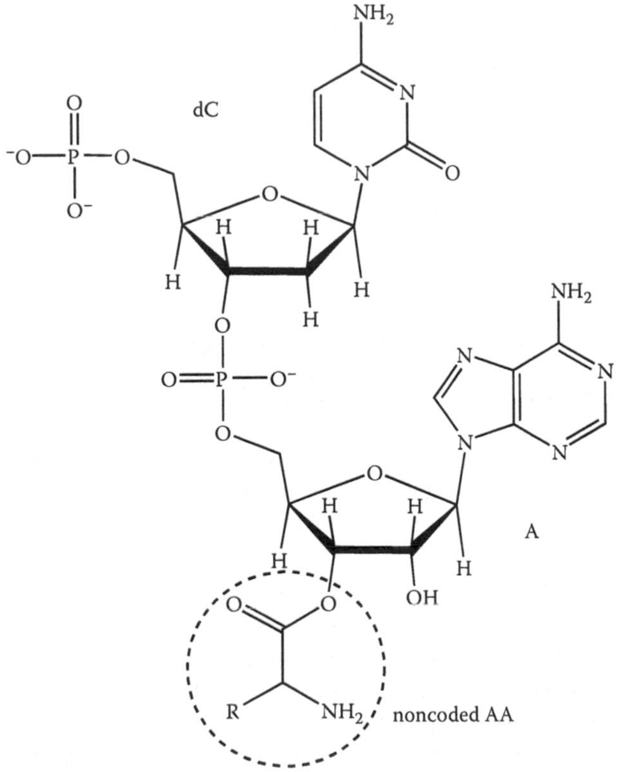

FIGURE 4.8 The noncoded amino acid is coupled to pdCpA to obtain the construct pdC-pA-AA (AA denotes a noncoded amino acid). This is then used by T4 RNA ligase as a substrate that is ligated to the tRNA CCA terminus. The tRNA is subsequently truncated to yield the usual CCA sequence at the 3′ end with the bound noncoded amino acid.

organisms or cells. Loading of the nonsense tRNA with the desired noncoded amino acid can be accomplished by chemical synthesis or enzymatically, using T4 RNA ligase and the noncoded amino acid coupled to pdCpA as a substrate (Figure 4.8). In the latter case, the tRNA is truncated by two nucleotides at the 3′ end to yield the usual CCA sequence at the 3′ end after the ligation of the pdCpA construct. Finally, the expression of the protein with the incorporated noncoded amino acid is carried out in vitro or in vivo, for example, in Xenopus oocytes, by using modified mRNA and acylated nonsense tRNA. The application of the nonsense-suppressor technique has already yielded hundreds of different noncoded amino acids incorporated into many tens of different proteins, and it is expanding fast.

4.4 NONRIBOSOMAL BIOSYNTHESIS OF PEPTIDES

Many peptides are coded in the DNA and are synthesized in the same way as proteins. This synthesis and its regulation are subject to the same rules and procedures as described earlier for the proteins. Very often a larger polypeptide chain is

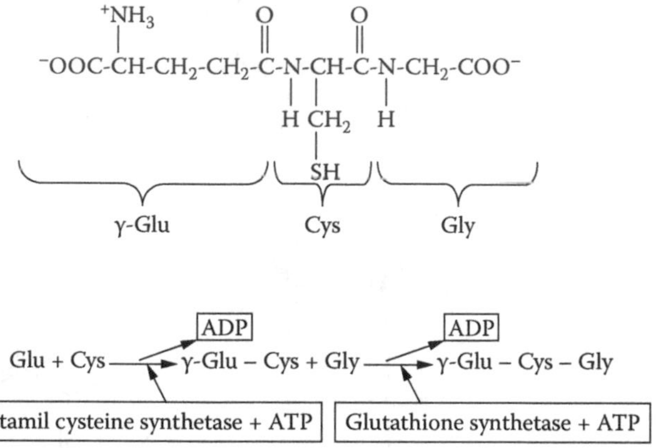

FIGURE 4.9 Gluthation synthesis: The structure of glutathion is shown in the upper panel. Below is the scheme for the synthesis of gluthation catalyzed by γ-glutamylcysteine synthetase and glutathion synthetase coupled to ATP hydrolysis.

synthesized, which is subsequently cleaved in a series of different functional peptides (see Chapter 5). However, some peptides are synthesized nonribosomally with the help of enzymes. The requirement for enzymatic rather than ribosomal biosynthesis arises from the nonstandard structure of these peptides, which includes nonstandard bonding or nonstandard constituents. A well-studied example is that of glutathione (Figure 4.9). This peptide serves as the principal antioxidant agent in eukaryotes and in Gram-negative bacteria, but it also participates in biosynthetic reactions (the formation of leukotriens in mammals), in the detoxification conjugation of some drugs, and as a cofactor of glutathione peroxidase. The biosynthesis of glutathione proceeds in two enzyme-catalyzed steps (Figure 4.9). First, the formation of the dipeptide γ-glutamylcysteine is catalyzed by γ-glutamylcysteine synthetase. This is a dimeric enzyme composed of a catalytic and a regulatory subunit that require ATP for the action. In the next step, glutathione synthetase catalyzes the addition of Gly to γ-glutamylcysteine, again at the expense of ATP hydrolysis.

Glutathione synthetase is a multimeric enzyme (dimer in humans and tetramer in *E. coli*) composed of identical subunits. As shown in Figure 4.9, the peptide bond between Glu and Cys is unusual, involving the γ-carboxyl group and not a standard α-carboxyl group. The reason is probably the protection of glutathione against degradation. Many nonribosomally synthesized peptides are found in bacteria and fungi. These organisms use large multifunctional enzyme complexes called nonribosomal peptide synthetases for the synthesis of peptides with an unusual structure, frequently composed of nonstandard amino acids and also of other non-amino-acid components. The nonribosomal peptide synthetases are the modular enzymes capable of many more different catalytic conversions than ribosomes, including the transformation of the L-amino acids into the D-enantiomers and the cyclization of the product peptide. They can also incorporate in the peptide structure some non-amino-acid-like organic molecules, for example, lactate, oxazoles, and thiazoles. Two well-studied examples are valinomycin and gramicidin. Both are designed by microorganisms for defense, and function as antibiotics. They are ionophores that incorporate in the hostile cell membrane, making it permeable to ions. The toxic action by ionophores is associated with the destruction of the transmembrane gradient of the ions. Valinomycin, which is obtained from different Streptomyces strains, is specific for K^+ ions that are incorporated into its cyclic structure, typically consisting of three L-Val, six D-Val, and three L-lactates. The bonds between amino acids are standard peptide bonds, but the three bonds between D-Val and L-lactate are ester bonds.

Gramicidin, produced by *Bacillus brevis*, is not cyclic and consists of 15 amino acids, eight in the l-, six in the D-configuration, and Gly. The N-terminus and C-terminus are modified for protection against degradation by peptidases, the first by formylation and the second by the addition of ethanolamine. Gramicidin is synthesized by four multimodular peptide synthetases that comprise 16 modules with 56 different catalytic domains with various catalytic activities, including the formation of a peptide bond and the conversion L-amino acids into D-amino acids. The gramicidin functions as a channel in the membrane consisting of two gramicidin molecules, forming a large helical structure with a hole in the center that allows the transport of H^+, Na^+, and K^+ ions.

4.5 DETERMINATION OF THE POLYPEPTIDE SEQUENCE

The protein sequence provides valuable information for determining the higher levels of protein structure as well as exploring the relationship between proteins on the basis of the sequence similarity. Sequence determination can be undertaken from three sources: directly from the protein polypeptide chain, from the DNA sequence of the nucleotides in the identified gene that codes for the protein under study, or from the sequence of the corresponding mRNA. Each approach has its advantages and disadvantages.

When starting from the protein, the first step in the classical approach is usually to determine the protein amino acid composition, which helps in the subsequent procedure. This is achieved by the hydrolysis of the peptide bonds in 6 M hydrochloric acid at 100 to 110°C for 24 h, or more. Subsequently, the released amino acids are separated and identified using chromatography (ion exchange or hydrophobic), and their amount is determined spectrophotometrically using reagents that give a colored derivative. Ninhydrin or fluorescanine can be used for this purpose (see Chapter 1 for the reactions). In the next step, the N-terminal and C-terminal amino acids can be determined. The N-terminal amino acid can be specifically labeled by several reagents. Most commonly, 1-fluoro-2,4-dinitrobenzene (Sanger's reagent), dansyl chloride, or phenylisothiocyanate (Edman's reagent) are used (see Chapter 1). After degradation, the labeled amino acid is identified in the same way as described earlier. For the labeling of the C-terminal amino acid, no reliable reagent is available. However, it is possible to use a digestion of the C-terminus by a carboxypeptidase performed for different time intervals. The C-terminal amino acid can be identified by an analysis of the time-dependent release of amino acids. The polypeptide chain is sequenced by Edman degradation. This is a set of the reactions (Figure 4.10) started by the action of phenylisothiocyanate on the N-terminal

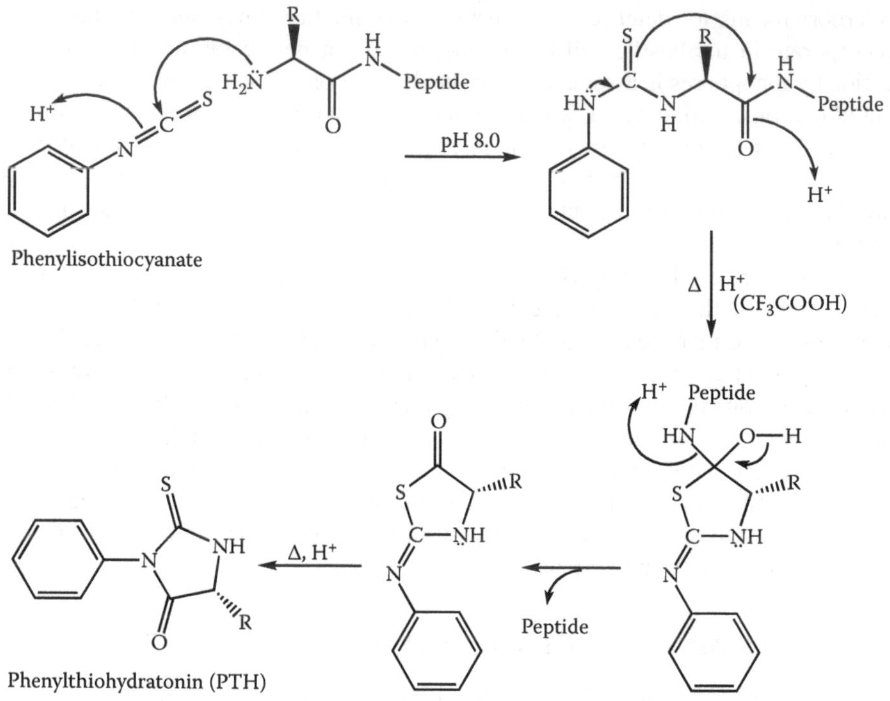

FIGURE 4.10   Scheme of Edman degradation. See text for details.

amino group in the polypeptide, and it results in the released, labeled N-terminal amino acid and the shortened polypeptide chain. By repeating the procedure, amino acids can be successively cleaved and analyzed one by one from the N-terminus to the C-terminus. In practice, the sequencing of an amino acid up to 50 peptides long can be achieved. Because the efficiency of the Edman reaction is around 98%, the error after 50 turns of the Edman cycle becomes substantial, and the sequencing of polypeptides longer than 50 amino acids is not reliable. Therefore, longer polypeptide chains must be cleaved into shorter fragments, and each fragment should be separately sequenced.

The standard procedure, after determining the composition of the amino acid and both terminal amino acids, would start by breaking the disulfide bridges, by oxidation with performic acid, or by reduction with dithiotreitol. Oxidation will yield cysteic acid residues that cannot be combined back to the disulfide bridge. The reduction results in two cysteine residues that must be further modified, for example, by acetylation with iodoacetate, to prevent any disulfide-bridge recovery. The polypeptide chain is then cleaved into fragments by a suitable endopeptidase, for example, trypsin. The amino acid's composition can be a clue to which proteolytic enzyme should be used for the cleavage because endopeptidases are specific, and most will cleave the peptide bonds near to the specific amino acid. The fragments are purified and each is subjected to Edman degradation and thus sequenced. The protein is then cleaved by another endopeptidase or another reagent (cyanogen bromide is frequently used) with different specificity in order to obtain a different set of fragments. After sequencing the fragments, the order of the fragments can be determined by the alignment of both sets of overlapping fragments. If the N-terminal and C-terminal amino acids of the protein are known, this information may help to identify the first and the last fragments in the line.

Disulfide bonds are determined by a separate cleavage of the polypeptide chain, this time without any disruption of the disulfide bonds. After separation of the peptides, these are compared with the set of peptides obtained using the same enzyme but after the cleavage of the disulfide bonds. The comparison reveals the regions were the disulfide-bond-forming Cys resides. A determination of the sequence by the described method is time consuming. It took Sanger and coworkers over 10 years to disclose the sequence of the first polypeptide insulin, composed of only 51 amino acids. Today, Edman degradation is automated and is performed using machines called sequenators. This greatly simplifies and accelerates the procedure. Another method to deduce the sequence directly from the protein is mass spectrometry (MS). To analyze the sequence of the polypeptide, tandem MS, also called MS/MS, is used. The principles of this method are given in Chapter 18.

Protein sequence can also be obtained by determining the corresponding DNA or mRNA sequence. Sequencing a protein from DNA is possible if the gene of the protein is known, including the intron and exon positions. If this is not the case, it is convenient that a short section of the protein sequence, usually 10 to 20 amino acids long, be sequenced by one of the methods described earlier. This fragment is translated into the corresponding DNA fragment, which is synthesized and used to hybridize with the mRNA coding for the protein. In this way, mRNA can be isolated and translated into the complementary DNA (cDNA), which can be amplified in a polymerase chain reaction (PCR). Using this procedure, a sufficient amount of cDNA can be obtained for subsequent sequencing.

DNA sequencing, usually by the Sanger method, is very fast and relatively simple. The Sanger method uses the DNA to be sequenced as a template, a short oligonucleotide primer that is radioactively or fluorescently labeled, DNA polymerase, four deoxyribonucloetides in the triphosphate form, and dideoxyribonucleotide triphosphates to stop the growth of the complementary DNA strand. All these components are mixed in order to achieve the growth of the complementary DNA strand on the cDNA template. At least four separate runs are needed. In each run, only a small amount of one type (A, T, C, or G) of the dideoxyribonucleotide triphosphates is added. This nucleotide lacks the hydroxyl group at position 3′, and when incorporated, the growth of the DNA strand is stopped. In this way, a series of polynucleotide fragments is obtained, ranging from the 5′ beginning of the strand to the position of each nucleotide added in the form of dideoxyribonucleotide. If, for example, the dideoxyribonucleotide triphosphate added is adenin, in this run the fragments with a length from the 5′ origin to the position of each adenin in the sequence will be synthesized. Fragments of all four runs are separated by gel electrophoresis, with each run being in a separate column. The mobility of the fragments in the gel during electrophoresis corresponds to their length. Since the primers are labeled, the position of each fragment in the gel can be easily detected. From the developed gel, the sequence of the DNA can be read directly. Although quick, this method has some drawbacks. It cannot detect modifications in the sequence that occur posttranslationally, for example, those resulting from proteolytic cleavage, and is not suitable for the detection of disulfide bridges. These must still be detected by classical protein sequencing. A very large number of protein sequences have been resolved, and they are available today in databases (Swiss-prot and many others). A resolved sequence is cornerstone information for an assessment of the higher levels of protein structure and is used to classify proteins into families, predict their function, follow their evolution on a molecular level, and undertake large proteomic studies. Some of these studies are covered in subsequent chapters.

FURTHER READING

1. Nelson, D. L. and Cox, M. M. 2008. Lehninger Principles of Biochemistry, 5th edition, W. H. Freeman, New York (Chapters 3 and 8).
2. Ronneberg, T. A., Landweber, L. F., and Freeland, S. J. 2000. Testing a biosynthetic theory of the genetic code: Fact or artifact? PNAS 97, 13690–5.
3. Korostelev, A., et al. 2006. Crystal structure of a 70S ribosome-tRNA complex reveals functional interactions and rearrangements. Cell 126, 1065–77.
4. Steward, L. E. and Chamberlin, A. R. 1998. Protein engineering with nonstandard amino acids. Methods Mol. Biol. 77, 325–54.
5. Budisa, N. 2006. Engineering the Genetic Code. Wiley-VCH, Weinheim, Germany.
6. Finking, R. and Marahiel, M. A. 2004. Biosynthesis of nonribosomal peptides. Annu. Rev. Microbiol. 58, 453–88.
7. Steen, H. and Mann, M. 2004. The abc's (and xyz's) of peptide sequencing. Nature Reviews Mol. Cell Biol. 5, 699–711.

5 Posttranslational Modifications

Matjaž Zorko

CONTENTS

During or after biosynthesis, parts of the backbone in many proteins are altered in a process called posttranslational modification. This includes shortening of the amino-acid chain by the proteolytic cleavage and, in some cases, the alteration of the order of the obtained fragments, the attachment of additional functional groups and chemical structures to the termini of the backbone and to amino-acid radicals, and, in some rare cases, even an apparent prolongation of the sequence. The vast majority of modifications are catalyzed by specific enzymes, but a few of them can happen spontaneously, without the help of enzymes. The number of possible modifications is huge—the number is estimated to exceed 80—including acetylation, acylation, amidation, glycosylation, carboxylation, methylation, disulfide-bond formation, hydroxylation, phosphorylation, sulfation, ADP-ribosylation, and many others. In this chapter, we will limit our review to only the most important modifications.

5.1 PROTEOLYTIC CLEAVAGE

This is a modification of the polypeptide chain in which parts of the chain are cleaved off by proteolytic enzymes and the chain is consequently shortened. The modification can serve various purposes, but one type of this modification is associated with the targeting of proteins. All proteins that are synthesized to be excreted from the cell, and most of those produced to enter specific organelle, contain a signal peptide. Membrane proteins can also contain a signal peptide; this can be as short as three amino acids or longer, containing over 50 amino acids, and is typically composed of several different regions. Out of these regions, one is a sequence of hydrophobic amino acids, and, in another, there are normally amino acids with a basic character. After translocation, the signal peptide is removed from the protein by the proteolytic action of signal peptide peptidases, which are transmembrane aspartyl proteases. Details about signal peptide cleavage are given in Chapters 7 and 24 and will not be further elaborated here.

Another purpose of proteolytic cleavage is to achieve a proper three-dimensional structure in the protein or peptide to enable specific interactions with other biomolecules, for example, receptors. The process is common in peptide hormones. An example of this type of reshaping of the peptide is the processing of the hormone insulin. This is synthesized in pancreatic β-cells as a 110-amino-acids-long preproinsulin that undergoes two proteolytic modifications (Figure 5.1). First, a signal peptide of 23 amino acids is cleaved off by the signal peptide peptidase after introducing the preproinsulin into the endoplasmic reticulum. The resulting proinsulin is stored in secretory granules and is further cleaved. The internal peptide of 33 amino acids is removed by proprotein convertase 1 and 2: the first being aspartyl, and the second seryl protease. Cleavage occurs after two pairs of basic amino-acid residues: -Arg-Arg- at the C-terminus of the chain B and -Lys-Arg- at the C-terminus of the excised peptide C. Both these basic pairs are subsequently removed by carboxypeptidase. The resulting active form of the hormone is released into the blood circulation system. It consists of two polypeptide chains, the shorter chain A and the longer chain B, joined by two –S-S- bridges, and is only recognized by an insulin receptor as such. After denaturation, proinsulin can regain the native conformation in the renaturation process, but not insulin, the two chains of which are separated if denatured.

Another example is the proteolytic conversion of preproopiomelanocortin, which is a precursor of nine physiologically active shorter peptides (Figure 5.2). The precursor polypeptide chain is 265 amino acids long and is synthesized in the central nervous tissue, to be processed in different regions of the brain. Each region can produce a different mixture of hormones. Even in the pituitary, the composition of the products differs in the various lobes. Here, a number of specific proteases are involved. Specific cleavage points are the pairs of basic residues -Lys-Arg-, -Arg-Lys-, and -Lys-Lys-, but cleavage also depends on the three-dimensional structure of the precursor polypeptide. Proteolytic cleavage is also characteristic of the processing of proteases. These enzymes catalyze the degradation of peptides and proteins, and their uncontrolled action could be harmful, if not lethal, for the cells. For this reason, the digestive proteolytic enzymes that degrade proteins in

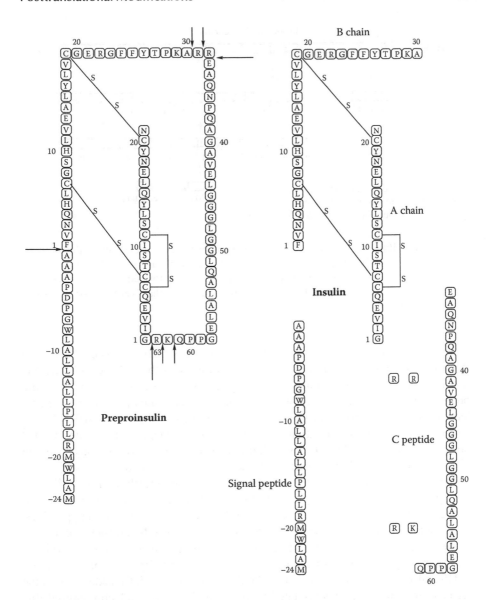

FIGURE 5.1 Insulin maturation by proteolytic cleavage. First, the signal peptide is cleaved off. The resulting proinsulin is then cleaved at two points to remove C peptide, which is further cleaved to release four basic amino acids, two at each terminus. The cleavage sites are indicated by arrows.

the intestinal tract of vertebrates may become active only after secretion. Proteases such as trypsin or carboxypeptidase are thus kept inactive in their zymogene form until they are delivered to the site where they are intended to function, and are activated by a proteolytic cleavage in situ. This activation is described in more detail in Chapter 21.

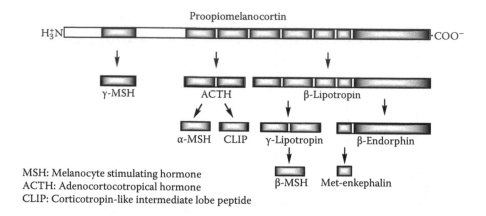

MSH: Melanocyte stimulating hormone
ACTH: Adenocortocotropical hormone
CLIP: Corticotropin-like intermediate lobe peptide

FIGURE 5.2 Processing of preproopiomelanocortin by proteolytic cleavage in the pituitary. Nine different hormones can be obtained from the single precursor peptide. However, in each of three lobes of the pituitary, only some of the cleavages are performed. Some of the cleavage products undergo additional modifications such as amidation of the C-terminus or acetylation of the N-terminus.

Proteolytic processing is very extensive in some RNA viruses. They insert their RNA into the host cell, where it can function directly as the mRNA molecule and is used as a template to produce a long polypeptide chain containing all the viral proteins. The polypeptide is subsequently split into individual proteins by the proteases. Most of the required proteases are also encoded by the viral RNA, and some of these proteases are already functional as the integral part of the original polypeptide in order to be able to start the sequence of proteolytic cleavages without the help of the host cell proteases. In all the examples described so far, parts of the polypeptide chain were removed, but they did not subsequently join in a different arrangement. In the processing of concanavalin A, a plant lectin, an initial glycosylated precursor is first deglycosylated and then cleaved into smaller polypeptides, which subsequently reanneal over a period of 10–27 h. The N-terminal sequencing of the precursors confirmed that the intact subunit of the functional concanavalin A was formed by the reannealing of two fragments since the alignment of the residues 1–118 and 119–237 was reversed in the final form of the lectin. Consideration of the tertiary structure of the glycosylated precursor and the mature lectin showed that the entire series of processing events could occur without significant refolding of the initial translational product. Proteolytic events included the removal of a peptide from the surface of the precursor molecule that connected the N- and C-termini of the mature protein. It seems that the cleavage and the formation of a new peptide bond occur simultaneously in a transpeptidation event rather than in two consecutive, independent steps consisting of a cleavage and subsequent religation. This processing is functionally important, establishing the carbohydrate-binding activity of the lectin.

Another example of the similar processing of proteins is inteins. These are segments of protein within longer polypeptides that are able to excise themselves and anneal the remaining regions, called exteins, with the peptide bond. The process is also called protein splicing because of the analogy with RNA splicing, and inteins are sometimes called protein introns. Inteins that are a hundred to several hundred amino acids long have been found in all three domains of life: eukaryotes, bacteria, and archaea. The initial polypeptide consisting of N-extein and C-extein with intein in between is not functional but becomes active after the excision of the intein. Protein splicing is a self-proteolytic process in which Ser and Asn residues neighboring a His residue are operating. Most of the inteins found so far incorporate an endonuclease domain called homing endonuclease, which is involved in the propagation of the intein-coding DNA. The enzyme breaks the DNA inside the gene that is homologous to the corresponding extein. The break in the DNA is repaired by a recombination mechanism in which the intein-coding DNA sequence is inserted. It seems that the only function of the inteins is to provide the mobility of the intein-coding DNA sequences, which are also called the parasitic genetic elements.

5.2 MODIFICATION OF THE BACKBONE TERMINI

5.2.1 ALTERATIONS AT THE N-TERMINUS

The initiating amino acid in all proteins is Met, which is formylated in prokaryotes. In almost all prokaryotic polypeptides, the formyl group is subsequently removed. In many proteins, both prokaryotic and eukaryotic, the Met residue is also cleaved by the enzyme methionine aminopeptidase. The specificity of the enzyme is associated with Met residues and also depends on the residues proximal to Met. Small Met-neighboring residues favor the cleavage, but Pro in the third position might inhibit the process. Sometimes not only Met but also the adjacent residue is removed.

The acetylation of the N-terminus is a widespread and much conserved modification in eukaryotes:

$$H_3C-\overset{\overset{\displaystyle O}{\|}}{C}-\overset{\displaystyle H}{N}- \tag{5.1}$$

Almost 90% of human cytosolic proteins are modified in this way. Enzymes that perform a transfer of the acetyl group from the acetyl-CoA to the α-amino group of the N-terminal amino acids are called N^{α}-acetyltransferases. It appears that the enzymes are associated with the ribosome, and the modification occurs cotranslationally. This is the reason why only cytosolic proteins are found to be acetylated at the N-terminus, since the signal peptide is removed from the N-terminus after the translocation of the noncytosolic proteins to endoplasmic reticulum.

The functional consequences of acetylation are still poorly understood. Actin and tropomyosin form functional actin filaments only when acetylated. Acetylation also protects proteins from digestion by aminopeptidases and might affect the degradation rate of proteins via the ubiquitination-dependent proteasome system. N-terminally acetylated proteins are also resistant to Edman degradation. Besides acetylation, other modifications, by adding methyl, formyl, glucuronyl, fatty acyl, and other groups to the N-terminal amino group, are known. The attachment of a fatty acid to the N-terminus has been identified for many cytoplasmic proteins. The fatty acids most commonly used in this modification are myristilyc acid and palmitoylic acid:

$$H_3C-(CH_2)_{12}-\overset{\overset{\textstyle O}{\|}}{C}-\overset{\textstyle H}{\underset{}{N}}-$$

myristoylation

(5.2)

$$H_3C-(CH_2)_{14}-\overset{\overset{\textstyle O}{\|}}{C}-\overset{\textstyle H}{\underset{}{N}}-$$

palmitoylation

Fatty-acid-associated proteins can be attached to the lipid bilayer of the membrane and are usually peripheral membrane proteins. However, when cleaved off the fatty-acid chain, they become freely soluble in the cytoplasm. Myristoylation and palmitoylation are mediated by the enzymes myristoyl and palmitoyl transferases, which share many common features but which also have some important differences. Myristoylation is a cotranslational process because myristoyl transferase interacts with the ribosome and is able to transfer the myristoyl group from the myristoyl-CoA to the N-terminal Gly during polypeptide chain assembly. The reaction is favored when amino acids at the second and fifth positions are small and uncharged. Palmitoylation occurs posttranslationally in cytosole by a transfer of the palmitoyl group from palmitoyl-CoA to N-terminal Cys followed by two aliphatic amino acids. An important difference is that palmitoylation is reversible. The palmitoyl group can be removed by acyl-protein thioesterases that regulate the palmitoylation status of various proteins, particularly those participating in signal transduction. Besides the association with the membrane, the acylation of the N-terminus is important in protein–protein interactions, cell signaling, and protein trafficking.

5.2.2 ALTERATIONS AT THE C-TERMINUS

The main modifications of the C-terminus are amidation, prenylation, and the attachment of the glycosylphosphatidylinositol (GPI) anchor. Amidation is common in peptides, particularly peptide hormones. The main step in making an amidated C-terminus is the action of the enzyme peptidyl-glycine-α-amidating monooxygenase, which catalyzes the introduction of the oxygen to the terminal Gly residue. The resulting hydroxylated product is unstable and is decomposed. This step is also assisted by the enzyme; it is called peptidylamidoglycolate lyase.

$$
\begin{array}{c}
\overset{O}{\underset{\parallel}{C}}-\overset{H}{\underset{\mid}{N}}-CH_2-\overset{O}{\underset{O^-}{C}} \\
\downarrow \\
\overset{O}{\underset{\parallel}{C}}-\overset{H}{\underset{\mid}{N}}-\underset{OH}{CH}-\overset{O}{\underset{O^-}{C}} \\
\downarrow \\
\overset{O}{\underset{\parallel}{C}}-NH_2 + HC-\overset{O}{\underset{O^-}{C}} \\
\underset{O}{\parallel}
\end{array}
\qquad (5.3)
$$

As in many other oxygenation reactions, the presence of vitamin C is required to maintain the proper oxidation state of copper, a constituent of the enzyme. Amidation is functionally important, enabling the full activity of many peptides, including substance P. Amidated peptides are more stable due to protection against exopeptidases acting at the C-terminal side.

Prenylation is a posttranslational modification comprising the attachment of an isoprenoid to the C-terminal Cys residue. Two related isoprenoids are used, farnesyl diphosphate or geranylgeranyl diphosphate, both arising from the same biochemical pathway that produces cholesterol. The farnesyl and geranylgeranyl groups are branched, partly nonsaturated lipids, the first consisting of 15 and the second of 20 carbon atoms. Some prenyltransferases recognize the CaaX box (Cys-aliphatic-aliphatic-any C-terminal residue), and others just the C-terminal Cys. When the lipid is attached to Cys in the CaaX box, the first three amino acids are removed so that Cys becomes a C-terminal amino acid. Targeting cellular membranes and the mediation of protein–protein interactions are well-known biological functions associated with these lipid anchors. Prenylation is a modification characteristic of the small GTPases, and because these proteins are involved in cellular signaling and tumorogenesis, they are of great medical interest.

The GPI anchor is a complex glycolipid consisting of a single phospholipid spanning the membrane and a head group consisting of a phosphodiester-linked inositol, to which a glucosamine is linked, a linear chain of three mannose sugars linked to glucosamine, and an ethanolamine phosphate linked to the terminal mannose. They act to anchor proteins to the cell membranes by inserting two fatty acid chains into the lipid bilayer. In fungal cells, GPI anchors are also used to covalently link proteins to cell-wall glucans. The proteins to be attached to the GPI anchor contain signal peptide, leading them to endoplasmic reticulum, where they are attached on the luminal side of the reticulum to the membrane with a short C-terminal hydrophobic domain (10 to 20 amino acids long). This is followed by a short hydrophilic sequence and the attachment site for the anchor addition. These general features of the GPI signal are conserved in the evolution from mammals to the lower eukaryotes, for example, protozoa. The attachment of the GPI anchor requires a series of reactions in which the protein is cleaved from the anchoring peptides and transferred to the GPI anchor. The process is catalyzed by a transamidase complex of at least five proteins. The biosynthesis of the GPI moiety occurs in the ER, and the complete GPI anchor is fully assembled prior to attachment to the protein. After the attachment process is accomplished, the protein is transferred via vesicular transport out of the cell, where it remains attached to the plasma membrane or the cell wall.

The only function of the GPI anchor seems to be the attachment of otherwise soluble proteins to the cell exterior. A well-known example is acetylcholinesterase, an important enzyme in synaptic signal transduction, which is anchored to the basal lamina in the synapse cleft in a high concentration. Anchored proteins, including acetylcholinesterase, can be released by the controlled action of phospholipases, particularly phospholipase C. The GPI-linked proteins are thought to be preferentially located in lipid rafts.

5.3 THE ATTACHMENT OF GROUPS TO AMINO-ACID RADICALS

5.3.1 Glycosylation

Carbohydrates are attached to specific amino-acid residues in a majority of the eukaryotic proteins via the N-glycosydic or O-glycosydic bond. There is no common role of the carbohydrates attached to the proteins. It has been shown that some proteins do not fold properly if not previously glycosylated; the stability of some secreted proteins is decreased if the glycans attached to the Asp are removed, and protein–ligand as well as protein–protein interactions are also affected by glycosylation. This plays a role in cell-to-cell interactions, including bacteria—host-cell recognition and the defense mechanisms of the immune system. Although several mechanisms of glycosylation are known, they all share at least three common features: (i) the transfer of carbohydrates to the protein is an enzyme-catalyzed process; (ii) the donor of the carbohydrate molecule is a nucleotide-activated sugar; and (iii) the attachment of the carbohydrate is site specific.

In N-glycosylation, a complex branched-chain oligosaccharide commonly containing 14 monosaccharides is transferred in endoplasmic reticulum from the carrier alcohol dolichol to an Asn residue. In eukaryotes the precursor oligosaccharide is normally composed of 3 glucoses, 9 mannoses, and 2 N-acetylglucosamines. The acceptor Asn is a part of the tripeptide glycosylation sequence Asn-X-Ser, Asn-X-Thr, or Asn-X-Cys, where X is any amino acid, except Pro. Interestingly, Asn is not glycosylated in all available glycosylation sequences in the protein, indicating that there must be additional determinants besides the sequence to specify the glycosylation site. The linkage is achieved by the oligosaccharyltransferase, catalyzing the formation of the N-acetylglucosamine-β1-Asn N-glycosydic bond:

(5.4)

After the attachment of the precursor oligosaccharide, the proper folding of the protein is achieved, and three glucose residues are removed. This is followed by the transport of the protein to the Golgi, where the oligosaccharide can be further processed. In some cases additional monosaccharides are added, and in others some of the monosaccharides are further removed. The processing is variable, producing a heterogeneous population of the carbohydrates on the same type of the protein. There are three major types of N-linked saccharides in vertebrates: high-mannose oligosaccharides, complex oligosaccharides, and hybrid oligosaccharides. Furthermore, in the lower eukaryotes, additional types are known (Figure 5.3). Functionally, N-glycosylation appears to be associated with proper folding, excretion, and the ability of the glycosylated protein to interact with other molecules. However, glycosylation is not always necessary for these processes. Many proteins deprived of glycosylation would properly fold, and a number of proteins are excreted without being glycosylated.

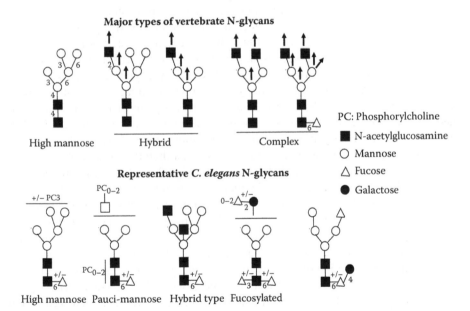

FIGURE 5.3 Comparative overview of the major types of vertebrate N-glycan subtypes and some representative *C. elegans* N-glycans. Top panel: Vertebrate diversification in the Golgi apparatus generates high-mannose, hybrid, and complex N-glycan subtypes. Most cell-surface and secreted N-glycans are of the complex subtype. Vertical arrows indicate the locations of branch formation in diversification, not all of which occur on a single N-glycan. Bottom panel: The main classes of *C. elegans* N-glycans include high-mannose (up to Man9GlcNAc2), paucimannosidic and hybrid type. Note that this scheme only includes some of the documented structures. (Free source: Berninsone, P. M. (2006). Carbohydrates and glycosylation, WormBook ed. The C. elegans Research Community, WormBook, doi/10.1895/wormbook.1.125.1, freely accessible at http://www.wormbook.org.)

Several types of O-linked glycosylation occur predominantly in the Golgi apparatus. The addition of N-acetyl-galactosamine to Ser or Thr residues with the enzyme
UDP-N-acetyl-d-galactosamine:polypeptide N-acetylgalactosaminyltransferase, followed by other carbohydrates, is the most common. The attachment to Ser is shown
in Equation 5.5; the site of a possible prolongation of the carbohydrate chain is indicated by the arrow.

$$
\tag{5.5}
$$

protein backbone

The process is characteristic for proteoglycans, which involves the addition of
glycosaminoglycans to an initially nonglycosylated core protein. The main functions of this modification are assistance in the secretion of the glycosylated protein into the extracellular matrix, where it participates in cell adhesion or acts
as a component of mucosal secretion. The function of glycosylation can also be
to improve protein solubility. Between the rare O-glycosylated proteins in blood
are the immunoglobulins of the type IgA and IgD. In collagen, glycolysation by
galactose characteristically occurs at hydroxy-Lys; Lys must thus be first modified by hydroxylation to serve as a glycosyl acceptor. The sugar that is also frequently used in glycosylation is fucose. It is important in several physiological
processes, such as selectin-mediated leukocyte endothelial adhesion, ABO blood-
type antigen determination, and Notch receptor signaling. Fucose modification is
catalyzed by fucosyltransferases, of which four groups are known in mammals.
The substrate is always GDP-fucose, which is transferred to the Ser or Thr in the
epidermal growth-factor-like repeats (EGF-like repeats), or the thrombospondin
type-1 repeats, defined by the specific pattern of three disulfide bonds. In the case
of EGF-like repeats, the O-fucose may be further elongated to a tetrasaccharide
by the sequential addition of N-acetylglucosamine (GlcNAc), galactose, and sialic
acid, and for thrombospondin, repeats may be elongated to a disaccharide by the
addition of glucose. Both of these fucosyltransferases have been localized to the
endoplasmic reticulum, which is unusual for glycosyltransferases, most of which
function in the Golgi apparatus.

The addition of N-acetylglucosamine to Ser or Thr residues has also been
observed, and it seems to be important in proteins participating in signal
transduction. Glycosylation alternates with phosphorylation at the same site.
Thus, if phosphorylation occurs, glycosylation does not, and vice versa. Both
N-acetylglucosamine and phosphate can be selectively removed by the processes
that control signal transduction. Finally, the attachment of mannose to Trp in
thrombospondin repeats should be mentioned; this is an exception in glycolysation
because the sugar is linked to a carbon of Trp rather than to a nitrogen of Asp or
an oxygen of Ser or Thr.

$$(5.6)$$

The role of this modification is not well understood; however, it seems to occur only in excreted proteins.

5.3.2 Phosphorylation

In eukaryotes, a large number of proteins are phosphorylated by the transfer of a phosphate group from the ATP to the hydroxyl group of Ser, Thr, or Tyr. In prokaryotes, the additional phosphorylation sites are the basic residues of His, Arg, and Lys. The reaction is catalyzed by kinases. Phosphorylation is a reversible modification, dephosphorylation being controlled by enzymes called phosphatases. Phosphorylation controls the activity of many proteins by switching them on and off via conformational changes. These are usually a consequence of the changes in the hydrophobicity/hydrophilicity profile of the phosphorylated protein. The introduction of the hydrophilic and charged phosphate group to the hydrophobic part of the protein structure, for example, would prevent hydrophobic interactions with hydrophobic partners within the protein molecule or in other molecules, and will promote new interactions on the basis of charge and polarity. An example is the phosphorylation of the p53 tumor-suppressor protein. It has over 18 different phosphorylation sites in the N-terminal domain that can be selectively phosphorylated by a number of kinases and dephosphorylated by phosphatases. Phosphorylation alters the conformation of p53 and thus affects its interaction with other proteins, for example, Mdm2, which transports p53 to the cytosole, where it is degraded by the proteasome. In this way, phosphorylation decreases the rate of p53 degradation and alters the effect of the protein on the transcription of the selected genes to mRNA. This can have a dramatic impact on p53-promoted tumor suppression. It is clear that the phosphorylation of p53 and other proteins has to be strictly controlled. Complex signal transduction cascades therefore control the active state of kinases and phosphatases.

The specificity of kinases and phosphatases is of great interest. In the cyclic-AMP-dependent kinases catalysis, the phosphate group is transferred to the Ser residue in the sequence Arg-Arg-X_n-Ser, where X can be any amino acid, and n is between 0 and 2. In tyrosin-kinases, the Tyr in the sequence Lys/Arg-X_5-Asp/Glu-X_3-Tyr is phosphorylated. However, in many kinases, the specificity is not entirely understood. Furthermore, the sequence seems not to be a sufficient determinant of the specificity since model peptides with the sequences just given are not readily phosphorylated, probably due to the lack of a proper three-dimensional structure.

Phosphorylation-related activation/deactivation of the protein function occurs to fully folded proteins in most cases in the cytoplasm. In this, phosphorylation differs from the majority of posttranslational modifications that mostly occur in endoplasmic reticulum and Golgi.

5.3.3 HYDROXYLATION

The hydroxylation of selected amino-acid residues in proteins is an oxidation reaction catalyzed by enzymes hydroxylases in which oxygen is introduced into the recipient molecule and a hydroxyl group is formed. The best known is the hydroxylation in Pro and Lys in collagen. The hydroxylation of Pro occurs at the C^γ atom, and 4-hydroxy-Pro (Hyp) is obtained. In some cases, Pro may be hydroxylated at the C^β atom, resulting in 3-hydroxy-Pro. Lys is hydroxylated on its C^δ atom, forming 5-hydroxy-Lys (Hyl). Three very large, multisubunit enzymes—prolyl 4-hydroxylase, prolyl 3-hydroxylase, and lysyl 5-hydroxylase—are needed to catalyze hydroxylation reactions. For the structure of 4-hydroxy-Pro, see Chapter 1, Figure 1.6. Molecular oxygen, iron, and α-ketoglutarate are required for the oxidation, as is ascorbic acid (vitamin C), to maintain iron in its oxidized state. Both hydroxylated amino acids are very important for collagen stability (see Chapter 26): hydroxy-Pro for the hydrogen bonds stabilizing the procollagen triple helix, and hydroxy-Lys for the crosslinking of the procollagen building blocks into the collagen fiber and for the attachment of the glycosyl groups. Also known is the hydroxylation of Asp, and the resulting β-hydroxy-Asp was considered important in the interaction of the modified proteins with Ca^{2+}, but another important role might be a negative regulation of protein fucosylation. The hydroxylation of Asp is related to the γ-carboxylation of Glu (described in the following text) and is found, for example, in vitamin-K-dependent coagulation plasma proteins.

5.3.4 γ-CARBOXYLATION OF GLU

In some proteins that bind the Ca^{2+} ion, Glu is additionally carboxylated. This is particularly true for blood-coagulation proteins and proteins involved in the calcification of bone. The obtained γ-carboxy-Glu has two neighboring carboxyl groups, suitably positioned to interact with Ca^{2+}:

$$(5.7)$$

This modification is the product of γ-glutamyl carboxylase, an oxidoreductase attached to the membrane of the endoplasmic reticulum. The enzyme requires HCO_3^-, oxygen, and the reduced form of vitamin K, which is oxidized while Glu is carboxylated in the coupled reactions. The mechanism of the catalysis is complex, coordinating the transformation of all four substrates. The process starts by entering the precursor protein containing a signal peptide and a γ-carboxylation recognition site, which binds to the enzyme after the cleavage of the signal protein. The carboxylated preprotein, which is still inactive, is transferred to Golgi; the proprotein is cleaved by proconvertase; and the matured, fully active protein is secreted from the cell. Although no obvious consensus sequence prevails in the carboxylation recognition sites of vitamin-K-dependent proteins, this site is best defined by a Z-F-Z-X-X-X-X-A motif, where Z is an aliphatic hydrophobic residue (Ile, Val, Leu), F is Phe, A is Ala, and X is any amino acid.

5.3.5 ATTACHMENT OF LIPIDS

Besides the attachment of fatty acids and isoprenoid to the termini of the protein backbone, described earlier, lipids can also be attached to the side chain of Cys, either in soluble proteins such as G-proteins or in transmembrane proteins such as seven-transmembrane receptors. The purpose of the modification is to provide the protein with the membrane-attachment site. In most cases the fatty acid transferred from the acyl-CoA to the protein Cys in the transesterification reaction is a palmitoylic acid, but other fatty acids can also participate. For the moment, the motif or the sequence determining which Cys residues in the protein will be targeted by the acyl transferases is not clear.

5.3.6 DISULFIDE BOND FORMATION

The formation of the disulfide bond between two Cys thiol groups is a common posttranslational modification, both in prokaryotes and eukaryotes. Disulfide bonds are found more readily in the secreted proteins. The formation of the disulfide bond is an oxidation reaction:

$$
\begin{array}{c}
\text{protein backbone} \\
| \\
(CH_2)_2 \\
| \\
SH \\
+ \\
SH \\
| \\
(CH_2)_2 \\
| \\
\text{protein backbone}
\end{array}
\xrightarrow{\text{oxidation}}
\begin{array}{c}
\text{protein backbone} \\
| \\
(CH_2)_2 \\
| \\
S \\
| \\
S \\
| \\
(CH_2)_2 \\
| \\
\text{protein backbone}
\end{array}
+ 2H^+ + 2e^-
\qquad (5.8)
$$

In eukaryotes this takes place in the endoplasmic reticulum, where the environment is not so reductive as in most other organelles and cytosole. In prokaryotes the reaction occurs along the passage of the protein across the membrane. A disulfide bond is typically referred to as an –S-S- bridge, denoted by, for example, the Cys26-Cys84 disulfide bond, or the 26-84 disulfide bond, or simply as C26-C84. The formation of the disulfide bond by the thiol-disulfide exchange reaction is catalyzed by disulfide oxidoreductases. These enzymes are different in prokaryotes, where they are called the Dsb (from disulfide bond formation) family of the disulfide oxidoreductases, and in eukaryotes, where they are called protein disulfide isomerases, but all share the motif Cys-X-X-Cys in the active center. One of the Cys residues in the active center has a much lower pKa, of around 3.5, than normal (around 8.5, see Chapter 1, Table 1.1), and upon dissociation, the thiolate ion S⁻ is obtained, which catalyzes the –S-S- bond formation. Proteins in the native state will always have specific Cys residues joined in –S-S- bonds. As the number of Cys residues in the protein increases, the number of possible disulfide bonds increases factorially; however, only one bond pattern corresponds to the native state. The number of ways (N) of forming p disulfide bonds from n Cys residues is given by the formula

$$N = \frac{n!}{p! \cdot (n-2p)! \cdot 2^p} \tag{5.9}$$

For ribonuclease A, containing eight Cys residues that make four disulfide bonds, N is 105. When several possibilities exist, the thiol–disulfide exchange makes it possible to establish proper disulfide bonds by disulfide reshuffling (see Chapter 6 for details). The role of disulfide bonds in proteins and peptides is to stabilize the structure. As calculated, each –S-S- bridge can contribute up to 17 kJ mol⁻¹ of stabilization energy when the bond is optimal in terms of stereochemistry. Stabilization is a result of favorable enthalpy contribution due to bond formation, destabilization of the unfolded state due to lowering of its entropy, and increased potential for the formation of the hydrophobic core of the folded protein. Disulfide bonds also contribute to stability by keeping together multichain proteins, such as immunoglobulines (see Chapter 25) and peptides, as in insulin (see Figure 5.1).

5.3.7   SULFATION

The Tyr hydroxy group is sulfated in membrane and secreted proteins, including adhesion proteins, G-protein-coupled receptors, coagulation factors, serine protease inhibitors, extracellular matrix proteins, and peptide hormones. The reaction is catalyzed by tyrosylprotein sulfotransferase in Golgi apparatus. Two types of this trans-Golgi membrane-bound enzyme are known in practically all tissues throughout the plant and animal kingdoms. The donor of the sulfate group is adenosine 3'-phosphate 5'-phosphosulfate in a sulfotransferase reaction. The obtained sulfo-Tyr (Equation 5.10) is very stable and can be identified after protein/peptide decomposition.

protein backbone

CH$_2$

O

O$=$S$=$O

O$^-$

(5.10)

No specific sulfation site sequence has been identified, but it seems the target Tyr resides between three and four acidic amino acids. The role of sulfation is in strengthening protein–protein interaction by the introduction of a negatively charged highly polar group. As shown, sulfated Tyr residues are directly involved in protein–protein contact, in the recognition of ligands by the receptors, in optimal proteolytic processing, and in the proteolytic activation of extracellular proteins.

5.4 MULTIPLE POSTTRANSLATIONAL MODIFICATIONS

Many proteins are modified at multiple sites by a whole set of different posttranslational modifications that, each separately and several in concert, fulfill complex physiological tasks. In the maturation of collagen (see Chapter 26 for the structure and role of the molecule), for example, at least five modifications are involved. After the synthesis of the preprocollagen peptide chains, these are sent to the endoplasmic reticulum where the signal peptide is cleaved and a procollagen is obtained. The next step is the hydroxylation of the selected Pro and Lys residues, as described in Section 5.3.3, followed by the hydroxylation of hydroxy-Lys by galactose or disaccharide glucosylgalactose (in some rare collagens the glycosylation site is Thr). This is a nice example of how posttranslational modifications are not functionally independent—glycosylation would not be possible without a precedent hydroxylation of Lys. Different types of collagen show different glycosylation patterns. After this step, a triple-helix procollagen can be assembled. Here, hydroxy-Pro provides an important number of hydrogen bonds that stabilize the triple helix. Procollagen is transferred to Golgi to be secreted by exocytose. Outside the cell the procollagen is further cleaved, and the obtained tropocollagen is assembled into the fibrils and fibers by crosslinking reactions. The major role in these reactions is played by hydroxy-Lys, which interacts with other hydroxy-Lys and His residues from the neighboring tropocollagens. In the fibril organization, glycosylated hydroxy-Lys plays an important role and also determines the diameter of the fibers. As seen, each modification has a specific role, and some of them are closely related. Moreover, in some proteins, the same amino-acid residue is potentially a target of several, usually competitive, modifications. Lys can be acetylated, methylated, hydroxylated, ubiquitinated, neddylated, sumoylated, and biotinylated—some of these modifications have been discussed earlier.

While all modifications except hydroxylation occur at the ε-amino group and must therefore compete with each other, hydroxylation proceeds at a different site (C5). However, a bulky modifying group attached to ε-nitrogen would make recognizing the modified Lys with the hydroxylase very difficult. Multiple modifications of the same protein molecule can have an agonistic or antagonistic effect. These effects can be observed on the level of the modifications where one modification promotes or prevents the other modification, or on the level of the effect where two modifications can increase or oppose the effect of each other. It is thought that multisite modifications on the protein constitute a complex regulatory program that transduces molecular information to and from the signaling pathways, controlling the protein function in this way.

5.5 NONENZYMATIC MODIFICATIONS

All the modifications described so far were enzymatically catalyzed and therefore under strict control. There are some spontaneous chemical reactions of proteins known, but most of them will occur in physiological conditions very rarely and at a slow rate. There are some exceptions: glycation is the binding of sugar molecules such as glucose and fructose to the protein, and can be exogenous or endogenous. Although exogenous glycation is important in, for example, the food industry and includes the binding of sugar molecules to the proteins during food processing, we will limit our interest to endogenous glycation that occurs mainly in the bloodstream. Fructose and galactose have higher potentials for glycation than glucose, but all of them will very slowly bind to amino-acid residues containing a hydroxyl or amino group. These interactions can start a series of other spontaneous conversions of proteins and can lead to advanced glycation end products, some of which are reactive enough to damage tissues and which can contribute to chronic diseases such as diabetes mellitus, cardiovascular diseases, Alzheimer's disease, cancer, and others. Patients with diabetes have elevated concentrations of glucose in the bloodstream and are at higher risks than healthy people. The glycation status with these patients is monitored via the glycated hemoglobin level.

Another spontaneous modification that will inevitably occur in aerobic organisms is the oxidation by molecular oxygen and by other oxidants arising from metabolic processes, such as peroxides and free radicals. Sulfur atoms in Cys and Met are the most sensitive to oxidation; however, the general reductive environment in the cell, maintained by a high level of the glutathione, tends to protect, particularly Cys, from excessive oxidation. In Cys, oxidation by oxygen leads to the formation of a disulfide bond (see Equation 5.8). Oxidation with stronger oxidants such as peroxides produces sulfonic acid:

$$
\begin{array}{c}
\text{protein backbone} \\[4pt]
\rule{3cm}{1pt} \\[2pt]
| \\
(CH_2)_2 \\
| \\
SO_3H
\end{array}
\qquad (5.11)
$$

In Met, the first step of oxidation is the sulfoxide, which can be further converted to sulfone:

$$
\begin{array}{ccccc}
\text{protein backbone} & & \text{protein backbone} & & \text{protein backbone} \\
| & & | & & | \\
CH_2 & & CH_2 & & CH_2 \\
| & \xrightarrow{\text{oxygen}} & | & \xrightarrow{\text{oxygen}} & | \\
CH_2 & & CH_2 & & CH_2 \\
| & & | & & | \\
S & & S{=}O & & O{=}S{=}O \\
| & & | & & | \\
CH_3 & & CH_3 & & CH_3 \\
\end{array}
\tag{5.12}
$$

In the cell, this reaction is reversed by the enzyme Met sulfoxide-peptide reductase.

FURTHER READING

1. Anraku, Y., Mizutani, R., Satow, Y. 2005. Protein splicing: its discovery and structural insight into novel chemical mechanisms. IUBMB Life 57, 563–574.
2. Linder, M. E. and Deschenes, R. J. 2007. Palmitoylation: policing protein stability and traffic. Nature Rev. Mol. Cell Biol. 8, 74–84.
3. Bradbury, A. F. and Smyth, D. G. 1991. Peptide amidation. Trends Biochem. Sci. 16, 112–115.
4. Ikezawa, H. 2002. Glycosylphosphatidylinositol (GPI)-anchored proteins. Biol. Pharmaceut. Bull. 25, 409–418.
5. Berninsone, P. M. 2006. Carbohydrates and glycosylation, WormBook, ed. The C. elegans Research Community, WormBook, doi/10.1895/ wormbook.1.125.1, accessible free at http://www.wormbook.org.
6. Luo, Y. and Haltiwanger, R. S. 2005. O-fucosylation of notch occurs in the endoplasmic reticulum. J. Biol. Chem. 280, 11289–11294.
7. Castellino, F. J., Ploplis, V. A., and Zhang L. 2008. γ-Glutamate and β-hydroxyaspartate in proteins. Methods Mol. Biol. 446, 85–94.
8. Furie, B., Bouchard, B. A., and Furie, B. C. 1999. Vitamin K-dependent biosynthesis of γ-carboxyglutamic acid. Blood 93, 1798–1808.
9. Moore, K. L. 2003. The biology and enzymology of protein tyrosine O-sulfation (Minireview). J. Biol. Chem. 278, 24243–24246.
10. Yang, X.-J. 2005. Multisite protein modification and intramolecular signaling. Oncogene 24, 1653–1662.

Folding of Proteins

6 Protein Folding

Matjaž Zorko

CONTENTS

After being synthesized, proteins are subject to the complex process of folding into their compact, functional structure. In terms of chemical formality, folding is a transition from the fully extended, unfolded conformation, designated by U, into the tightly packed, native folded conformation, designated by F. This is, in its simplest mechanism, in accordance with the equation

$$U \underset{}{\overset{K_{eq}}{\rightleftharpoons}} F \tag{6.1}$$

characterized by

$$K_{eq} = \frac{[F]}{[U]} = \frac{k_f}{k_u} \tag{6.2}$$

where $[U]$ and $[F]$ are the equilibrium concentrations of the protein in the unfolded and native folded conformations, respectively, and k_u and k_f are the corresponding rate constants. As pointed out by Levinthal in the mid-1960s, the process of folding should not be random. A random folding is too slow and is not in accordance with the observed protein folding times. Taking a small protein consisting of 100 amino acids as an example, and assuming only three different conformations for each residue (which is a gross underestimation!), the total number of different conformations is 5×10^{47}. If each conversion takes only 10^{-13} s, the total time to scan randomly through all possible conformations would be 10^{27} years—much longer than the age of the Universe. This inconsistency is often referred to as Levinthal's paradox. The observed folding times for small proteins made up of approximately 100 amino acids are around 0.01 s. It thus appears that folding is a directed process that occurs by a multistep, progressive stabilization of various intermediate states. The resolved pathways of protein folding and the computer simulations of this process confirm this scenario. However, this does not contradict the fact that the apparently two-step processes in accordance with Equation 6.1 have frequently been observed for small proteins with fewer than 100 amino acids, but rather speaks for the possibility that the folding in these cases is too fast to detect the intermediate states. With larger proteins that fold in a timescale of seconds and even minutes, several intermediate states are readily observed. To understand the mechanism of folding, all the steps of the process should be resolved, and the kinetic constants for each transition should be determined.

6.1 STABILITY OF THE PROTEIN STRUCTURE AND DENATURATION

Most of the current knowledge about the stability of the protein structure and the folding processes has come from studying protein denaturation and subsequent renaturation. Proteins in the compact, folded conformation exhibit secondary, tertiary and, in some cases, also quaternary structures, as described in previous chapters. The primary structure of the protein and the proper environment are usually sufficient to ensure the functional, native conformation of the proteins. At all levels higher than the primary structural level, it is predominantly weak interactions that operate, assisted by covalent disulfide bridges if they are present. However, by changing the protein environment, these bonds can be broken, resulting in protein denaturation. This denaturation is, by definition, a loss of all the levels of protein structure except the primary structure because only the peptide bonds and sometimes the disulfide bridges remain intact. Denaturation can be achieved by increasing the temperature, changing the pH to either extreme, increasing the pressure, or adding denaturants, that is, chemicals such as guanidinium chloride or urea (see Equation 6.3).

$$\text{guanidinium chloride} \qquad \text{urea} \qquad (6.3)$$

Sometimes the reduction of the disulfide bridges or the removal of the cofactors will also lead to denaturation, and the stability of the protein structure can also be affected by mutations, for example, by the site-specific replacement of any of the crucial residues or by truncating the sequence. With small proteins, denaturation is a two-state process, described by Equation 6.1, and is usually fully reversible. However, with larger proteins, particularly those composed of several domains, the process is multistep and not always reversible. It can, though, sometimes be rendered reversible by applying a special regime of renaturation; for example, when denatured by heating, reversibility can be achieved by carefully controlled slow cooling, by denaturants, by a slow decrease of the denaturant concentration, or by dialysis. However, many proteins—not only the large ones—aggregate and precipitate during denaturation. A typical reversible denaturation–renaturation curve is shown in Figure 6.1. Note

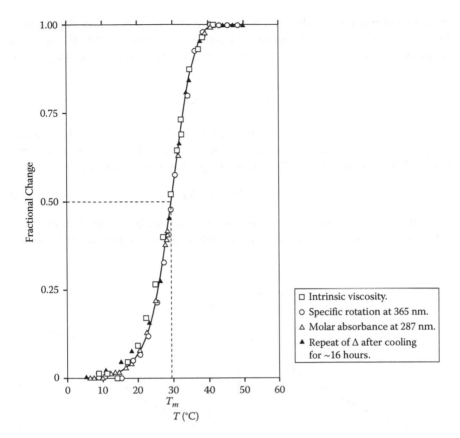

FIGURE 6.1 Typical denaturation–renaturation curve for the two-state system. The temperature-induced unfolding of bovine ribonuclease A in HCl-KCl (pH 2.1, ionic strengths 0.019) is measured by the increase in viscosity (squares), by the decrease in optical rotation at 365 nm (circles), and in UV absorbance at 287 nm (open triangles). Measurements of the second heating after cooling for 16 h is shown by closed triangles. (From Ginsburg, A. and Carroll, W. R. 1965. Intermediates in protein folding reactions and the mechanism of protein folding. Biochemistry 4, 2159–74. With permission.)

that the midpoint of the curve corresponds to the "melting temperature" T_m, where an equal amount of the native and unfolded forms of the protein are present. Several methods can be used to follow denaturation or renaturation. Absorbance at 287 nm and near-UV fluorescence change with the progress of denaturation, reflecting the increased exposure of aromatic residues of Trp and Tyr to the solvent. Circular dichroism detects changes in the proportion of the ordered secondary structure (helices, β-structure) and unordered structure during denaturation. The intrinsic viscosity and NMR chemical shift are also affected by denaturation. See Chapters 9, 11, and 12 for the principles of most of these methods.

In thermal denaturation, energy is applied to the protein directly by heating. When the thermal energy exceeds the energy of the bonds that sustain the protein structure, these bonds are broken and the protein starts to denature (see Figure 6.1). The native protein conformation is stable in a temperature interval that is very dependent on the protein's structure. Thermodynamic treatment can be applied to the reversible denaturation described by Equations 6.1 and 6.2. The free energy of folding is given by Equation 6.4:

$$\Delta G^o = -RT \ln K_{eq} \qquad (6.4)$$

This value should be negative if the folded conformation is to be stable. The free-energy change is made up of the enthalpy contribution (ΔH^o) and the entropy contribution ($T\Delta S^o$), according to the fundamental equation of thermodynamics:

$$\Delta G^o = \Delta H^o - T\Delta S^o \qquad (6.5)$$

hhAs illustrated in Figure 6.2, for the case of lysozyme, both contributions do not differ very much and are almost equally dependent on temperature for the folded and

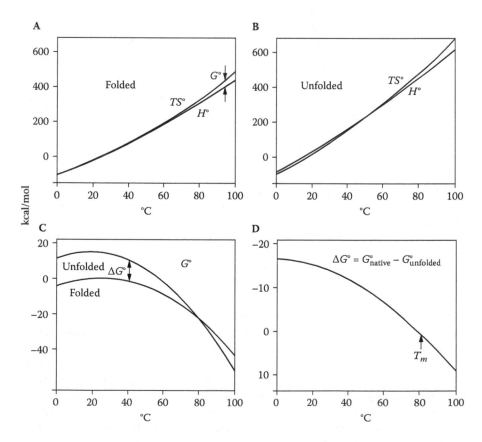

FIGURE 6.2 Thermodynamic parameters for the folded and unfolded conformations of lysozyme at various temperatures. See text for an explanation. (Reproduced from Creighton, T. E. 1993. Proteins Structure and Molecular Properties, 2nd ed., W. H. Freeman, New York, p 298, Figure 7.20, PV. With permission.)

unfolded form of the protein (Figure 6.2A and B). This results in rather small, negative values of ΔG^o for the folding, which increases with increasing temperature until it becomes equal to zero at T_m (Figure 6.2D). The difference between the free energies of the folded and unfolded states is thus usually small: in the case of lysozyme, it is not greater than 16 kcal mol^{-1} (67 kJ mol^{-1}; see Figure 6.2C). The whole system, including the environment (i.e., the solution), must be taken into the account. The protein-stabilizing forces represent the sum of the hydrophobic interactions, van der Waals interactions, hydrogen bonds, and electrostatic interactions in the folded state, while the destabilizing forces are related to the greater conformational entropy and the solvation of nonpolar residues in the unfolded state. Enthalpy increases with temperature due to the disruption of the bonds that stabilize the protein structure, while entropy increases with temperature because the highly ordered structure of the protein becomes progressively less ordered. As explained in Chapter 2, the entropy contribution of water, which becomes increasingly more organized around the denatured state, can partly compensate for the contribution of the protein. The organized packages of water around the protein are called "icebergs," and with increasing temperature, these icebergs melt, thus contributing to the increase in the entropy of the system. The melting of the icebergs also contributes significantly to the specific heat C_p of the protein's denatured state. The temperature dependence of ΔH and ΔS is related to C_p, and can be combined, according to Equation 6.5, into the Gibbs–Helmholtz equation for the dependence of ΔG of the denaturation on the temperature. The free energy of the denaturation at any temperature T, ΔG_T, is thus

$$\Delta G_T = \Delta H_{T_m}\left(1 - \frac{T}{T_m}\right) - \Delta C_p\left[T_m - T + T\ln\left(\frac{T}{T_m}\right)\right] \qquad (6.6)$$

As is clear from Figure 6.2D, the net stability of the proteins decreases both at higher and lower temperatures. Proteins should, therefore, be expected to unfold at both high and low temperatures. The unfolding of proteins as a result of an increase in the temperature is a well-known phenomenon; however, unfolding at low temperatures can only be observed when it occurs in an accessible temperature range, above the freezing point of water, which is rarely the case. Nevertheless, the decreased stability of proteins at low temperatures has been observed in solutions with antifreeze additives. This low-temperature unfolding has thermodynamic characteristics that are the opposite of those observed at high temperatures; that is, the heat is released during the unfolding at low temperatures but is taken up during the unfolding at high temperatures. The thermal unfolding of proteins can be recorded by differential scanning calorimetry, with a comparison made against a reference buffer solution. Denaturation is an endothermal reaction, accompanied by a large absorption of heat, which is required for the disruption of the bonds. This can be recorded, and so the values of T_m and of the enthalpy of unfolding can be determined. Assuming a two-state denaturation process (Equations 6.1–6.6), the ΔG for the unfolding can be determined in different conditions (including the standard condition) from Equation 6.6, and K_{eq} can be calculated from Equation 6.4. The value of K_{eq} can also be obtained by determining the equilibrium values or the fractions of $[U]$ and $[F]$ using spectroscopic methods. The values of ΔG^o for unfolding in water are usually in the range 15–45 kJ mol^{-1}, yielding values of K_{eq} for the unfolding of 10^{-4}–10^{-7}. In other words, for these proteins, it is 10 thousand to 10 million times more probable that they are in the folded state rather than in the unfolded state in standard conditions.

The effects of denaturants on the stability of proteins can be explained in terms of preferential interactions of the denaturants with aqueous interfaces and protein surfaces. In general, denaturants tend to increase a protein's solubility and to interact preferentially with protein surface by replacing/excluding water from this interaction. A substance that interacts more favorably with the protein's surface than with water potentially acts as a denaturant, and one that is excluded from the protein surface by water stabilizes the protein's folded state. Both guanidinium chloride and urea have a complex effect on proteins and on water in the surrounding solution, predominantly due to their hydrogen-bond-forming potential. These two denaturants can build strong hydrogen bonds with the carbonyl oxygen of the polypeptide backbone and the polar amino-acid side chains, but they can also affect hydrophobic interactions inside the protein core by reorganizing the structure of water. The relatively strong interaction of the denaturants with aromatic residues (the ion–π interaction) has also been observed. The hydrogen bonds between the denaturants and water are comparable to water–water hydrogen bonding, but they tend to disrupt water's structure because of unfavorable geometry. By doing this, guanidinium chloride and urea decrease hydrophobic interactions between nonpolar amino acid residues by around 30%.

The general effect of these denaturants is the shift in equilibrium between the protein's folded and unfolded states toward the unfolded state. Because the unfolded protein offers more specific and nonspecific binding sites to denaturants in comparison to the folded protein, denaturants can act as ligands, showing a higher overall affinity toward the unfolded conformation than to the folded one. In the case of guanidinium chloride, counter ions also play an important role in denaturation. Their effect is in accordance with their solvation potential, described by the Hoffmeister series. Thus, guanidinium thiocyanate is a more potent denaturant than chloride, while guanidinium sulfate does not show a denaturating effect at all but instead it stabilizes the protein's folded structure.

Extreme values of pH promote the unfolding of proteins. There are three effects operating in this process. The first is related to the ionizable side chains that are hidden in the interior of the folded protein in a nonionized form. His and Tyr are good examples of this. The interior His can be protonated at a pH below six, but only in contact with water, which occurs when the protein unfolds. At a low pH, the protonation of an interior His stabilizes the unfolded protein form because of the negative effect of the charge on the hydrophobic interaction required for the refolding. The same is true for Tyr at pH >10. The second effect of an extreme pH is the disruption of the salt bridges that stabilize the protein structure. Each salt bridge can contribute to the stability of proteins with an energy of up to approximately 20 kJ mol^{-1}. The stronger the interaction between the charged groups in the salt bridge, the more extreme the change of pH that is needed for salt-bridge disruption. In a grossly simplified way, this can be explained in terms of the competition between the binding of the groups in the salt bridge to the proton or to each other. The third effect is the electrostatic repulsion of the protein surfaces at low pH, where the protein has a net positive charge because of the protonation of Asp and Glu, and at a high pH, where the protein has a net negative charge because of the deprotonation of Lys, Tyr, and Arg. There is no direct evidence of the contribution of this effect to denaturation, but the fact that proteins are the least soluble at an intermediate pH corresponding to the isoelectric point corroborates the involvement of charge in the protein's stabilization.

6.2 FOLDING PATHWAYS

6.2.1 INTERMEDIATES IN THE FOLDING PATHWAYS

Folding studies are undertaken by observing the refolding of the unfolded protein that is "returned" to simulated native conditions in vitro. The hydrogen-exchange method and spectroscopic methods, including absorbance, fluorescence, circular dichroism, small-angle x-ray scattering, NMR, and others are used to monitor protein-structure perturbations during folding. Since folding is usually very fast, these methods are combined with those that follow fast kinetic events, such as stopped-flow and temperature-relaxation methods. In order to assess the importance of a particular amino-acid residue for the stability of the protein, site-directed mutagenesis is frequently included. As discussed earlier, the two-state model of unfolding in accordance with first-order kinetics is valid for many small proteins (Equation 6.1). In refolding, more complex kinetics is often encountered, which is usually represented by Equation 6.7:

$$U \rightleftharpoons I \rightleftharpoons F \qquad (6.7)$$

where I represents an intermediate state. Note that several intermediate states along the pathway of the refolding can exist, and that Equation 6.7 is only the simplest representation of a more-than-two-state refolding model. Alan Fersht studied the refolding of two small proteins—the 64-residues-large chymotrypsin inhibitor 2 (CI2) and the 110-residues-large ribonuclease barnase—in detail (see Figure 6.3 for the structures). The CI2 consists of only one helix and a β-sheet composed of five strands, organized in both parallel and antiparallel configurations. All the peptidyl-Pro bonds are in a favorable trans configuration and no disulfide bridges are present. CI2 shows a simple two-state kinetics, in accordance with Equation 6.1, with a half-life of 13 ms at room temperature. Small deviations from the first-order kinetics were attributed to the minor cis–trans transitions of the peptidyl-Pro peptide bonds. Simulation studies revealed that particular patches in the α-helix and the β-sheet are important when forming the initial interactions in the core of the structure. The nucleation of the core is followed by a condensation around the core, followed by a collapse into the final native structure. The patches that promote nucleation are not present in the denatured state but emerge only during the transition state.

The results obtained with CI2 appear to be representative of the refolding of small proteins or small protein domains. The refolding of barnase is another story, however. It is a multimodal protein containing a central five-stranded β-sheet and three helices. In this case, no disulfide bridges are present. The multistate kinetics of the refolding is basically in accordance with Equation 6.7, but at least two major

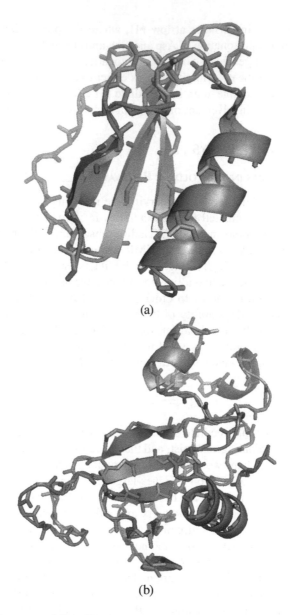

(a)

(b)

FIGURE 6.3   Structure of CI2 (PDB id 1COA) and barnase (PDB id 1BNI). Pictures were generated by the Pymol computer program. Side chains and hydrogen atoms were omitted for clarity. Presented as cartoons are α-helices and β-strands.

intermediate states are present. Kinetic and thermodynamic studies of the refolding of wild-type barnase and several mutated forms, as well as computer-simulation studies have revealed that the helices are formed first and very quickly. The simulations also indicate that the helices do not completely disappear in the denatured state; however, they do show dynamic transitions in which they are rapidly built and destroyed. The central part of the β-sheet is also formed relatively quickly, while the sheet periphery folds later. The main hydrophobic core of barnase, consisting of the β-sheet and one of the helices, is thus a nucleus around which the complete structure is built sequentially in subsequent steps. It seems that the helix-supporting β-sheet is not absolutely necessary since the structure also folds in the truncated barnase, which lacks the sequence of this helix. The initially formed β-sheet is planar, and it becomes curved only after establishing the interactions that maintain the tertiary structure. It thus seems that not the initial but rather the later folding events are rate-limiting, occurring with a half-life of around 30 ms at room temperature.

6.2.2 THE MOLTEN GLOBULE

In general it is believed that, during their folding, most proteins pass an intermediate state with the structural and thermodynamic characteristics somewhere in between the unfolded and native state. This intermediate state is called the molten globule. Whether or not a molten globule is a real intermediate or just a concept in understanding the folding mechanism is still a matter of debate. The properties that are usually ascribed to the molten globule are

- The dimensions and the average content of the secondary structure are near to that of the native state.
- The side chains of the amino acids are in homogeneous surroundings, and the tertiary structure is largely absent.
- The values of enthalpy are very close to that of the unfolded state.
- The structure is very dynamic with motions on a timescale longer than nanoseconds.
- The hydrophobic core is formed but packed very loosely.

In most cases the structures corresponding to the characteristics of the molten globule were observed during denaturation rather than during folding. In α-lactalbumin, for example, the molten-globule-like structure was detected using circular dichroism studies during denaturation by a decrease of the pH and by the introduction of moderate concentrations of guanidinium chloride. Similar structures were also observed after the addition of low concentrations of alcohol or fluoroalcohol and denaturing salts such as sodium perchlorate. During thermal denaturation, the molten globule state is seldom detected. Thermodynamically, folding is accompanied by a large entropy and free-energy decrease due to the formation of a more stable but, at the same time, more ordered structure. The folding process can be represented by the so-called folding funnel (see Figure 6.4) in which a gradual reduction of entropy and free energy results in it finally achieving the lowest native energy state. The molten global structure is positioned somewhere in the middle of the funnel, and the whole folding process can be regarded as a two-step procedure in accordance with Equation 6.7, in which the intermediate state I is a molten globule. The funnel in this figure is in accordance with the two-state folding mechanism and is mainly a result of hydrophobic collapse.

In a more complex multistep folding mechanism, the energy landscape should show several local energy minima, separated from the deepest minimum of the

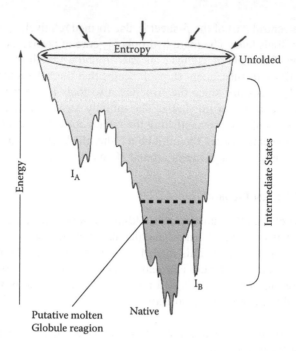

FIGURE 6.4 A scheme of a folding funnel for the folding mechanism comprising two inter-mediate states (I_A and I_B). (Modified from Clark, *P. L.* 2004. Protein folding in the cell: reshaping the folding funnel. Trends Biochem. Sci. 29, 527–534. With permission.)

native state by partly elevated passages corresponding to the activation energies. The parameter Q was introduced to measure the progress in the folding along the travel of the polypeptide chain from the top to the bottom of the folding funnel. Q is the ratio between the number of native pairwise contacts at any stage of the folding process and the total number of pairwise contacts in the native state:

$$Q = \frac{(\text{number of native pairwise contacts})_{\text{in state i}}}{(\text{total number of pairwise contacts in native state})} \tag{6.8}$$

The value of Q approaches zero at the top of the funnel and gradually increases dur-ing the folding process. A molten globule is characterized by a Q of around 0.27 and a transition state of roughly twice this value. The transition state is different from the transition states of ordinary chemical reactions. First, several pathways that can usually be observed for the folding of proteins will generate a number of different transition states. Additionally, proteins are large, and a huge number of bonds are involved in the transition states that dynamically oscillate between a large number of different conformations. The major transition state is thus a collection of structures with the highest free energy.

It is interesting to compare the molten globule with the intermediates observed in kinetic studies. An example is the folding of cyt *c*, which, as found by Colón and coworkers, folds sequentially at the neutral pH. Three intermediate states (I_1, I_2, I_3)

can be kinetically detected, according to Equation 6.9:

$$U \rightleftharpoons I_1 \rightleftharpoons I_2 \rightleftharpoons I_3 \rightleftharpoons F \tag{6.9}$$

It seems that the last intermediate, I_3, in some of the refolding conditions corresponds to the molten globule state. Similar to the case of cyt c, apomioglobin also shows a multistep folding pattern with at least two intermediate states. The second, later-stage intermediate seems to be analogous to the molten globule structure. In these two examples and in some other proteins, the molten globule appears relatively late during the folding procedure. In lysozyme, which shows a very complex folding mechanism (see Figure 6.5), in most of the refolding conditions, the molten globule stage has not been detected. However, two of the intermediates seem to be closely

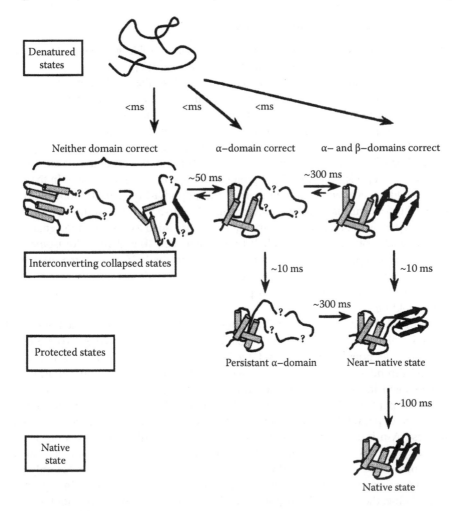

FIGURE 6.5 Suggested lysozyme refolding pathway. (From Radford, S. E. and Dobson, C. M. 1995. Insight into protein folding using physical techniques: studies of lysozyme and α-lactalbumin. Phil. Trans. R. Soc. Lond. B348, 17–25. With permission.)

related to the molten globule, one at the initial stage of the refolding and the other in a late stage of the process. The former apparently shares several characteristics with the similar initial-stage molten globule detected in α-lactalbumin folding. This shows that a molten globule can appear at different stages of the refolding and more than just once during the refolding pathway. Alternatively, particularly with small proteins, a molten globule stage can sometimes not be detected. The physiological relevance of the molten globule state attracts attention since it seems that native and nonnative states can coexist in the cells. The nonnative partially denatured states seem to appear transiently in some processes, for example, during the transloca-tion of proteins across membranes (see Chapter 7), in chaperone-assisted folding or refolding (see the following text), and during amyloidogenesis (see Chapter 29).

6.2.3 SLOW STEPS IN PROTEIN FOLDING

Two distinct processes can substantially slow down the folding process: the isomer-ization of the peptide bond preceding the Pro residue (peptidyl-Pro peptide bond) and a proper formation of the disulfide bonds. As described in Chapter 1, the pep-tide bonds in peptides and proteins are preferentially in the trans configuration. In general, the ratio between the cis and trans configurations is approximately 1:1000. However, the prolyl-peptide bond is an exception, with a cis:trans ratio of approxi-mately 1:3. In the native protein structure, the configuration of each peptidyl-Pro peptide bond is defined. A study of around 1500 proteins revealed at least one cis peptidyl-Pro peptide bond in 43% of proteins in the native conformation. In the pro-cess of folding, a proper configuration of the peptidyl-Pro peptide bond must be achieved before this part of the sequence can fold into a native structure. The transi-tion from the cis to the trans configuration is accompanied by an activation energy of around 85 kJ mol^{-1}, resulting in a half-time of between 10 and 100 s for isomeriza-tion at room temperature. This is much slower than the half-time of the folding of small proteins that do not require this isomerization, which is usually measured in milliseconds. Isomerization of the peptidyl-Pro bond can be a rate-limiting step in protein folding.

In unfolded proteins, peptidyl-Pro bonds can slowly isomerize, as demonstrated by studies of short peptides lacking the defined secondary and tertiary structures. A heterogeneous population of the unfolded protein is thus obtained, in which some peptidyl-Pro bonds are in the correct, and some in the incorrect, configuration. As demonstrated in the case of ribonuclease A, molecules with the correct conforma-tion of the peptidyl-Pro bond refold quickly (in less than a second), and those with the incorrect peptidyl-Pro bond configuration refold much more slowly (in several minutes). A mixture of the quickly and the slowly refolding processes makes the study of peptidyl-Pro bond isomerization difficult. Kinetic studies using a "double-jump" procedure (successive unfolding and refolding) seem to be reliable only in a limited number of cases; therefore, other methods must also be used. One approach to validating the role of the peptidyl-Pro bond isomerization in the folding pro-cess is to perform the folding experiment in the presence of prolyl isomerase, the enzyme that catalyzes the cis–trans isomerization of the peptidyl-Pro bond. The other approach is to introduce a mutation in the polypeptide chain, replacing the Pro residue that is involved in the cis peptidyl-Pro bond in the native conformation. In both cases the slowly folding phase should disappear. However, the interpretation of these experiments is not always straightforward. In mutation studies, for example, the residue that is introduced to replace Pro might affect the stability of the protein and thus the rate of folding. In proteins with several Pro residues in the sequence, prolyl isomerase accelerates the formation of the correctly, and also the incorrectly, configured peptidyl-Pro bonds in the unfolded protein, thus increasing the fraction of the protein that folds slowly. This was observed, for example, in the folding of carbonic anhydrase, which has 15 trans and two cis peptidyl-Pro bonds in the native conformation. Prolyl isomerases were found in both prokaryotes and eukaryotes. The well-known and thoroughly studied examples of these enzymes are cyclophilins, FKBP (FK506-binding protein), and parvulins. The first two are larger, and share some structural features, particularly the active site, being associated with the β-sheet and forming a hydrophobic pocket. They are not homologous proteins, and have a different substrate specificity. Cyclophilins are rather unspecific regarding the residue preceding Pro in the sequence, but FKBP shows a much higher affinity for substrates with hydrophobic residues preceding Pro. Thus, Leu and Phe are particularly favorable at this position, while replacing Leu or Phe with Glu decreases the efficiency of the enzyme by about 1000 times. Parvulins, which are found in *E. coli*, are rather small (only 92 amino acids) and resemble FKBP in their specificity; however, they show a lower affinity for the substrates, probably due to the smaller active site pocket.

An interesting feature of prolyl isomerases is their autocatalytic folding. This was demonstrated with the human FKBP12 containing seven trans peptidyl-Pro peptide bonds, and also with parvulins. Domains similar to prolyl isomerases have been found in some larger proteins. One such example is the *E. coli* trigger factor. This is a 48-kDa large protein that includes the FKBP-like domain with prolyl isomerases activity. It is associated with ribosomes and, as it seems, it assists in the cotranslational folding of the nascent polypeptide chain. The trigger factor might also have a protective role in the cell. Dimers of the trigger factor are very resistant to thermal unfolding and function as heat-shock proteins, responding to stress damage to the cell. A similar protective role for prolyl isomerases has also been suggested for cyclophilin Cys-40 and FKBP52 that bind to the heat-shock protein Hsp90.

The isomerization of the peptidyl-Pro bond has been extensively studied in ribonuclease A, which is probably the most studied enzyme so far. It contains four Pro residues, two in the cis and two in the trans configuration in the native form (see Figure 6.6). A very complex pattern of cis–trans isomerization exists in a multistep sequential folding mechanism in which all 16 possible isomeric states of the peptidyl-Pro bonds are included. It has been shown that, in some transitions between the unfolded state via several intermediate states to the final native state, the isomerization of the peptidyl-Pro bonds represents the rate-limiting process.

Another process that was initially considered as potentially slow is the proper formation of disulfide bridges. Proteins containing several Cys residues in the sequence can theoretically form a number of different disulfide bridges; however, in the native fold, disulfide bridges are precisely defined, and folding in the cells always results in a single, well-defined disulfide bond pattern. Investigating disulfide bond formation

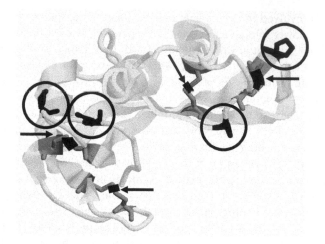

FIGURE 6.6   Structure of ribonuclease A (PDB id 3RN3) with the indicated four SS bridges (arrows) and four Pro (circles) related to the peptide bond cis–trans transitions. The picture was generated by Rasmol computer program.

during the folding process is possible by chemically modifying the free Cys residues, for example, by iodoacetate, and by subsequent identification of their positions in the protein sequence by sequencing the obtained peptides after proteolysis. The identification of the modified Cys residues has recently become much quicker as a result of the use of mass spectrometry. Alternatively, intermediate disulfide bonds can be trapped by decreasing the pH; this procedure does not require any chemical modifications. In ribonuclease A, there are eight Cys residues that could theoretically generate 28 different disulfide bonds (Figure 6.6). Ribonuclease can be totally denatured by disrupting the existing four native disulfide bonds (between the Cys residues 26–84, 40–95, 58–110, and 65–72). After refolding, these disulfide bonds regenerate spontaneously and accurately in a mildly reductive environment provided by the presence of a thiol reagent, such as mercaptoethanol. If no thiol reagent is present, folding can still be achieved, but the proper protein conformation is not regained since the activity of the enzyme corresponds only to a fraction of the normal activity.

The probability of randomly forming a proper disulfide pattern in ribonuclease is 1 in 105. Thiol reagents catalyze the disulfide exchange reaction (Figure 6.7A), which prevents the protein structure from remaining in the initially randomly formed disulfide bonds. Although disulfide bonds substantially stabilize native protein conformation, they do not determine this conformation. In the physiological conditions in the cells, the role of the thiol reagent could be played by the oxidized form of glutathione. However, establishing the correct disulfide bonds is so important during protein folding that specialized enzymes have evolved. These enzymes are protein disulfide isomerases (PDIs), and they occur in eukaryotes in the endoplasmic reticulum and in prokaryotes in the periplasm. A PDI efficiently catalyzes the disulfide exchange reaction shown in Figure 6.7; however, it also has other functions (the binding of Ca^{2+} and being a constitutive domain of some other proteins).

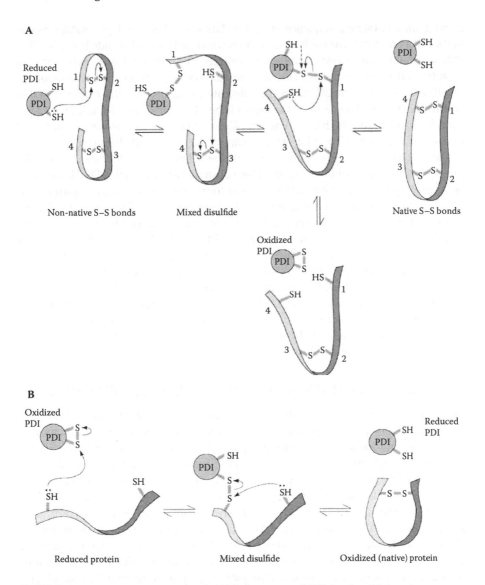

FIGURE 6.7 The role of protein disulfide isomerase (PDI) in the thiol exchange reaction (A) and in de novo disulfide bond formation (B). (From Voet, D. and Voet, J. G. 2004. Biochemistry, 3rd edition. John Wiley & Sons, Hoboken, NJ. With permission.)

The concentration of PDI in the endoplasmic reticulum is very high, up to 1 mM, which is similar to the concentration of glutathione in both the oxidized and reduced forms. The presence of glutathione makes the disulfide bonds in the endoplasmic reticulum very stable. Human PDIs are a family of enzymes related to thioredoxin, a protein that functions in a way that is analogous to a PDI, but in the cytosole. These PDIs consist of two identical subunits, each containing almost 500 amino acids organized in two pairs of the homologous domains a-a′ and b-b′. The domains a and a′ incorporate the sequence –Cys-Gly-His-Cys-. The two Cys residues can be in the reduced form (–SH) or both Cys can be oxidized into a disulfide bond (–S-S–). PDI is catalytically active in both the oxidized and reduced forms; however, this is in two different processes. The reduced PDI participates in the catalysis of the disulfide bonds' rearrangement in the substrate protein, while the oxidized PDI is able to catalyze the formation of the disulfide bonds in the de novo synthesized polypeptide chain (Figure 6.7B). The catalytic Cys residues are situated in the middle of the nonpolar region of the PDI, which enables the binding of the substrate protein in the unfolded form. It is believed that all the subunits of a PDI are required for the productive high-affinity binding of the substrate protein to the PDI. However, in order to keep the PDI active in the oxidized form, other proteins are required. One such protein is the membrane-associated flavoprotein named Ero1, which is largely conserved in eukaryotes. It contains two essential disulfide bonds included in a disulfide relay comprising a substrate protein, a PDI, an Ero1, and an FAD. This relay is needed for electron transfer from the newly formed disulfide bond to the FAD, which eventually passes the electron to an oxygen. Recently, other proteins similar to Ero1 have been identified, for example, flavoprotein Erv2. In prokaryotes, a similar disulfide relay exists in periplasm, consisting of a PDI named DsbA, a membrane protein DsbB, a quinone cofactor, and a cytochrome oxidase. Since the formation of the disulfide bond is associated with the oxidation of Cys—SH groups, PDI-assisted folding is sometimes referred to as oxidative protein folding.

6.3 ASSISTED FOLDING—CHAPERONES AND CHAPERONINS

Many proteins can only fold with the assistance of other proteins and protein complexes called molecular chaperones. The term chaperone is frequently used in a very broad sense and includes any molecule or structure that helps in protein folding. In this sense, prolyl isomerases, protein disulfide isomerases, and other proteins that assist in folding are also called chaperones. Complex multi-subunit structures that are also involved in assisted folding are large supramolecular machines that bind unfolded or partially folded proteins and drive their folding by the active process accompanied by the hydrolysis of ATP. These barrel-like structures were termed chaperonins. However, this terminology is not always strictly obeyed. Early observations related chaperones to the cell's stress response; this was due to their increased abundance following heat shock. Consequently, they were termed heat-shock proteins (Hsp), and are differentiated by their molecular mass—Hsp100, for example, denotes a heat-shock protein with a molecular mass of approximately 100 kDa.

Protein folding in vivo is not the same as protein folding in vitro. The concentration of proteins and other structures in the cytoplasm and endoplasmic reticulum is extremely high, including the large number of nascent protein polypeptide chains that are synthesized in parallel in many copies. This offers ample possibilities for the interactions and aggregation of the nascent proteins. Additionally, in in vitro folding, a complete protein polypeptide chain is exposed to a change of environment in order to collapse into the folded conformation. During protein synthesis, the N-terminal part of the polypeptide chain gradually emerges from the ribosome with a rate that is much slower than the average rate of folding. Therefore, local folding that started in the absence of other parts of the polypeptide chain and interfered with by interactions with other proteins could result in nonfunctional folding. To avoid this, specific proteins are used to protect the polypeptide chain from premature folding. In prokaryotes, this function seems to be associated with the trigger factor, discussed earlier, which is not only an efficient prolyl-isomerase but also shows a strong affinity for the unfolded polypeptide chain and interacts with the ribosome. In eukaryotes, a similar function was assigned to the nascent-polypeptide-associated complex (NAC). This heterodimeric protein not only protects the polypeptide chain from misfolding but also participates in targeting the protein to different organelles, such as endoplasmic reticulum and mitochondria. However, the NAC is not homologous to the trigger factor and shows no prolyl-isomerase activity. Two other systems also have a strong affinity for the nascent polypeptide chain, particularly toward its short hydrophobic segments. In prokaryotes this is the DnaK–DnaJ system, and in eukaryotes its homologous counterpart is Hsp70–Hsp40. The observed interactions with the nascent polypeptide chain, ribosomes, and trigger factor in prokaryotes, as well as with the NAC in eukaryotes, imply the function of these systems in the early phases of folding. Both systems assist protein folding via a cycle coupled to ATP hydrolysis. The DnaK–DnaJ cycle starts by binding the substrate protein to the 70-residues-large consensus sequence named the J domain in DnaJ. A substrate protein is then transferred from the DnaJ to the DnaK. This protein contains three major domains: the substrate-binding domain, the C-terminal α-helical domain, and the ATP-binding domain. The substrate-binding domain comprises a cavity that recognizes the hydrophobic amino acid residues in the substrate protein. The ATP-binding domain shows ATPase activity. The hydrolysis of the ATP induces a conformational change that drives the conformational alteration of the other two domains. The C-terminal domain functions as a cap that, upon hydrolysis of the ATP, closes the active-site cavity in the substrate-binding domain. After transferring the substrate protein to the DnaK, the DnaJ dissociates. This, together with substrate binding, provokes the conformational change of the DnaK, which strongly increases ATPase activity. The ATP is rapidly hydrolyzed, and the C-terminal domain closes the substrate-binding cavity with the substrate protein trapped inside. All this happens cotranslationally during protein synthesis, so that the entrapped part of the nascent polypeptide chain is protected from false folding. When the complete protein chain is eventually synthesized, a dimeric protein called GrpE is bound to the DnaK, which decreases the affinity of the DnaK for ADP by 200 times, and increases the affinity for ATP by 5000 times. This causes the exchange of ADP for ATP and a conformational change, resulting in opening of the cap of the substrate-binding cavity and release of the nascent protein. The system is now ready to start another cycle.

In eukaryotes, the ADP/ATP exchange and the release of the substrate protein are mediated by other nucleotide exchange factors, such as BAG-1 and HspBP1, while the roles of DnaJ and DnaK are played by Hsp40, and Hsp70, respectively. DnaK/Hsp70, DnaJ/Hsp40, and GrpE represent families of proteins with many related members. Frequently, DnaK/Hsp70 is regarded as chaperones, and the other players in the cycle are called cochaperones. The released nascent polypeptide chain can be subsequently left on its own to fold in the native form, or it can be transferred to another system that further assists folding. The next step in assisted folding is fulfilled by chaperonins. Two types of chaperonins are known: type I (group I) is found in prokaryotes, mitochondria and chloroplasts, and type II (group II) in archaea and eukaryotic cytoplasm. The best-known chaperonin family is the type-I GroEL chaperonin of *E. coli*. It is composed of two sets of seven identical subunits organized in two rings with the cavity inside. GroEL requires a cochaperonin GroES for the function. The structure of the GroEL complex is shown in Figure 6.8. Each GroEL subunit is composed of three domains: equatorial, intermediate, and apical. The equatorial domains that function as ATPases form the basis of the ring structure and keep both rings together. The intermediate and apical domains make up the walls and roof of the cavity. The apical domain has a hydrophobic inner surface adapted for the binding of partly unfolded proteins with exposed hydrophobic

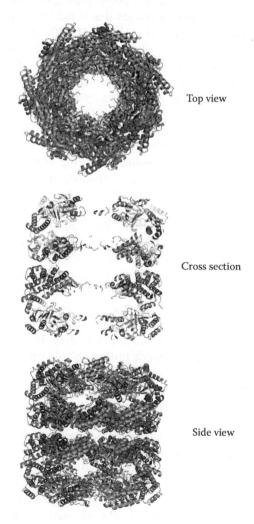

Top view

Cross section

Side view

FIGURE 6.8 The structure of GroEL (PDB id 1KP8) generated by Pymol computer program. Top view and shown cross section reveal the inner chambers of the protein complex.

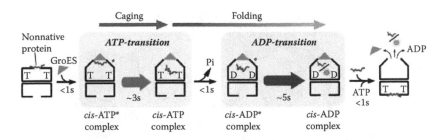

FIGURE 6.9 GroEL cycle containing two slower steps, one after binding of GroES to GroEL, and another after ATP hydrolysis. (From Ueno, T., Taguchi, H., Tadakuma, H., Yoshida, M., and Funatsu, T. 2004. GroEL mediates protein folding with a two successive timer mechanism. Mol. Cell 14, 423–434. With permission.)

residues. Proteins folded in the native structure and fully unfolded proteins do not tend to bind to the apical domain; it seems that the conformation of the protein that is able to bind resembles the molten globule structure. The apical domain also contains interaction sites for the binding of GroES, which is composed of seven identical subunits and functions as a cap to close the cavity in GroEL. The apical and intermediate domains are subjected to major conformational change after ATP hydrolysis catalyzed by the equatorial domain. This results in a twofold increase of the cavity volume and modification of the nature of the ring surface that is exposed to the cavity lumen from the hydrophobic to the more hydrophilic. The intermediate domain seems to be the transmitter of the conformational change from the equatorial to the apical domain.

Similar to the DnaK–DnaJ system, the GroEL–GroES system functions in cycles (Figure 6.9). The cycle starts with the binding of the substrate protein to the rim of the cavity on one ring of the GroEL complex; this ring is now named the cis-ring. Parallel to this, the ATP is bound to the equatorial domains of the same ring. Different researchers obtained different results regarding the sequence of these two events. In some models, the cis-ring is preloaded with ATP when the substrate protein binds (Figure 6.10), and in others the substrate is bound to the empty cis-ring and the ATP binds later. In the next step, GroES is bound to the same rim, and the substrate protein is transferred into the ring cavity. According to recent studies by Rye and coworkers, the binding of the ATP induces a conformational change in the cis-ring, accompanied by the forced unfolding of the substrate protein. Only now the ATP is hydrolyzed, causing a major conformational change, which by rotating the apical and the intermediate domains of the ring proteins by approximately 90°, uncovers the hydrophilic surfaces of the ring proteins and exposes them to the lumen of the cavity. The folding of the protein proceeds during this stage, and this is followed by the binding of a new molecule of the substrate protein to the rim of the trans-ring, accompanied by the binding of the ATP, which represents the start of the new cycle. Only now follows the release of the folded (or partly folded) protein, GroES, and ADP from the cis-ring. It is estimated that the folding of only 10% to 15% of the proteins in *E. coli* strongly depends on GroEL. These proteins are called class-III proteins, and they may undergo several GroEL cycles, being sometimes

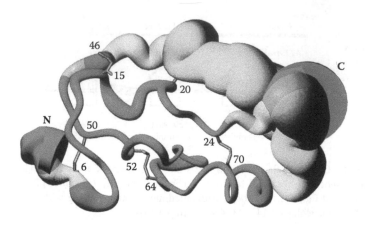

FIGURE 6.10 Flexibility of the structure of a marginally stable nematode anticoagulant protein C2 (NAPc2) is revealed from the apparent thickness of the backbone, particularly in the segments near the C-terminus. Residues in this region show great flexibility due to considerable conformational freedom, which does not allow for the precise determination of the coordinates of the atoms. (Reproduced from Duggan, B. M., Dyson, H. J., and Wright, P. E. 1999. Eur. J. Biochem. 265, 539–548. With permission.)

in contact with GroEL for several minutes. Class-II proteins pass only one GroEL cycle (this lasts about 10 s), and the folding of class-I proteins is GroEL independent. Eukaryotic chaperonins (class-II chaperonins) are not known in such detail as GroEL. They also consist of two rings, and the basic mechanism of action is similar to GroEL; however, their rings are composed of a larger number of subunits (usually eight or nine), and they do not require a separate cap of the GroES type. As shown earlier, chaperones and chaperonins protect de novo synthesized proteins against a premature false folding and assist in the folding of many proteins by dynamic modifications to their folding environment. In addition, chaperons participate in other important physiological tasks, such as the repair and clearance of misfolded or damaged proteins after cell stress. As discovered recently, chaperones are also present extracellularly, where they may have an important role in amyloid formation and degradation.

6.4 PROTEINS DESIGNED TO BE MARGINALLY STABLE

Proteins in their native fold are considered to be in their lowest energy state (see Figure 6.4) and thus stable. However, the native state is in most proteins very flexible. The atoms in a protein undergo a small-scale vibration of bond lengths and angles that occur at frequencies between 6×10^{12} s^{-1} and 10^{14} s^{-1}. Conformational change, which may be provoked by very small local deviations of lengths and bond angles, such as in hemoglobin, can spread over a large area in the protein, accompanied by a substantial modification of the protein structure. Still larger movements can occur in large protein domains that are flexibly linked to the protein core by hinge regions. In antibodies, for example, the domains rotate relative to each other in 10^{-8} to 10^{-7} s. Protein flexibility, therefore, involves movements of very different magnitudes on a time and space scale. In some proteins, parts of the structure can be extremely flexible. However, this does not represent a disadvantage for these proteins in terms of low stability but rather enables great adaptability of the structure during binding to different partners. An example of a marginally stable protein is a nematode anticoagulant protein C2 (NAPc2). As seen in Figure 6.10, large parts of this molecule are extremely flexible so that segments near the C-terminus practically lack the information on the secondary structure, as revealed by NMR studies undertaken by Peter Wright and coworkers. This reflects a high degree of motion on a nanosecond scale for this region and a high degree of conformational heterogeneity comparable to that in unfolded proteins. Such behavior has functional reasons. NAPc2 binds to the coagulation factor Xa with an inhibition constant of 2.4 nM. This tight binding is achieved by multiple weak interactions between the two proteins. This is possible only by the substantial adaptation of the inhibitor structure, to provide a large number of favorable contacts with the target protein. Without the extreme flexibility of the NAPc2 structure, such tight binding would not be possible. Additionally, NAPc2 also binds to other molecules, particularly to the tissue factor–factor VIIa complex, for which it requires the presence of the factor Xa. It was suggested that the flexibility of the NAPc2 structure is also advantageous in this process. In a similar way to NAPc2, some other proteins also exhibit a large structural flexibility. Further examples of such proteins are the transcription factors. The flexibility of their structure is pertinent to their function, associated with the binding to a large repertoire of different partner molecules.

FURTHER READING

1. Pain, R. H. (Ed.). 2000. Mechanisms of Protein Folding. 2nd ed. Oxford University Press, Oxford.
2. Fan, D. J., Ding, Y. W., Pan, X. M., Zhou J. M. 2008. Thermal unfolding of Escherichia coli trigger factor studied by ultra-sensitive differential scanning calorimetry. Biochim. Biophys. Acta 1784, 1728–34.
3. Tu, B. P. and Weissman, J. S. 2004. Oxidative protein folding in eukaryotes: mechanisms and consequences. J. Cell Biol. 164, 341–46.
4. Hiniker, A. and Bardwell, J. C. A. 2004. Disulfide relays between and within proteins: the Ero1p structure. Trends Biochem. Sci. 29, 516–19.
5. Horwich, A. L., Fenton, W. A., Chapman, E., Farr, G. W. 2007. Two families of chaperonin: physiology and mechanism. Annu. Rev. Cell. Dev. Biol. 23, 11–45.
6. Stan, G., Lorimer, G. H., Thirumalai D., and Brooks B. R. 2007. Coupling between allosteric transitions in GroEL and assisted folding of a substrate protein. PNAS 104, 8803–8.
7. Ueno, T., Taguchi, H., Tadakuma, H., Yoshida, M., and Funatsu, T. 2004. GroEL mediates protein folding with a two successive timer mechanism. Mol. Cell 14, 423–34.
8. Lin, Z., Madan, D. and Rye, H. S. 2008. GroEL stimulates protein folding through forced unfolding. Nature Struct. Mol. Biol. 15, 303–11.

7 Intracellular Sorting of Proteins

Gunnar von Heijne

CONTENTS

7.1 CELLULAR COMPARTMENTS

Cells are not bags of randomly mixing molecules but are highly organized entities with rich internal structures. On a gross level, a cell can be viewed as being divided into a number of compartments separated by lipid bilayer membranes, but even within a given compartment complex, macromolecular assemblies help to maintain the shape and function of the whole.

A rudimentary compartmentalization is present already in bacteria. In Gram-positive bacteria, the inner (or plasma) membrane sets the cytoplasm apart from the outside world and is in its turn surrounded by a thick peptidoglycan cell wall. In Gram-negative bacteria such as Escherichia coli, the cell wall is replaced by a second glycolipid membrane—the outer membrane—and the periplasmic space between the inner and outer membranes contains a thin peptidoglycan layer that shapes the cell and helps it withstand assaults such as osmotic shock.

Eukaryotic cells have a much more interesting inner life, sporting scores of membrane-delimited intracellular compartments: the nucleus, the compartments along the exo- and endocytic pathways (endoplasmic reticulum, Golgi apparatus, trans-Golgi network, endosomes, lysosomes), mitochondria, peroxisomes, and—in plants—chloroplasts. Each of these compartments has its own specific complement of proteins, most of which are encoded in the nuclear genome and need to be imported into the organelle from the cytosol.

Although the mechanisms of protein sorting and import differ in their molecular details between the different organelles, there are a couple of common themes. Proteins destined for a particular compartment share one or a few kinds of targeting signals, often linear stretches of amino acids that are identifiable already in the primary structure. Most import machineries further require that the polypeptide is unfolded during transit across the limiting membranes. Finally, import, being a unidirectional process, is driven by the hydrolysis of high-energy compounds such as ATP or GTP or by an electrochemical membrane potential.

In this chapter, we will discuss the most important protein-sorting pathways, starting with the secretory pathway and moving on to mitochondrial and chloroplast protein import, peroxisomal protein import, and protein trafficking in and out of the nucleus.

7.2 PROTEIN SECRETION AND MEMBRANE PROTEIN ASSEMBLY IN BACTERIA

Gram-negative bacteria can sort proteins between five different compartments: the cytoplasm, the inner membrane, the periplasm, the outer membrane, and the extracellular milieu. The cytoplasm is the default location: proteins devoid of a targeting signal or hydrophobic transmembrane segments remain in this compartment.

Sorting of water-soluble proteins to the periplasm generally requires the presence of an N-terminal signal peptide (Figure 7.1A). The typical signal peptide is a stretch of some 15-25 amino acids composed of three distinct regions. The n-region contains one or a few positively charged amino acids, the central h-region is defined by a contiguous uncharged stretch of mainly hydrophobic amino acids, and the c-region harbors the cleavage site for the signal peptidase I enzyme. Since periplasmic proteins need to be translocated across the inner membrane in an unfolded conformation, the presence of a signal peptide is not always sufficient to ensure efficient translocation but the protein must also not fold too quickly in the cytoplasmic environment.

Protein translocation across the inner membrane is posttranslational, that is, the protein is fully made and released from the ribosome before it engages the secretory machinery in the inner membrane. As an insurance against premature folding, most periplasmic proteins bind to the cytoplasmic chaperone SecB as they come off the ribosome and therefore remain unfolded during their short stay in the cytoplasmic compartment.

SecB hands over the unfolded polypeptide to the SecAYEG translocation machinery (or translocon) in the inner membrane (Figure 7.1B). SecA is an ATPase that cycles on and off the membrane, while the SecYEG complex forms a translocation channel across the inner membrane. The signal peptide interacts first with SecA and is then transferred to a binding site inside the SecYEG complex, triggering the opening of the translocation channel. Using ATP hydrolysis as the energy source, SecA then drives the nascent chain through the SecYEG channel into the periplasm. Finally, signal peptidase I cleaves the signal peptide from the translocated protein, which then folds in the periplasm.

High-resolution structures are available for all the main players in this pathway (Figure 7.2). SecB is a chaperone with two shallow hydrophobic grooves on its surface

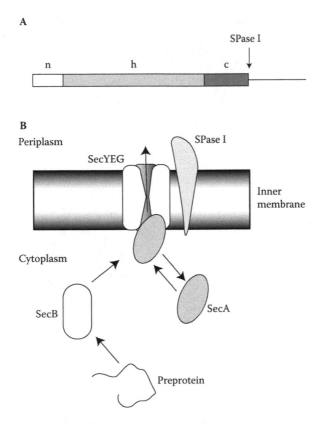

FIGURE 7.1 (A) The canonical signal peptide. The short n-region is positively charged and 1–5 residues long, the h-region is hydrophobic and 7–15 residues long, and the c-region harbors the cleavage site for signal peptidase I and is 5–6 residues long. (B) The posttranslational pathway for protein translocation across the inner membrane in *E. coli*. The preprotein is first bound to the SecB chaperone, then delivered to translocon-bound SecA, and finally translocated through the SecYEG translocon in an ATP-driven process. The SecA ATPase can cycle between a translocon-bound and free state. Finally, the signal peptide is cleaved from the mature protein by signal peptidase I.

that are thought to bind segments of unfolded proteins. SecA is a complex molecule composed of five separate domains: the wing domain (WD), the nucleotide-biding domain (NBD), the preprotein-biding domain (PBD), and two ATPase regulatory domains (IRA1 and 2). Somehow, ATP hydrolysis in the NBD domain can drive the unfolded polypeptide through the SecYEG translocon, but the details of how this works are still unclear.

The only high-resolution structure available for the SecYEG complex is of the channel in a closed state. In this state, the channel is obstructed from the periplasmic side by a "plug domain" that is thought to swing out when the signal peptide binds. This would open an hourglass-shaped channel through the protein, constricted near its center by a ring of hydrophobic residues. Possibly, the diameter of the ring can adapt to the size of the translocating polypeptide, forming a tight seal around it that

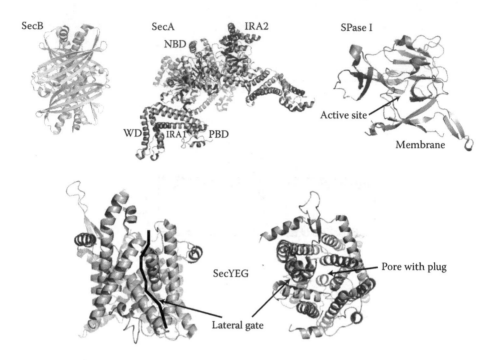

FIGURE 7.2 Structures of E. coli SecB (PDB code 1QYN), SecA (PDB code 2FSG), the catalytic domain of signal peptidase I (PDB code 1B12), and the SecYEG translocon from an archaea (PDB code 1RH5) as seen in the plane of the membrane (left) and from the cytoplasm (right). The wing domain (WD), nucleotide-binding domain (NBD), preprotein-binding domain (PBD), and two intramolecular regulator of ATPase domains (IRA1 and 2) are indicated on SecA. The active site and the membrane-facing side are indicated on SPase I. The lateral gate and the pore (with the plug domain in the middle) are indicated on the SecYEG translocon.

prevents ions from leaking across the channel. There is a putative "lateral gate" between transmembrane helices 2 and 7 in SecY that is thought to open up toward the surrounding membrane to give hydrophobic segments in the translocating chain free passage into the lipid bilayer where they will form transmembrane α-helices (see Chapter 33).

Signal peptidase I protrudes from the periplasmic side of the inner membrane, and its active site is located near the membrane–water interface region. The c-region of the signal peptide binds in an extended conformation to an exposed β-strand in the enzyme, positioning the scissile bond near a catalytic dyad where the hydroxyl of a Ser residue performs the nucleophilic attack, and the ε-amine of a Lys residue serves as the general base (in eukaryotic signal peptidases, the Lys residue is generally replaced by a His).

7.2.1 THE TAT SECRETION MACHINERY

Bacteria (and the thylakoid membrane in chloroplasts) have a second translocation machinery for periplasmic proteins, the so-called Twin-Arginine Translocation

(TAT) machinery. The TAT translocon is unique in that it transports folded proteins across the inner membrane and can even handle oligomeric protein complexes, thereby allowing certain proteins devoid of TAT signal peptides to "piggyback" on proteins that have such signals. Typically, the TAT machinery is used by proteins that need to pick up various cofactors in the cytoplasm before moving to the periplasm, and hence must fold prior to translocation.

The TAT signal peptides are similar to the Sec signal peptides but have a distinct twin-arginine (-Arg-Arg-) motif in the n-region. Their h-regions also tend to be less hydrophobic than the h-regions of the Sec signal peptides. The detailed workings of the TAT machinery are unknown, but it is believed that the TatA protein can assemble into multimeric channels of varying stoichiometry to adapt to substrate proteins of different sizes.

7.2.2 OUTER- AND INNER-MEMBRANE PROTEINS

The SecYEG translocon is also used by proteins destined for both the outer and inner bacterial membranes. Similar to periplasmic proteins, outer membrane proteins are made with an N-terminal signal peptide and are translocated posttranslationally into the periplasmic space where they are taken care of by a set of periplasmic chaperones. The integration into the outer membrane involves the BamA (also called Omp85 or YaeT) complex present in the outer membrane, but how this complex ensures the proper assembly of the outer membrane proteins is unknown.

Lipoproteins—periplasmically exposed proteins anchored to the inner or outer membranes via an N-terminally attached lipid—are diverted from the main SecYEG translocation pathway at the level of signal-peptide processing. Rather than being cleaved by signal peptidase I, lipoproteins receive their lipid anchor when cleaved by the lipoprotein signal peptidase (signal peptidase II), which transfers a lipid onto a Cys residue situated immediately downstream of the cleavage site. Unless provided with a "Lol avoidance" signal (an Asp residue next to the lipid-modified Cys), lipoproteins interact with the LolCDE complex in the inner membrane, are handed over to the LolA periplasmic chaperone, and their lipid anchor is integrated into the outer membrane with the help of the LolB lipoprotein.

In contrast to periplasmic and outer membrane proteins, inner membrane proteins are generally inserted cotranslationally into the membrane. Most inner membrane proteins lack a cleavable signal peptide; instead, targeting to the SecYEG translocon is ensured by the most N-terminal transmembrane segment acting as a targeting signal. Transmembrane segments are typically both longer and more hydrophobic than signal peptides, and can bind to the so-called signal recognition particle (SRP) as they emerge from the ribosome. In bacteria, SRP is composed of the Ffh protein and a 4.5S RNA molecule, while the eukaryotic SRP has a more complex structure.

SRP, together with its receptor FtsY, targets the ribosome/nascent-chain complex to the SecYEG translocon. The ribosome then binds to the translocon, SRP/FtsY are released, and the growing polypeptide is fed into the SecYEG channel. Periplasmic domains in the membrane protein are translocated through the SecYEG channel, while hydrophobic transmembrane α-helices are moved laterally from the channel into the surrounding lipid bilayer through a "lateral gate" in the channel wall (Figure 7.2).

7.2.3 DEDICATED SECRETION SYSTEMS

Gram-negative bacteria harbor specialized secretion machineries that allow them to secrete proteins through the outer membrane into the medium or even inject proteins directly into target eukaryotic cells. These machineries can be classified in at least three groups, called type I–III secretion systems. Briefly, type I secretion systems are independent of the SecYEG machinery and are composed of an inner-membrane ATPase coupled via an adapter protein to TolC, a long smokestack-like molecule that forms a continuous tube through the periplasm and the outer membrane. The substrate protein is fed into and pushed through the TolC tube using energy derived from ATP hydrolysis by the inner-membrane ATPase.

Type II secretion systems act on substrates that are translocated into the periplasm by the SecYEG or TAT machineries. The secretion machinery itself couples ATP hydrolysis in the cytoplasm, via an inner membrane component and periplasmic coupling proteins, to a multimeric, ring-shaped outer-membrane secretion channel.

Type III secretion systems are highly complex assemblages with a basal body built from two large ring-like structures, one in the inner and one in the outer bacterial membrane, connected by a periplasmic cylindrical structure. The periplasmic cylinder continues outside the cell in the form of a hollow needle-like structure with a tip that can bind to and penetrate the target membrane. Substrate proteins are thought to be pushed through the needle into the target cell. The whole machinery is powered by ATP hydrolysis in the cytoplasmic part of the complex.

Finally, some proteins carry their own dedicated outer-membrane secretion pore into the periplasm. Such autotransporters are translocated into the periplasm by SecYEG, whereupon their translocator domain inserts into the outer membrane, forming a large β-barrel. The effector domain then translocates through the β-barrel pore to the outside of the cell, where it may be cleaved off from the transporter domain.

7.3 PROTEIN TARGETING TO THE SECRETORY PATHWAY IN EUKARYOTIC CELLS

The secretory (or exocytic) pathway leads from the endoplasmic reticulum (ER) through the Golgi apparatus and the trans-Golgi network to the plasma membrane (with a side-branch to the lysosome). The endocytic pathway brings proteins and lipids back from the plasma membrane into endosomes and onwards to the lysosome, or recycles them to the plasma membrane via the trans-Golgi network. In polarized cells, proteins and lipids can be shipped across the cell by transcytosis between the two separated plasma-membrane domains. Trafficking between compartments in the exo-, endo-, and transcytotic pathways is carried out by vesicular transport, and will not be further discussed here.

The entry point to the secretory pathway is the ER. In animals and plants, protein translocation into the ER is cotranslational, while in fungi both co- and posttranslational mechanisms exist. There are many similarities between protein translocation across the inner bacterial membrane and across the ER membrane, and many of the components are homologous between prokaryotes and eukaryotes. The lumen of the ER hosts systems for protein folding and quality control, and misfolded proteins can be dislodged from the ER back into the cytoplasm for degradation.

Just as in bacteria, the cotranslational pathway in eukaryotic cells depends on the SRP and an SRP-receptor in the ER membrane. The eukaryotic SRP is more complex than its bacterial counterpart, and not only binds to signal peptides or transmembrane segments emerging from the ribosome but also slows down translation of mRNAs encoding secreted and membrane proteins. Translational slowdown is thought to give the ribosome more time to dock to an empty translocon while ensuring that the ensuing translocation reaction will still be cotranslational.

Signal peptides in eukaryotic proteins look much like their bacterial counterparts but are always recognized by the SRP. The translocon (called the Sec61 complex) in the ER is also similar to the bacterial SecYEG complex, as is the signal peptidase. After signal-peptide recognition, the SRP docks to its receptor on the ER membrane and hands over the ribosome/nascent chain complex to the Sec61 translocon in a process orchestrated by interdependent GTP-hydrolysis reactions on the SRP and the SRP receptor. Polypeptide translocation through the Sec61 channel is driven by chain elongation on the ribosome and possibly also by the binding of BiP, a lumenal chaperone of the Hsp70 family, to the nascent chain as it emerges on the lumenal side of the membrane.

Besides signal-peptide cleavage, most secreted proteins are also covalently modified by glycosylation as they enter the ER. Glycosylation is catalyzed by oligosaccharide transferase, an enzyme that is colocalized with the Sec61 translocon and that scans the nascent polypeptide as it emerges from the translocon channel. The substrate for oligosaccharide transferase is a short motif in the nascent chain: Asn-X-Thr-Y or Asn-X-Ser-Y (where X and Y can be any amino acid except Pro).

Asn-linked oligosaccharides not only make proteins more soluble and robust but also function in the quality-control system in the ER. Quality control depends on cycles of glucosylation/deglucosylation reactions on the Asn-linked oligosaccharides, from which the protein can escape only by attaining a properly folded structure. Proteins that get trapped in misfolded states are eventually shipped back across the ER membrane in a poorly understood process and are finally targeted for degradation by the cytoplasmic proteasome system.

Integral membrane proteins destined for the different compartments of the secretory pathway (including the plasma membrane) are inserted cotranslationally into the ER membrane by the SRP/Sec61 system, as described earlier for bacterial inner-membrane proteins.

7.4 PROTEIN TARGETING TO MITOCHONDRIA

Proteomics studies indicate that mitochondria contain on the order of 1000 different proteins, yet only a handful (13 in human mitochondria) are encoded in the mitochondrial genome and synthesized inside the organelle. The remaining 99% are encoded in the nuclear genome and must be imported from the cytoplasm.

The mitochondrial protein import machinery is surprisingly complex. So far, the outer mitochondrial membrane has been shown to contain two protein complexes involved in protein import and assembly—the TOM and SAM complexes—and two distinct import machineries—the TIM23 and TIM22 complexes—that are present in the inner mitochondrial membrane. In the matrix, one finds the PAM import-motor complex, the presequence peptidase (MPP) and other proteases, and chaperones required for folding of the imported proteins. In addition, the intermembrane space between the outer and inner membranes contains components such as the "small TIM" chaperones and the MIA complex. In total, more than 30 different proteins, many of which are encoded by essential genes, have been identified in these complexes so far.

Posttranslational protein imported into isolated mitochondria can be reconstituted in vitro, and full-length precursors of mitochondrial proteins bound to chaperones such as Hsp70 and Hsp90 can be detected in the cytoplasm. Posttranslational import seems to be the rule, and it is still an open question whether there exist mitochondrial proteins that can only be imported cotranslationally.

Most proteins located in the mitochondrial matrix are targeted for import by an N-terminal mitochondrial targeting peptide—a stretch of residues that can fold into an amphipathic α-helix with one hydrophobic and one positively charged face. Negatively charged and helix-breaking residues (Asp, Glu, Pro) are largely absent from the targeting peptides. The Tom20 receptor in the outer membrane has a hydrophobic groove into which the amphipathic α-helix can bind, positioning the precursor protein for transfer into the general import pore of the TOM complex. Proteins with internal targeting sequences are initially bound by another receptor, Tom70, before transfer to the general import pore; the critical sequence characteristics of internal targeting sequences are not well understood.

The precise mechanism that drives precursor proteins across the TOM channel is not entirely clear. The prevailing model posits that there are binding sites for the targeting peptide both on the cytoplasmic (cis) side and the intermembrane-space (trans) side of the outer membrane, and that the latter have a higher binding affinity for the targeting peptide. The N-terminal targeting peptide would therefore be pulled across the TOM complex by a series of interactions with increasingly strong binding sites and finally be positioned in the intermembrane space, ready to be transferred to the TIM23 complex.

The TOM and TIM23 complexes interact with each other via the Tim21 protein, and this protein may also help release the targeting peptide from the TOM complex. Upon insertion into the TIM23 channel, the positively charged targeting peptide is believed to be driven across the inner membrane by the membrane electrical potential ($\Delta\psi$). The remainder of the precursor protein is pulled into the matrix by an ATP-powered motor, the PAM complex. The central component of the PAM complex is mitochondrial Hsp70, an ATP-dependent chaperone that binds the incoming, unfolded polypeptide and prevents its backsliding in the import channel. The binding of multiple Hsp70 molecules to the preprotein eventually leads to complete translocation into the matrix, where, after removal of the targeting peptide, the protein finally folds with the help of additional chaperones.

Some membrane proteins located in the inner membrane are made with a typical N-terminal targeting peptide, followed by a single hydrophobic transmembrane segment. Such proteins use the TIM23 complex both for translocation of the matrix-localized domain across the inner membrane and the lateral insertion of the transmembrane segment into the inner membrane.

The TIM22 complex in the inner membrane is specialized for another class of inner-membrane proteins, mostly belonging to the class of mitochondrial metabolite carrier proteins (e.g., the ADP/ATP carrier). These proteins have multiple transmembrane segments and are targeted to the TOM complex by internal targeting sequences recognized by the TOM70 receptor. Upon transfer into the intermembrane space, they bind to the TIM9/10 chaperone complex (the "small TIMs"), are delivered to the TIM22 complex, and are finally inserted into the inner membrane.

Proteins can also be targeted to the inner membrane via TIM23-mediated import into the matrix, followed by insertion into the membrane from the matrix side. The insertion step is carried out by the so-called Oxa1 translocon in the inner membrane, a protein with homology to the translocon-associated protein YidC found in *E. coli* and other bacteria that can insert certain small inner-membrane proteins independently of the SecYEG machinery.

Proteins are delivered into the intermembrane space by either of two routes: TIM23-mediated insertion of a transmembrane precursor form into the inner membrane, followed by proteolytic removal of the membrane anchor by an intermembrane-space protease, or direct import through the TOM channel driven either by the formation of a disulphide-bonded complex with the Mia4 protein in the intermembrane space or by the irreversible association with proteins already present in the intermembrane space. The first route is a simple extension of the TIM23 pathway discussed earlier. In the second, the imported protein is diverted from the TOM–TIM pathway and trapped in the intermembrane space by virtue of its affinity for other proteins in that compartment. An important player in the intermembrane space is the MIA complex that can trap small cysteine-rich proteins by forming intermolecular disulphide bonds and then release them in a properly folded state.

Many proteins in the outer mitochondrial membrane have a β-barrel structure, as do bacterial outer-membrane proteins. In both cases, membrane insertion is mediated by evolutionarily related outer-membrane protein complexes: the BamA complex in bacteria (see section 7.2) and the SAM complex in mitochondria. The Tim9/10 chaperone complex binds the β-barrel proteins as it exits the TOM channel and hands it over to the SAM complex for insertion into the outer membrane.

7.5 PROTEIN TARGETING TO CHLOROPLASTS

Being an organelle with three membrane systems—the outer envelope, the inner envelope, and the thylakoid membrane—the chloroplast presents a particular challenge for protein sorting. As for mitochondria, only a small number of proteins are encoded in the chloroplast genome, and the great majority have to be imported post-translationally from the cytoplasm and routed to the correct subcompartment.

The chloroplast-targeting peptides, called transit peptides, contain a high proportion of hydroxylated amino acids and few, if any, negatively charged residues. They have no discernible secondary structure preferences, and may in fact be random coils in solution.

In contrast to mitochondria, only one outer membrane (TOC) and one inner membrane (TIC) translocation complex have been identified to date. Docking of the precursor protein and translocation across the outer and inner membranes depend on GTP and ATP hydrolysis, and Hsp70 chaperones in the intermembrane space and the stroma may drive separate TOC and TIC translocation reactions.

A large number of components in the TOC and TIC complexes have been isolated, but their functions remain largely conjectural. Toc75, a large β-barrel protein, is likely to be the major import channel in the TOC complex. Toc75 has significant sequence similarity to the bacterial BamA protein involved in mediating the insertion of β-barrel proteins into the outer bacterial membrane. The bacterial BamA, the mitochondrial SAM, and the chloroplast TOC complexes thus appear to have a common evolutionary origin.

In the TIC complex, Tic20 has detectable sequence similarity to the channel subunits Tim23 and Tim17 in mitochondria and may form part of the TIC channel, possibly together with Tic110.

After import into the stroma, many proteins are further routed to the thylakoid. The pathways for protein import into the thylakoid are basically the same as those involved in protein translocation across the inner bacterial membrane. Proteins destined for the lumen of the thylakoid are made with a bipartite presequence composed of an N-terminal transit peptide followed by a bacterial-type Sec or TAT signal peptide.

Thylakoid membrane proteins in most cases use either the SRP/YidC pathway or a "spontaneous" insertion pathway in which no protein components have been identified so far.

7.6   PROTEIN TARGETING TO PEROXISOMES

Peroxisomes are surrounded by a single membrane and lack an organellar genome. Protein import is posttranslational, and both folded proteins and oligomeric protein complexes can be imported.

Two kinds of peroxisomal targeting signals are known, called PTS1 and PTS2. PTS1 is a highly conserved C-terminal tripeptide –Ser-Lys-Leu (-SKL). Proteins with a PTS1 are bound by the soluble cytoplasmic receptor Pex5, which in turn binds to receptors on the surface of the peroxisome. The nature of the translocation step is unknown, but it is thought that the Pex5 receptor accompanies the cargo protein into the peroxisome and then recycles out of the organelle.

PTS2 signals are less well defined than PTS1 and are found at the N-terminus of some peroxisomal proteins. The cytoplasmic PTS2 receptor is called Pex7, and is also thought to accompany the cargo protein into the peroxisome.

7.7   NUCLEAR PROTEIN TRAFFICKING

In eukaryotic cells, the processes of transcription and translation are separated by the nuclear membrane. This necessitates an intense trafficking of proteins and RNA into and out of the nucleus. Transport is mediated by the nuclear pore complex

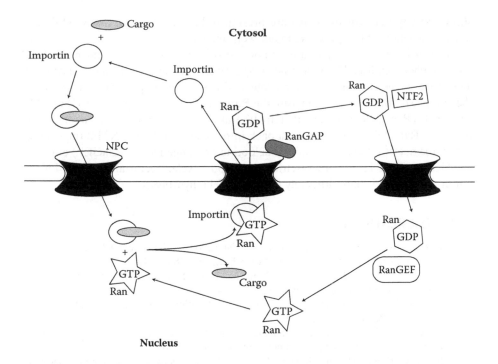

FIGURE 7.3 Ran-dependent import of cargo protein into the nucleus. Cargo proteins equipped with a nuclear localization signal bind to an importin-αβ dimer in the cytosol. The cargo–importin complex then transits through the nuclear pore complex (NPC). In the nucleus, RanGTP displaces the cargo from importin-αβ; the RanGTP–importin complex is reexported to the cytosol while the free cargo remains in the nucleus. In the cytosol, the RanGTP–importin complex meets the RanGTPase-activating protein (RanGAP), inducing the conversion of RanGTP to RanGDP, which lacks affinity for importin-αβ. RanGDP then binds the import factor NTF2 and translocates back into the nucleus. Finally, RanGTP is regenerated from RanGDP by the action of the Ran guanine-nucleotide exchange factor (RanGEF) that catalyzes the exchange of GDP for GTP (which is present at high concentration in the nucleus).

(NPC), a huge complex of ~100-MDa molecular weight composed of multiple copies of ~30 different polypeptides. It has been estimated that a single NPC can maintain a transport rate of up to 1,000 cargo proteins per minute. Proteins smaller than ~40 kDa can diffuse freely through the NPC, but larger proteins and protein complexes require active transport for entry or exit.

Cargo protein recognition and transport through the NPC is carried out by a family of proteins collectively called karyopherins ("importins" and "exportins") that act together with the Ran GTPase.

A simplified view of the transport cycle is shown in Figure 7.3. Cargo proteins contain nuclear localization signals (NLSs). The classical NLS is composed of one or two closely spaced stretches of four to five basic amino acids (Lys or Arg) and binds to an importin-αβ dimer in the cytoplasm. The cargo–importin complex can partition into and cross the NPC by virtue of its affinity for so-called FG-repeats, short Phe-Gly rich structures that are present in high numbers in many NPC proteins. The details of this process are not well understood.

Once inside the nucleus, the cargo–importin-αβ complex is dismantled by RanGTP and NUP50/CAS/RanGTP that bind, respectively, to importin-β and importin-α. This releases the cargo inside the nucleus and primes the importins for reexport to the cytoplasm. Back in the cytoplasm, the Ran GTPase-activating protein (RanGAP) triggers GTP hydrolysis on Ran, thereby liberating the importins and generating free RanGDP. RanGDP is finally brought back into the nucleus by the NTF2 import factor. In the nucleus, the Ran guanine-nucleotide exchange factor (RanGEF) converts RanGDP to RanGTP, thereby resetting the import cycle. Vectorial transport into or out of the nucleus is thus ultimately driven by GTP hydrolysis.

FURTHER READING

1. Bohnert, M., Pfanner, N., and van der Laan, M. (2007). A dynamic machinery for import of mitochondrial precursor proteins. FEBS Lett. 581, 2802–2810.
2. Cook, A., Bono, F., Jinek, M., and Conti, E. (2007). Structural biology of nucleocytoplasmic transport. Annu. Rev. Biochem. 76, 647–671.
3. Neupert, W. and Herrmann, J. M. (2007). Translocation of proteins into mitochondria. Annu. Rev. Biochem. 76, 723–749.
4. Rapoport, T. A. (2007). Protein translocation across the eukaryotic endoplasmic reticulum and bacterial plasma membranes. Nature 450, 663–669.
5. Ruiz, N., Kahne, D., and Silhavy, T. J. (2006). Advances in understanding bacterial outer-membrane biogenesis. Nat. Rev. Microbiol. 4, 57–66.
6. Stewart, M. (2006). Molecular mechanism of the nuclear protein import cycle. Nature Rev. Mol. Cell. Biol. 8, 195–208.
7. Wickner, W. and Schekman, R. (2005). Protein translocation across biological membranes. Science 310, 1452–1456.

Protein Degradation

8 Protein Turnover

Matjaž Zorko

CONTENTS

For the cell, the correct balance of protein synthesis with protein degradation is essential. Half-lives of proteins vary from several seconds to over 100 days. As observed, half-lives of proteins are related to the structure of their N-terminus. Proteins with N-terminal Met, Ser, Ala, Thr, Val, or Gly are long-lived, having usually half-lives greater than 20 h. Those with N-terminal Phe, Leu, Asp, Lys, or Arg are rapidly degraded, having half-lives of few minutes or less. This is called the N-terminal rule. Another rule is the PEST rule. Proteins rich in Pro, Glu, Ser, and Thr (one letter abbreviations are P, E, S, T) are more short-lived as other proteins. Proteins accumulate modifications via nonenzymatic and enzyme-catalyzed reactions, including partial proteolysis and becoming less active with aging. Protein degradation is needed to remove damaged or unwanted proteins and to recycle amino acids. Two main protein-degrading systems exist. Around 90% of all proteins are degraded by proteasomes, large supramolecular proteolytic structures. They are found in eukaryotes and prokaryotes and degrade mainly defective and partly unfolded proteins. Lysosomes represent another system for protein degradation in eukaryotes. These organelles containing different hydrolytic enzymes, including proteases, are mainly intended to degrade extracellular proteins, membrane proteins, and proteins with very long half-lives.

8.1 PROTEIN DEGRADATION BY PROTEASOMES

Lysosomes were thought to be the main site of protein degradation in cells until, in 1970s and early 1980s, a ubiquitin-dependent protein degradation coupled to ATP hydrolysis was discovered and related to the large protein complex called proteasome. Work accomplished by a number of researchers culminated by elucidating the

main features of the ubiquitin proteasome system and resulted in the Nobel Prize to
Hersko, Ciechanover, and Rose in 2004.

8.1.1 PROTEASOME STRUCTURE

Proteasome consists of the 20 S core complex to which two types of the regulatory
cap structures of 19 S or 11 S can be attached. The core complex is a cylindrical
structure composed of four stacked rings of proteins, two α-rings, and two β-rings
(Figure 8.1). Seven α-type proteins form each of the two α rings at the ends of the
structure, and seven β-type proteins compose each of the two central β-rings. In
eukaryotes, there are seven different but closely related α-subunits, and the same
number of related but not identical β-subunits. In prokaryotes, all α- and β-subunits
are identical. Inside the core complex, there are three cavities separated by narrow
passages. Three of the β-subunits function as proteases, each with different substrate
specificity, determined by the substrate residue recognized by the protease P1 position (the first residue at the N-terminal site of the peptide bond to be cleaved). One
subunit has a chymotrypsin-like activity, with Tyr and Phe preferentially bound to
P1. Another resembles trypsin, accommodating the basic amino acids Lys or Arg in
the position P1, and the third's P1 recognizes the acidic residue such as Glu or Asp.
Though similar in specificity, they are not Ser-proteases.

The role of the nucleophile-attacking group is associated with the Thr hydroxyl
group. All three proteases are synthesized in the inactive proenzymes and are activated by the cleavage near the N-terminus, making Thr the N-terminal residue. The
active centers of all three types of proteases are situated in the luminal space of the
proteasome. When the core complex of the proteasome is not associated with the
caps, it seems to be closed. N-terminal domains of α-subunits act as a gate that opens
after the interaction of the core complex with the cap. The association of two 19 S
caps with a proteasome core complex, each to one α-ring, results in the formation
of the 26 S protreasome. The atomic structure of the 26 S proteasome has not been
resolved yet. The cap consists of 20 subunits organized in two rings. The base ring

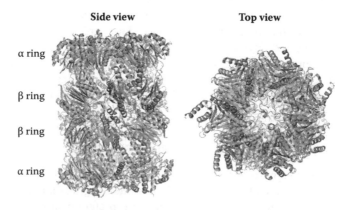

Side view   **Top view**

α ring

β ring

β ring

α ring

FIGURE 8.1   Two views on the 20 S core of the yeast proteasome. (PDB 1JD2, picture generated by the Pymol computer program.)

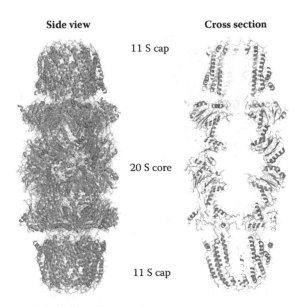

Side view   Cross section

11 S cap

20 S core

11 S cap

FIGURE 8.2 Two views on the 20 S core of the yeast proteasome capped on both sides with the 11 S Trypanosome cap. Cross section reveals cavities in the structure. (PDB 1FNT, picture generated by the Pymol computer program.)

interacts with the α-ring of the core complex. It includes six ATPases of the AAA type (AAA stands for the ATPases associated with a variety of activities). The outer ring functions as a lid of largely unknown function and is composed of eight proteins. Between the lid and the base ring of the cap is situated at least one additional subunit with the deubiquitylating activity. This is an Zn^{2+}-dependent protease rather than a Cys-protease, as are numerous other deubiquitinylating enzymes in the cells. Higher eukaryotes such as humans use the 11 S regulatory cap. This is a dome-shaped structure consisting of seven proteins named PA28. A similar structure with subunits called PA26 has been found in Trypanosome. The 11 S cap does not comprise a lid, and is a permanently open structure. Although PA28 and PA26 proteins are not closely related (they are only 14% identical), mixed complexes composed of, for example, rat 20 S core and two Tryponosome 11 S caps have been shown to be fully active. A construct of 20 S yeast core and 11 S Trypanosome caps is shown in Figure 8.2. A complex with the 20 S core associated at each end with different caps (one 11 S and one 19 S) can also exist.

8.1.2 TARGETING FOR DEGRADATION

In general, proteins selected for degradation are marked by the attachment of ubiquitin. Ubiquitin is expressed in the cells as polyprotein, and individual ubiquitins are released from this structure by the action of the deubiquitinating enzymes (see the following text). Ubiquitin is a small, highly conserved protein with the exposed C-terminal end (see Figure 8.3). The final three C-terminal residues are Arg-Gly-Gly. Terminal Gly is covalently attached to the Lys residue of the target protein. Three

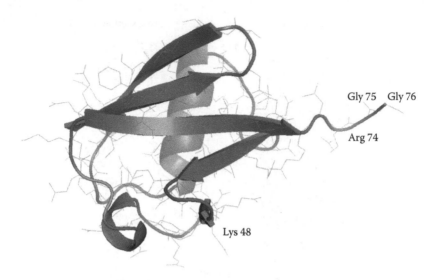

FIGURE 8.3 The structure of ubiquitin containing four β-strands and an α-helix. Note the exposed C-terminal end with Arg 74, Gly 75, and Gly 76. Lys 48, important for the formation of the polyubiquitin chain, is also shown. (PDB 1UBQ, picture generated by the Pymol computer program.)

enzymes and ATP are required for this procedure (see Figure 8.4). In the first step, ubiquitin is adenylated by ATP and is subsequently transferred to the Cys of the ubiquitin-activating enzyme E1 by forming a thioester bond. In the next step, ubiquitin is transferred to a Cys residue of the ubiquitin-conjugating enzyme (E2), from

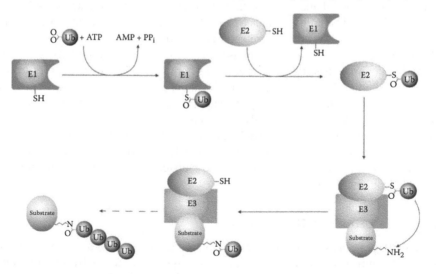

FIGURE 8.4 Tagging proteins for degradation. The substrate is a selected protein, Ub denotes ubiquitin, and enzymes required for ubiquitylation are denoted by E1, E2, and E3. For details see text. (Figure was created by Roger B. Dodd and was taken from Wikipedia [in public domain], http://en.wikipedia.org/wiki/Image:Ubiquitylation.png.)

where it is eventually transferred to the ε-amino group of the Lys residue in the target protein, forming an isopeptide bond. For this step, the enzyme ubiquitin-protein ligase (E3) is required. In humans, there is a single E1 enzyme, over 20 E2 enzymes, and many E3 enzymes, which specifically recognize different types of E2. There are two families of E3 known: one containing a HECT domain, and another containing a RING-finger domain (very similar to the zinc finger structure). RING-finger E3 transfer ubiquitin from E2 direct to the target protein, while HECT E3 contain a Cys residue that transiently accepts ubiquitin before it is transferred to the target. Target proteins comprise a short segment in the sequence (usually close to N-terminus) called ubiquitination signal, which recognized by HECT or RING-finger motif. At least in some proteins, partial unfolding of the target protein is required for the ubiquitination signal to become exposed. Following the described pattern, several ubiquitins are added to the condemned protein to form a chain of ubiquitins. Interubiquitin isopeptide bonds are formed between the terminal Gly of one molecule and Lys 48 of another. At least four ubiquitins are required to target a protein for degradation in the proteasome. However, not all proteins that are subjected to proteasome processing require ubiquitination. Unfolded and partly unfolded proteins can enter the proteasome without ubiquitin assistance. The same is true for the oxidized proteins and for some intrinsically unstable proteins incorporating the unstable regions (see Chapter 6). Well-studied examples of proteins subjected to ubiquitin-independent proteasomal degradation are ornithine decarboxylase and the cyclin-dependent kinase inhibitor p21Cip1. Ornithine decarboxylase incorporates several specific sites that are recognized by the proteasome. The recognition is assisted also by the ornithine decarboxylase accessory factor antizyme 1.

8.1.3 PROTEASOME MECHANISMS

The events in protein degradation by the proteasome are not known in detail. In the 26 S proteasome, the 19 S regulatory cap recognizes polyubiquitinated proteins, and the deubiquitylating enzyme cleaves the isopeptide bonds, enabling the recycling of ubiquitin. Six ATPases in the base ring of the cap may have more than one role. The essential one is to provide the conformational changes required to unfold the target protein. Although the process is not disclosed, it might be similar to that occurring in chaperonins whose structure is strikingly similar to that of the proteasome, although the constituent proteins are not related. Specific protein sequences (Gly-Ala repeats) have been identified that inhibit unfolding by the proteasome. Another role of the ATPases is to open the entrance into the core structure cavity. Both processes have been shown to be driven by ATP hydrolysis. The unfolded proteins can then enter the core and are proteolytically cleaved by Thr-proteases. The narrow entrance into the cavity restricts passage of the folded proteins, and these are thus saved from degradation.

The product of target protein degradation are short peptides, typically seven to nine residues long, that are released from the proteasome and further cleaved into individual amino acids by the peptidases in the cytosol. The proteasome core associated by the 11 S regulatory cap can accept nonubuquitinated unfolded proteins for degradation. The main function of this cap is to open the gate into the proteasome

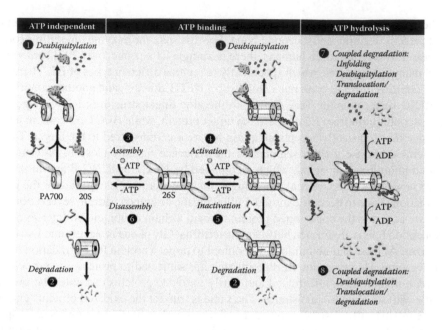

FIGURE 8.5 Different roles of ATP in proteasome action. The 26 S proteasome is composed of the 20 S proteasome and the 19 S regulatory complex (PA700). Purified PA700 is an ATPase but deubiquitylates polyubiquitylated proteins by an ATP-independent process (1). The 20 S proteasome can degrade certain unstructured proteins that promote gating (2). The assembly of the 26 S proteasome from 20 S proteasome and PA700 requires ATP binding but not hydrolysis (3). Activation of the 26 S proteasome may require a separate ATP-binding event (4). Removal of ATP promotes inactivation (5) and disassembly (6) of the 26 S proteasome. The 26 S proteasome deubiquitylates, by an ATP-independent mechanism, some polyubiquitylated proteins that are refractory to degradation (1). The 26 S proteasome degrades certain unstructured, nonubiquitylated proteins without ATP hydrolysis (2). The 26 S proteasome degrades both folded and unstructured polyubiquitylated proteins by a process that requires ATP hydrolysis, and it couples mechanisms required for degradation, including substrate unfolding (when needed), deubiquitylation, translocation, and degradation (7 and 8). (Figure taken from Liu, C.-W., et al. 2006. ATP binding and ATP hydrolysis play distinct roles in the function of 26 S proteasome. Mol. Cell 24, 39–50. With permission.)

core. ATP hydrolysis is not required for this process; the binding energy provided by the interaction of the 11 S cap with the N-terminal domains of the α-ring proteins is sufficient to induce the required conformational changes. However, ATP has an important role in the action of the proteasome. Proteasome processes that do not require ATP, those that depend on ATP binding but not hydrolysis, and those that require both binding and hydrolysis of ATP are reviewed in Figure 8.5.

8.2 PROTEIN DEGRADATION BY LYSOSOMES

Lysosomes are cell organelles containing digestive enzymes, including lipases, carbohydrases, proteases, and nucleases. They are all acid hydrolases with the pH optimum well bellow 7, corresponding to pH 4.5 to 4.8 maintained in the lysosome interior by the proton and chloride pumps. Proteases in the lysosomes are called cathepsins. There are 13 types of cathepsins known. Most of them are Cys-proteases, but some exceptions exist. For example, cathepsins A and G are Ser-proteases, and cathepsins D and E are Asp-proteases. Catalytic mechanisms of proteases are described in Chapter 21. Cathepsins mostly act as endopeptidases, and some show mixed endopeptidase and exopeptidase activity, but few are specific to the cleavage of the peptide bonds only at N-terminus (amino-peptidases) or C-terminus (carboxy-peptidases). In general, cathepsins have broad specificity corresponding to their role as general cleaners of the exogenous and membrane proteins brought into the lysosome by vesicular transport.

FURTHER READING

1. Voges, D., et al. 1999. The 26 S proteasome: a molecular machine designed for controlled proteolysis. Annu Rev Biochem 68, 1015–68.
2. Liu, C. W., et al. 2006. ATP binding and ATP hydrolysis play distinct roles in the function of 26 S proteasome. Mol Cell 24, 39–50.
3. Sharon, M., et al. 2006. Structural organization of the 19 S proteasome lid: insights from MS of intact complexes. PLoS Biol 4, e267.
4. Thrower, J. S., et al. 2000. Recognition of the polyubiquitin proteolytic signal. EMBO J 19, 94–102.

Introduction to Part II

Astrid Gräslund

It is sometimes stated that whenever an important new method becomes available for study in a certain scientific field, this gives rise to a whole new aspect of the field. This is certainly true for most chemical research areas. This section will cover some traditional and some relatively new methods available for studies of peptides and proteins. The methods concentrate on structure at different levels. It should be remembered that it was only a few decades ago that the first three-dimensional protein structures were presented, with a Nobel Prize going to John Kendrew and Max Perutz in 1962 for the crystallographic structures of myoglobin and hemoglobin, respectively—work that had been completed only a few years before. Their results emphasized the deep insight that the particular three-dimensional fold of a protein determines its unique function, a truth that nowadays we take for granted. Even if the structures were determined in the crystalline state, it soon became evident that they represented the true native state, since, for example, catalytic activity could be found also in crystalline enzymes.

It was, however, another big step forward when the three-dimensional structure of a protein in solution could be determined independently by another method, namely NMR (earning a Nobel Prize for Kurt Wüthrich in 2002). The NMR method also allowed protein dynamics on the nanosecond time scale to be studied in solution. Mass spectrometry is another important method for studies of proteins and peptides. In fact its first applications to proteins was also awarded the Nobel Prize in 2002. Nowadays, mass spectrometry is crucial in many studies where proteins and their modifications must be identified, and this method is deeply involved in modern proteomics studies. For these three methods mentioned here, a common denominator is that they were developed by physicists, first applied by chemists to small molecules. Great breakthroughs were then made when they could be applied to biomacromolecules. These and other important methods will be briefly described in the following section.

Introduction to Part II

Astrid Gräslund

9 Purification and Characterization of Proteins

Matjaž Zorko

CONTENTS

Characterization of a protein usually demands protein extraction and purification. Regardless of the source from which a protein is to be extracted, in order to follow purification efficiency, a procedure to identify the protein of interest must be known, or it should be developed. This procedure is based on some unique physical, chemical, or biological property of the protein. Very seldom do proteins incorporate a chromophore, such as hem in hemoglobins and cytochromes, which enables a spectroscopic determination of its presence in low concentration. Most often, a protein is identified via its specific interactions with other molecules. Enzymes can be detected by their enzymatic activity, which is usually specific enough to allow for the identification of a single enzyme type. Specificity can be enhanced by using inhibitors. Antibodies can be raised against the studied protein and used for its identification in the mixture of thousands of different proteins. All these approaches should also provide a quantitative measurement of the amount of protein.

The studied protein can originate from different sources, such as physiological liquids, tissues, cell lines, and microorganisms in which proteins normally reside or are expressed after genetic manipulation. In contrast to physiological liquids in which proteins are freely accessible, all other sources must be suitably treated to release the entrapped proteins. Tissues and cells can be disrupted mechanically by different types of homogenizers or by sonification. Chemical treatment is usually employed with cells that are protected by the cell wall. For example, microorganisms such as *Escherichia coli*, which are frequently used for the expression of proteins, are exposed to lysozyme or different mixtures of other enzymes in order to hydrolyze the bonds in the peptidoglycans of the cell wall. The obtained protoplasts can be subsequently disrupted by hypotonic shock, followed by the release of the proteins and other cell components. Detergents are used to release membrane proteins from the lipid bilayer. All these procedures must be used with care to preserve the protein structure intact. Some of the methods for the purification of proteins are discussed in the following text.

9.1 PRECIPITATION OF PROTEINS

Globular proteins that are not integrated into the membranes are usually soluble, while fibrilar proteins are essentially nonsoluble in water solutions. Solubility of globular proteins depends on several factors related to the protein structure and the composition of the solvent. Interactions with water molecules that keep the protein in solution are provided by the charged and polar residues on the protein surface. Water around the protein forms a cage composed of a layer of water molecules that are hydrogen bonded to the hydrophilic groups of the surface residues; however, the exchange of water molecules in the cage with the ones in bulk water is very fast. The solubility of proteins is highly dependent on their charge and the concentration of salts in the solution. All molecules of a protein carry the same net charge, which depends on the protein structure and on the pH of the solution. This causes electrostatic repulsion between protein molecules, which additionally prevents protein aggregation and increases protein solubility. In pH that equals the value of the isoelectric point, pI, the net charge of the protein is zero, and repulsion is lost. This is the reason why, at this pH, proteins are the least soluble.

Dependence of protein solubility on the concentration of the salts is complex. At low concentrations, salts increase the solubility of proteins because ions interfere with the electrostatic interactions between protein-charged groups, thus preventing aggregation. This is known as the salting-in effect. When the concentration of salt increases, it starts to affect the hydration layer of water molecules around a protein, generally by disrupting it. Ions withdraw water from the protein hydration layer, exposing in this way the hydrophobic parts of the protein surface, which tend to interact with each other, promoting protein aggregation. This is called the salting-out effect. The efficiency of different ions in this process is in accordance with the position of the ion in the Hofmeister series. Water structure breakers such as NH_4^+ and SO_4^{2-} are the most effective. The decrease of protein solubility is in accordance with the Cohn equation

$$\log S = B - KI \qquad (9.1)$$

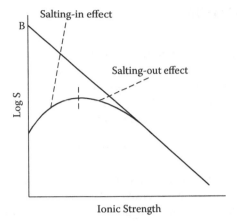

FIGURE 9.1 Schematic presentation of the solubility of the protein as a function of the ionic strength of the salt in solution. The curve part left of the peak value is in accordance with the salting-in effect, and the curve right of the peak represents the salting-out effect. The line corresponds to the Cohn equation (see Equation 9.1 in the text).

where S is solubility of the protein, B is idealized solubility, K is a salt-specific constant, and I is the ionic strength defined by the equation

$$I = \frac{1}{2} \sum_{i=1}^{n} c_i z_i^2 \qquad (9.2)$$

where c_i and z_i are the concentration and charge of the ions of the salt. The graphical presentation of the Cohn equation is shown in Figure 9.1.

The degree of the salting-out effect varies from protein to protein and is dependent on protein structure. Even closely related proteins can show substantial differences in solubility in the presence of salts. For example, myoglobin is, for several orders of magnitude, more soluble than hemoglobin in ammonium sulfate solution with ionic strength of 6 M. Because of this, the salting-out effect can be used for the fractionate precipitation of proteins. Ammonium sulfate, $(NH_4)_2SO_4$, is most often used for this purpose because of its efficiency, high solubility, biocompatibility, and low price. In solutions with pH values above eight, ammonium sulfate binds protons and shifts pH toward higher values, and is therefore usually replaced by sodium citrate. The fractionation of proteins is achieved by successive addition of the salt and by withdrawing the precipitated proteins after each portion of added salt. In this way, we can obtain a fraction that precipitates the studied protein with much less other protein species as in the starting homogenate. Proteins can be precipitated also by adding an organic solvent to the protein solution due to lowering of the dielectric constant. However, the organic solvents often act as denaturants, and their use as precipitating agents is very limited.

9.2   CENTRIFUGATION

The gravitational acceleration of the Earth (g) is too small to affect the Brownian motion of the proteins in solution. In the centrifuge, much greater acceleration of over 400.000 g can be created by rapid spinning of the protein solution. Three forces act on the protein or any particle that is in steady motion in the artificial gravitational field in the centrifuge: centrifugal force (F_c), buoyant force (F_b), and friction force (F_f). In the steady state, these forces are in equilibrium:

$$F_c + F_b + F_f = m\omega^2 r + m_o \omega^2 r + fv = 0 \qquad (9.3)$$

where m is a mass of a particle, m_o is a mass of the displaced solution, $\omega^2 \cdot r$ is radial acceleration, f is the translational friction coefficient, v is the velocity of the particle, F_c is the centrifugal force, F_b is the buoyant force (because of the displaced solution), and F_f is the friction force. The velocity v of the particle is

$$v = \frac{(m - m_o)}{f} \omega^2 r \qquad (9.4)$$

Note that v can have a positive value (the particle is moving away from the rotation axis in centrifuge) or negative value (the particle is approaching the rotation axis) depending on the relative value of m and m_o (or relative density of the particle in comparison to the density of the surrounding solution). The sedimentation coefficient s of a molecule is defined as its rate of sedimentation in a given centrifugal force (Svedberg equation):

$$s = \frac{v}{\omega^2 r} = \frac{M(1 - V_p \rho)}{Nf} \qquad (9.5)$$

where M is molecular weight, V_p is the partial specific volume of the particle (it is the inverse of the protein density), f is the translational fractional coefficient, ρ is the density of the solvent, ω is the radial velocity, and N is Avogadro's number. The unit of the sedimentation coefficient is named Svedberg (abbreviated by S) after the pioneer of ultracentrifugation, Theodor Svedberg, and is defined as 1×10^{-13} s. It is a characteristic of the molecule or particle sedimenting in the ultracentrifuge and is dependent on the mass of the particle, density of the particle (the partial specific volume of the particle, V_p, is a function of particle density), density of the solvent, and the shape of the particle that determines the friction coefficient f. In practice, the sedimentation coefficient is often used to characterize biological particles and macromolecules. For example, we speak of a 50 S ribosome subunit, 18 S rRNA, etc. In any centrifuge, a 10 S particle sediments twice as fast as a 5 S particle. Ultracentrifugation can be used to separate proteins and other biomolecules with different sedimentation coefficients, and also to determine their molecular masses (see Equation 9.5).

In the separation and purification of proteins by centrifugation, different approaches can be used. Differential centrifugation is often used to separate the cell homogenate into fractions, which are further used as a source of proteins under

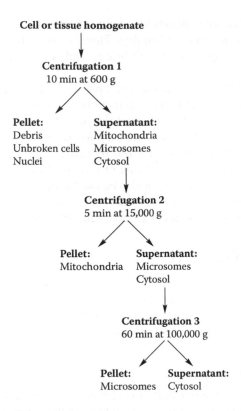

FIGURE 9.2 The scheme of differential centrifugation. Microsomes are small vesicles of disrupted endoplasmic reticulum. The obtained fractions are not pure since particles from different fractions overlap in size and mass. The mitochondrial fraction contains peroxisomes and lysosomes, but can also contain small nuclei. Small microsomes and ribosomes remain in the cytosole.

investigation. The cell homogenate is subjected to repeated centrifugations, each time removing the pellet and increasing the centrifugal force. An example of the procedure is shown in Figure 9.2. Proteins can also be separated by zonal centrifugation, in which a sample of protein mixture is applied on top of the solution in the centrifuge tube. During spinning, all proteins starting from the same point will separate into bands or zones according to their sedimentation coefficients. A density gradient can be applied to the solution in order to minimize the diffusion of the separating protein bands. In this procedure, the density of the gradient is usually kept lower in all parts of the centrifuge tube than the density of the proteins. Contrary to this, in isopicnic centrifugation, the density gradient is essential, and it is prepared so that the densities of the proteins to be separated lie within the lower and upper limit of the gradient densities. A sample is applied on the top of the gradient, and the proteins start to separate according to their sedimentation coefficients, as in zonal centrifugation. However, when the protein arrives at the gradient layer in which the density of the solution is equal to that of the protein, it stops to move (in this layer $m = m_o$, see Equation 9.4). Isopicnic centrifugation, in which proteins are separated by their density, demands very long centrifugation times because equilibrium is achieved only after one or two days. However, often the separation is already adequate during the approach to equilibrium, and centrifugation can be stopped after 8 or 10 h.

Determination of the molecular mass of a protein by centrifugation does not give very accurate results. It requires measurement of the sedimentation coefficient, but the values of the friction coefficient f and the partial protein volume V_p should be known. The last two parameters are not easy to determine. An approximate value of molecular mass can also be determined by comparing the sedimentation of the protein of interest with the sedimentation of a set of standard proteins of different molecular masses. Centrifugation can also be used to determine the number of polypeptide chains in the proteins, to follow the reorganization and dissociation of protein subunits, and to observe protein aggregation and interactions.

9.3   CHROMATOGRAPHY

Chromatography is a method to separate molecules according to their distribution between the mobile and stationary phases. For the separation of proteins, the following types of chromatography are most often used: gel filtration, ion exchange chromatography, hydrophobic chromatography, and affinity chromatography.

9.3.1   Gel Filtration

Gel filtration is also called size exclusion chromatography because it separates proteins on the basis of their molecular mass related to their size. The stationary phase packed into the column is a gel consisting of roughly spherical particles with pores of defined size. Gels for gel filtration are prepared from agarose, acrylamide, and dextran polymers. A suitable solution is applied as the mobile phase that carries the sample of proteins of different average diameter through gel beads. Proteins with sizes corresponding to the pores and smaller enter into the pores and are retained to different degrees by the gel, while those that are too large to be adsorbed into the pores pass through the gel together with the mobile phase. For each protein, a volume of the solution needed to elute the protein from the column, called the elution volume, V_e, can be determined. The values of V_e are roughly proportional to the logarithms of the molecular masses of the proteins. This makes it possible to estimate the molecular mass of an unknown protein by the calibration of the column with standard proteins of known molecular masses (Figure 9.3). More precise methods for determination of the molecular mass, such as mass spectrometry, have recently largely replaced gel filtration. However, gel filtration is still very much in use to separate small molecules and ions from proteins or to exchange the buffer solution in which a protein is dissolved.

9.3.2   Ion Exchange Chromatography

Proteins are charged depending on their isoelectric point, pI, and the pH of the solution, and can interact with a stationary phase that is an ion exchanger. Anionic

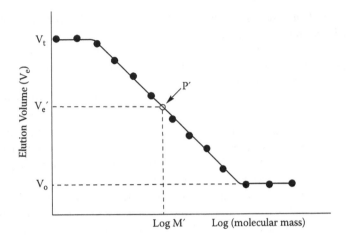

FIGURE 9.3 Determination of the molecular mass of an unknown protein (P′) by gel filtration. Closed points represent data of standard proteins, the open point represent data of an unknown protein with elution volume V_e' and molecular mass M'. V_t and V_o are the total volume of the column and the void volume, respectively, and represent the upper and lower limit of the molecular mass.

exchangers bind positively charged proteins with pI over 7, and cationic exchangers those with pI below 7. The stationary phase is made of cellulose, crosslinked dextranes, crosslinked agarose, or crosslinked acrylamide; more recently, polystyrene crosslinked with divinylbenzene has also been used. To these matrices, suitable acidic groups are attached to obtain a cationic exchanger, or basic groups to yield an anionic exchanger. A sample of proteins is applied to the column filled with the stationary phase at constant pH and low ionic strength. Proteins will interact with the ion exchange groups in the stationary phase in accordance with their pI and the number of interacting charged groups. Some proteins will form strong and some weak interactions with the stationary phase, and those that have the same net charge as the stationary phase will interact very weakly or not at all. In the next step, a mobile phase is changed by salt solution with gradually increasing ionic strength, resulting in the fractional release of the protein. Ions will displace proteins from the stationary phase: first, those interacting with the matrix only weakly, and subsequently, at higher ionic strength and also proteins that are attached more firmly. Ion exchange chromatography can be applied for the separation of proteins, peptides, amino acids, nucleic acids, and charged carbohydrates.

9.3.3 Hydrophobic Chromatography

Since some hydrophobic side chains remain at the protein surface also after folding, they can be used for separation, based on the hydrophobic interactions between proteins and the nonpolar stationary phase. Crosslinked agarose polymers with the attached hydrophobic hydrocarbon side chains of different lengths are used as the stationary phase. A protein sample is applied via a mobile phase of the high ionic strength achieved by the addition of NaCl or $(NH_4)_2SO_4$ in high concentration (1–2 M).

The presence of salts favors the hydrophobic interactions by attracting water. After the hydrophobic binding of proteins to the stationary phase, a mobile phase with gradually decreasing ionic strength is applied for the stepwise elution of the proteins. Fractionation is thus based on the hydrophobic properties of the proteins in the sample. A special type of this method is reversed-phase chromatography. Although this terminus has been used more generally for any chromatography with a nonpolar stationary phase, it is more specifically intended for chromatography that includes nonpolar solvents into the mobile phase. This procedure requires a stationary phase that is stabile and resistant to nonpolar solvents. Silica or synthetic polymers with the attached linear and hydrocarbon chains (C2 to C18) are usually employed. A phenyl ring containing chains can also be introduced. The protein sample is applied via the low ionic-strength aqueous mobile phase. In the next step, hydrophobically attached proteins or peptides are eluted by the mobile phase, in which the amount of a suitable nonpolar solvent such as acetonitrile, methanol, ethanol, tetrahydrofuran, 2-propanol, and others is gradually increased. The gradient of the nonpolar solvent can range from 5% to 80%. High concentrations of the nonpolar solvent can lead to denaturation of proteins; therefore, this procedure is more appropriate for the separation of peptides. In this case, the stationary phase with longer hydrocarbon chains such as C18 is used.

9.3.4 AFFINITY CHROMATOGRAPHY

In affinity chromatography, high-affinity, noncovalent interactions of proteins with specific ligands are utilized. A ligand that is specifically recognized by the protein of interest is bound to the stationary phase. Very different ligands can be used, including substrates, inhibitors, cofactors, hormones, antibodies, and oligonucleotides. A sample is applied in the mobile phase that is conditioned to enable optimal interaction between the ligand and the target protein. The ligand-bound protein is eluted by changing the conditions in order to decrease the affinity of the protein to the ligand, or by displacing the ligand attached to the stationary phase with the same ligand added to the mobile phase. An important advantage of the method lies in the specificity of the ligand–protein interaction, allowing the application of very crude samples such as complete-cell homogenates or lysates. It is not unusual to achieve a high-yield purification of the specific protein in only one step.

Affinity chromatography is very convenient for the purification of proteins that are expressed with the His tag. A sequence of His residues (at least six) can be introduced into the protein either at the C-terminus or at the N-terminus and can be used for specific interaction with the stationary phase containing metal chelates (usually with Ni or Co ions). Alternatively, proteins that cannot be directly captured because no suitable ligand is available can be expressed as fusion proteins with a protein that is easy to purify by affinity chromatography. Ligands for affinity chromatography must be attached to the stationary phase in such a way as to be able to favorably interact with the target protein. This means that the link between the ligand and the stationary phase must not interplay with protein–ligand interaction and must be stable to resist variable conditions during separation. Spacer molecules of different lengths are often used to flexibly connect the stationary phase matrix with the ligand. This facilitates the optimal adaptation of the ligand to the protein-binding site.

Technically, chromatography equipment can range from a simple column with suitable stationary and mobile phases that is operated manually to a very complex computer-assisted chromatographic system. The introduction of high-performance liquid chromatography (HPLC) improved the efficiency and resolution of peptide and protein separation and purification to a great extent. The efficiency of the chromatographic column of any of the earlier described types is dependent on the size of the gel particles. In general, the smaller the particles, the higher the gel capacity, and the more efficient the separation. However, small gel beads also mean small interbead spaces and much decreased flow of the mobile phase through the gel. In HPLC, the flow is maintained by the pump, which pushes the mobile phase through a tightly packed gel in a column by a pressure of several hundred bars. The whole system must be built to resist this very high pressure. Usually, the system is equipped with more than one pump and contains additional units such as a sample application unit, mixers, a detector to monitor separation by recording UV light absorption or fluorescence emission, a fraction collector, several valves to regulate flow, and a computer to control the separation parameters, and it performs the procedure automatically. Different types of chromatography represent the main set of methods for the preparative separation and purification of peptides and proteins, allowing large-scale purification.

9.4   ELECTROPHORESIS

Although preparative electrophoretic equipment exists, the main use of electrophoresis is in determining the purity of proteins and peptides after purification. The method is based on the movement of the charged proteins and peptides through a polyacrylamide gel exposed to the electric field. Different other electrophoretic matrices can be used, but polyacrylamide gel electrophoresis (PAGE) is today by far the most common electrophoretic method. The gel is prepared by the polymerization of acrylamide into the long chains of polyacrylamide crosslinked by N,N′-methylenebisacrilamide. The reaction is catalyzed by tetramethylethylenediamine and ammonium persulfate. The polymerization reaction can be controlled to give the gel pores of defined size that function as the molecular sieve that is able to slow down or exclude the movement of larger proteins through the gel. Thus, separation results from the combination of the charge and size effect. The shape of the protein can also affect the rate of migration through the gel. Samples are applied to the wells in the gel and, after the electric field is established, the proteins will move through the gel toward the positive anode if negatively charged, or toward the negative cathode if positively charged. Usually, PAGE is run in the buffer with pH of around 9, making most of the proteins negatively charged and forcing them to migrate toward the anode (see Figure 9.4).

When PAGE is performed with untreated proteins, it is called native PAGE. More often, proteins are pretreated with sodium dodecyl sulfate (SDS), which is a potent denaturing agent, and with a reducing agent such as mercaptoethanol to reduce disulfide bonds. This results in denatured proteins, which interact with SDS to obtain an additional negative charge for approximately every two amino acids. The charge is thus proportional to the molecular mass of the polypeptide chain and the intrinsic charges

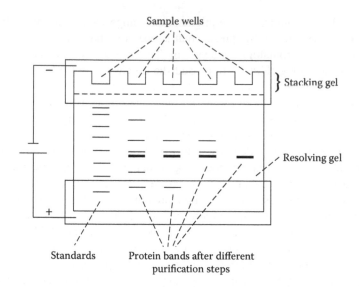

Sample wells

} Stacking gel

Resolving gel

Standards   Protein bands after different
purification steps

FIGURE 9.4 The scheme of polyacrylamide gel electrophoresis (PAGE). The first part of the gel just below the sample wells is a stacking gel with larger pores used to concentrate sample at the border between the stacking and the resolving gel, enhancing in this way the sharpness of the bands. In the first line are shown bands of standard proteins, and the other lines contain bands of sample after successive purification steps. The broad band is a protein that is purified, and other bands belong to those with impurities.

of the protein, and its shape does not influence the migration in the gel. Separation of proteins into the individual bands is dependent exclusively on the molecular mass, with larger proteins migrating more slowly than smaller ones. The resolution can be enhanced by forming a linear gradient of pores from the top to the bottom of the gel. After separation, the gel is stained to make protein bands visible, usually by applying Coomassie Brilliant Blue or silver nitrate. Both agents will stain proteins nonspecifically. Coomassie Brilliant Blue can detect protein bands of 50 ng, while silver staining is around 50 times more sensitive. Specific detection of proteins is possible by Western blotting. The procedure employs specific antibodies raised against the protein of interest, subsequently detected by the secondary antibodies linked to the enzyme producing a colored product after the addition of a suitable substrate. This identification technique is combined with the transfer of proteins from the gel to the nitrocellulose-supporting matrix. It is common to run molecular markers of known molecular mass in a separate lane in the gel in order to calibrate the gel and determine the molecular mass of unknown proteins by comparing the distance traveled relative to the marker.

Another electrophoretic method is isoelectric focusing. Electrophoresis is run in the gel in which a pH gradient is created by aliphatic ampholytic molecules. Proteins will migrate to the anode or to the cathode, depending on their charge. Traveling across the pH gradient, the charge of the protein will be modified. When in the region with pH equal to its isoelectric point, the migration of the protein molecule is stopped because the net charge of the protein is zero at this point and the molecule is not affected by the electric field. Isoelectric focusing thus separates proteins according to their isoelectric points, resolving proteins that differ in pI value by as little as 0.01. This method is also the basis of two-dimensional electrophoresis where it is combined with SDS-PAGE (see Chapter 20).

9.5 OTHER METHODS

Other methods can be included in the separation and purification of proteins, as well as in their identification and characterization. The one that has recently been shown to be most advantageous is mass spectrometry, which is discussed in Chapter 14. Dialysis of small molecules and ions through the pores of the semipermeable membrane can be used to remove ions from proteins after fractional precipitation of proteins or to change the buffer in the protein solution. It can also be used to purify proteins from low molecular contaminates and to quantify the association of small molecules with proteins. Ultrafiltration is an opposite procedure in which pressure is applied to thrust solvent and small molecules through the pores of the semipermeable membrane. It is effectively used to concentrate protein solutions.

9.6 PROTEIN PURIFICATION STRATEGY

No general strategy applicable for the purification of all proteins is available. Protocols to purify the unknown protein are usually based on protocols for the purification of similar proteins, if available, and on previous practical experience. Many times, the trial-and-error approach is an important part of the procedure to refine the protocol. After establishing the assay to detect the protein of interest, the next step is to choose the appropriate protein source. If these are tissues or cells, they have to be disrupted. The homogenate can be subjected to differential centrifugation to enrich the fraction in which the studied protein resides. At this stage, it is tempting to apply affinity chromatography if the ligand specifically interacting with the protein is known. This is more common with proteins expressed with the His tag or as a fusion construct in which the fused partner protein can be easily trapped. Normally, proteins are further treated by ammonium sulfate to obtain the fraction with the substantially smaller number of proteins that is subsequently dialyzed to remove ions and further separated by the appropriate chromatographic technique. SDS-PAGE is run after each purification step to follow purification efficiency. Usually, many purification steps are required to eventually obtain the purified form of the protein of interest, characterized by the single band in SDS-PAGE.

FURTHER READING

1. Wilson, K. and Walker, J. 2000. Principles and Techniques of Practical Biochemistry. Cambridge University Press, Cambridge, U.K.
2. Harrison, G. R., et al. 2003. Bioseparations Science and Engineering. Oxford University Press, Oxford, U.K.
3. Skoog, D. A. 2006. Principles of Instrumental Analysis, 6th ed. Thompson Brooks/Cole, Belmont, CA.
4. Zellner, M., et al. 2005. Quantitative validation of different protein precipitation methods in proteome analysis of blood platelets. Electrophoresis 26, 2481–9.

10 Crystallography and X-Ray Diffraction

Astrid Gräslund

CONTENTS

The development of techniques to determine three-dimensional (3-D) structures of biological macromolecules such as proteins has been absolutely necessary for us to reach the present level of understanding of their functions. The ordered arrangement of atoms to fill the space shows an architecture on the atomic level of the molecule, which is strongly correlated to function. This holds equally true for a structural protein, an enzyme, or a nucleic acid. Two major methods for 3-D structure determination on the atomic level of resolution are available today: x-ray crystallography, which has given rise to most 3-D structures of proteins reported up to now, and nuclear magnetic resonance (NMR), which can be applied to molecules in an aqueous solution. X-rays have to be used for diffraction because only they have the short wavelengths (order of Ångströms) that are of the same size as the atomic distances that are to be probed in a molecule. The NMR method is based on a different concept, probing interatomic distances in proteins by the magnetic interactions between neighboring nuclei. Electron diffraction is a third method in structural biology, but only a smaller number of structures have reached atomic resolution.

10.1 MAKING CRYSTALS

X-ray crystallography is based on diffraction of x-rays in an ordered material. The protein sample has to be prepared as a single crystal, where the molecules are ordered in unit cells. Each unit cell may contain one or more molecules in an ordered arrangement, and then the unit cells have to be packed so that all have the same orientation relative to the macroscopic dimensions of the crystal. Since protein molecules typically have irregular shapes, they are surrounded by disordered solvent molecules and make only a few direct molecular contacts among themselves. Making well-diffracting single crystals is a bottleneck in x-ray crystallography. Many procedures have been developed for crystallization, and conditions have to be varied over many parameters and wide ranges in order to produce well-behaving crystals that can be used for investigations.

In large structural biology laboratories, the techniques are often robotized so that a multitude of conditions in terms of pH, precipitant, etc., can be explored. One initial requirement for successful crystallization is that the protein has to be quite pure, ideally better than of 97% purity. Methods to crystallize vary, but the general principle is that a protein solution with a relatively low concentration of precipitant is made supersaturated by slowly removing the solvent, whereupon the crystals are formed by slow precipitation. A common method is called "hanging drop," in which the drop of protein solution slowly evaporates water to a larger reservoir of solvent containing a high concentration of precipitant (e.g., the salt ammonium chloride). One protein can crystallize into different crystal forms with different unit cells, depending on crystallization conditions. Although many crystals may form and look relatively fine under the microscope, it is not obvious which will be the best in terms of diffraction. This has to be studied at an x-ray source.

10.2   X-RAY SOURCES

The source of x-rays can be local, using tubes that produce them with a continuous background and some sharp spectral lines depending on the material of the tube. Nowadays, researchers often make use of large-scale radiation facilities called synchrotrons. These are large storage rings that contain electron bunches traveling close to the speed of light in a circular orbit. Electrons are first accelerated to high energies and then fed into a circular storage ring. The circular orbit is maintained via magnetic devices that force the electron beam to deviate from a linear course. As a result, the electron bunch emits electromagnetic radiation, so called "Bremsstrahlung" (German literally for brake radiation), with a very wide and continuous energy spectrum. The cutoff at the lowest wavelength or highest energy depends on the speed (energy) of the electrons. When they move with a speed close to the speed of light, they will have a cutoff wavelength in the x-ray region.

Originally, the "Bremsstrahlung" was a side effect observed by nuclear physicists using potent electron accelerators, but nowadays synchrotrons are built almost exclusively for the use of electromagnetic radiation ("the world's best lamp"). In large facilities, the radiation is extremely intense, and a narrow wavelength band can be selected to give monochromatic x-rays with a wavelength that can be varied if needed. A large European synchrotron is located in Grenoble, France. This facility is run as a joint effort by most European countries. The circumference of its synchrotron ring is 844 m, and wavelengths well below 1Å are available with high intensity. Experimental stations, so called beamlines, are built around the ring, and some of them are for the exclusive use for x-ray crystallography. A particular micro-focus beamline with a focus on the micrometer length scale has been constructed at

ESRF and can be used for extremely small protein crystals, also with micrometer dimensions.

10.3 THE DIFFRACTION EXPERIMENT

For structure determination, the protein crystal (typically with millimeter dimensions in favorable cases) is placed in a beam of x-rays. Because the wavelength of x-rays is on the order of Ångströms, it matches the distances between the atomic planes in the crystal, and a macroscopic x-ray diffraction pattern is produced. This is because some scattered x-ray waves recombine to add to one another's intensities, and some others subtract from each other. Figure 10.1 shows the principle of an experimental setup for x-ray diffraction and a schematic illustration of the principle

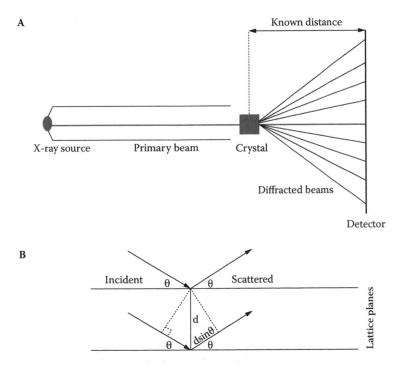

FIGURE 10.1 (A) A schematic setup for protein crystallography. The crystal is placed in the focus of an x-ray beam. The x-ray source can be an x-ray tube or, nowadays, a beamline in a synchrotron. The diffracted beams are recorded by the detector (originally a photographic plate, nowadays an electronic device) at a known distance from the crystal and form a diffraction pattern. (B) Illustration of Bragg's law, which is the basis of x-ray diffraction. In the crystal are depicted lattice planes that scatter the incident x-rays, which should be considered as sinusoidal waves with wavelength λ. The distance between two such planes is d (on the atomic Ångstrom scale). Diffracted beams from the two planes will travel in parallel but will have a pathlength difference of d sinθ. If d sinθ corresponds to 0.5 λ, the two scattered waves will cancel each other; if d sinθ corresponds to λ, the two waves will reinforce each other. By these variations, the scattered waves will give rise to a diffraction pattern with dark and light spots according to the different scattering planes in the crystal.

that links atomic distances to macroscopic diffraction angles, according to the physical law formulated by Bragg. All heavy atoms in the protein scatter the x-rays in all directions, but some atoms give rise to positive interference with each other. This will give a distinct spot in the diffraction pattern, and the appearance of this particular spot is linked to a particular set of atomic distances in the molecule.

Diffraction processes are governed by the 3-D arrangement of the atoms in the ordered molecules in the crystal, but the pattern of diffraction spots does not immediately give up this information. The diffraction pattern is related to the 3-D structure of the diffracting atoms by a mathematical relation called Fourier transformation. We come across Fourier relations in various fields of physics; the basic mathematical idea is that a sinusoidal wave can be described in two ways: either as a function of time or as a function of frequency, and the two descriptions give equivalent information.

In the present case, the two equivalent descriptions of diffracting objects are via the diffraction pattern (in the so-called phase space) or the 3-D structure in the real three-dimensional space. To link observations in the phase space to the 3-D space, we have to perform a Fourier transformation of the diffraction information, in order to obtain the desired map of positions in 3-D space for the atoms that helped to produce the diffraction pattern. The result is called an electron density map, to which one can fit the backbone and side-chain atoms of the studied protein. Only relatively do heavy atoms, not hydrogens, take part in the processes and will be mapped. Figure 10.2 illustrates how molecular structures are fitted into electron density maps.

10.4   THE PHASE PROBLEM

Besides producing the single crystals, and before performing the Fourier transformation, there is another practical problem that has to be solved before a successful protein structure determination can be made by x-ray crystallography. This is the so-called phase problem. At the beginning of protein crystallography, this was the major hurdle to overcome. The problem is one of insufficient information in the diffraction pattern. When digitizing the information in this pattern, we know the positions of the spots and their intensities, but in order to do the Fourier transformation properly, we need another parameter called phase, which is associated with each wave producing a diffraction spot. More information is therefore needed. In small-molecule crystallography, the problem could be solved by so-called direct methods, essentially guessing the phases and iterating until a consistent picture is obtained. For proteins, the additional information was originally provided by adding heavy metal ions such as gold or mercury to the protein, hoping that they would bind at a small number of well-defined binding sites. The heavy metal ions would then give rise to distinct changes in the diffraction pattern. In fact, at least two different types of heavy metal substitutions were needed to provide enough extra information to produce an electron density map and thus solve a 3-D protein structure. This original method of solving the phase problem (multiple isomorphous replacement, MIR) is still heavily used.

The phase problem can also be solved by making use of the knowledge of structural symmetries or by related and known protein structures (molecular

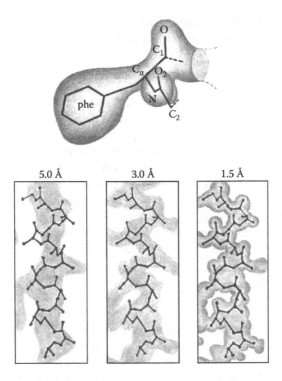

FIGURE 10.2 Fitting a molecular structure into an electron density map. Depending on the resolution, more and more details of the atomic structure can be visualized. (Upper part): With high resolution, side chains of amino-acid residues can be fitted such as, in this case, a phenylalanine side chain. (Lower part): Only below about 2 Å resolution is it possible to visualize protein side chains with any certainty. (Reprinted from Bränden, C. and Tooze, J. Introduction to Protein Structure, 2nd edition. With permission from Garland Sciences, New York. The upper part is adapted from A. Jones in Methods Enzym. Eds. H. W. Wyckoff, C. H. Hirs, and S. N. Timasheff 115B: 162, New York: Academic Press, 1985. With permission. The lower part is adapted from W. A. Hendrickson in Protein Engineering. Eds. D. L. Oxender and C. F. Fox, p. 11. New York: Wiley-Liss, 1987. With permission.)

replacement, MR). The MR method becomes more and more useful nowadays, as more and more 3-D structures are known and deposited in the data banks. It has also been possible to use specific physical properties such as the variable wavelength of the synchrotron beam. If diffraction patterns are produced using two different wavelengths, this also gives extra information (multiple anomalous dispersion, MAD). It is possible to combine the MAD technology with introduction of medium-heavy atoms such as selenium into recombinantly produced protein. This is done by allowing the protein to incorporate externally added selenomethionine instead of ordinary sulfur-containing methionine during recombinant production. When using more than one x-ray wavelength, it is enough with this single medium-heavy atom substitution to solve the phase problem. The phase problem has therefore become quite manageable for the majority of applications. Still, for the largest and most difficult structures of megadalton size, the phasing is certainly not trivial, and individual solutions have to be looked for, sometimes combining different structural biology methods and iterating toward a solution.

10.5 BUILDING THE STRUCTURE MODEL

The electron density map obtained by the successful solution to the phase problem and subsequent Fourier transformation has to be interpreted into a molecular structure. The quality of the original diffraction pattern determines a resolution limit. This goes back to the quality of the crystals used. The resolution limit of a diffraction pattern is given as a number of Ångström (the further out in the pattern one can detect distinct spots, the higher the resolution). A good protein structure with atomic resolution of side chains requires a resolution of 2 Å or less (see Figure 10.2); with resolution around 5 Å, only an overall shape of the molecule can usually be seen. For a map of better than 2 Å resolution, it is possible to fit pieces of the polypeptide chain and the side chains into the map. Interactive computer programs are used in this process.

10.6 REFINING THE STRUCTURE

After obtaining a first model of the molecule, it can be refined by continued calculations. The model is fitted by comparing its computed diffraction map with the real experimental one and then iteratively minimizing the difference. The final difference is measured by an R-factor, which is 0 for total agreement and 0.59 if there is no agreement. A structure is usually considered reliable if the R-factor is around 0.15. The structure is reported in terms of three space coordinates for each nonhydrogen atom. Hydrogens are typically added according to common wisdom. Besides the space coordinates themselves, there is one more parameter describing each determined atom, namely, the temperature factor, also called the B factor. The B factor carries information about the uncertainty in the determined coordinates. High B factors can signal that there is an error in the determined structure, but it can also be a signal for high mobility in the structure, for example, in dynamic loops at the surface of the protein. In many cases, regions of high disorder or mobility, such as at the N- or C-termini of a protein, or highly dynamic loops, are not even included in the 3-D structure since they are not well defined. It is interesting to note that such disordered/mobile segments can in many cases be easily studied by NMR if the molecules are not too large, and are often shown to have very high intrinsic mobility.

10.7 PROTEIN DATA BANKS

Tens of thousands of protein structures are now deposited in protein structure data bases such as The Worldwide Protein Data Bank (PDB), which by the end of 2008 had about 55,000 biomolecular structures stored. Among the largest and most impressive biomolecular structures reported to date are those of the whole ribosome, a huge protein/RNA complex of more than 2 MD, and a eukaryotic RNA polymerase in action, a similarly huge protein/DNA/RNA complex.

Structural genomic initiatives over the world have started to systematically crystallize and determine the structures of gene products with or without known function, such as from a particular gene in an organism. It turns out that not all proteins identified in this way can be crystallized. A likely explanation is that they do not have a well-ordered structure, or big enough parts of them are disordered so that they do not crystallize well. They may be considered to constitute a new class of proteins: intrinsically disordered proteins (IDPs). Possibly, there is often even a functional reason for the lack of structure: the protein in question requires one or more interaction partners in order to fold into a specific structure that may be associated with a particular function.

10.8   MEMBRANE PROTEINS

Membrane proteins constitute a particular problem in 3-D structure determination. Special solvents involving detergents or antibodies have to be used for the crystallization, and the results are often quite poor. However, the first breakthrough came in the 1980s when the high-resolution x-ray structure of a bacterial photosynthetic reaction center was reported. Now, the number of reported structures of membrane proteins exceeds 100, and massive research efforts are put into solving such structures. In this context, the field has moved forward by recent reports on the structure of a G-Protein-coupled receptor (GPCR, see chapter 23), the beta2-adrenergic receptor with 7 transmembrane helices linked by loop structures. One of these structures was recently solved by using a number of microcrystals in a "crystal droplet" in the microfocus beamline at the ESRF synchrotron facility. These microcrystals could only be found by systematically using the microfocus beam to scan the "crystal droplet" for spots that would diffract like a crystal should, and then adding the results together. Figure 10.3 shows the essential features of this membrane protein structure with well-resolved transmembrane helices.

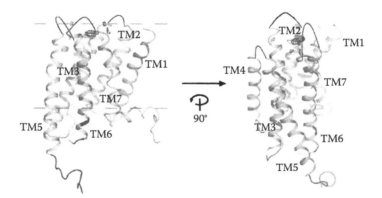

FIGURE 10.3   A cartoon representation of the crystal structure of the beta2-adrenergic receptor, the first G-protein-coupled receptor structure solved to atomic resolution. The transmembrane helices TM1–TM7 are shown, and their positions relative to a biomembrane are indicated. (Reprinted from Soren G. F. Rasmussen et al. (2007). Nature, 450: 383–387. Crystal structure of the human β_2 adrenergic G-protein-coupled receptor. With permission from Macmillan Publishers.)

FURTHER READING

1. Brändén, C and Tooze, J. 1999. Introduction to protein structure. 2nd edition. Garland
   Publishing Inc., New York, NY, USA.

11 Optical Spectroscopy

Astrid Gräslund

CONTENTS

11.1 CIRCULAR DICHROISM

Spectroscopy methods deal with the interaction between matter and electromagnetic radiation of varying wavelengths. The most common spectroscopy deals with UV and visible light absorption and can be used to characterize or quantify, for example, an aqueous solution of a protein. Aromatic amino acids typically absorb with a maximum wavelength around 260–280 nm due to their aromatic side chains. Trp has the largest ring system, and its relatively strong absorbance maximum is around 280 nm. Tyr and Phe have weaker absorbances, and their maxima are shifted toward shorter wavelengths. Absorbance measurements at 280 nm are often used for protein or peptide quantitation. Table 11.1 gives typical absorbance indices for Trp, Tyr, and Phe that can be used for this purpose. The peptide bond in a protein absorbs UV light at shorter wavelengths. In a wavelength range between 185 and 250 nm, the absorbance of a protein or peptide solution is dominated by peptide bond absorption. Studies of light absorption in this range therefore carry information about the peptide bond, its chirality, and its structure. Certain proteins also absorb light in the visible region, such as, for example, heme proteins, due to their conjugated prosthetic groups.

In a normal spectrophotometer, the absorbance in the 185–250 nm region can give information about protein concentration (e.g., if there are no aromatic residues in the sample), but if one makes use of polarized light in the experiment, then additional information about protein secondary structures may be obtained. This is due to the chirality of the peptide bond. The most common spectroscopy exploiting the chirality of the peptide bond is called circular dichroism spectroscopy (CD). CD experiments are often used to give an overall idea of protein structure in aqueous

TABLE 11.1

Absorbance Index and Quantum Yield of the Side Chains of Aromatic Amino Acids

	Absorbance Index (M^{-1} cm^{-1})	Quantum Yield
Trp	5400 (max 280 nm)	0.20
Tyr	1450 (max 275 nm)	0.14
Phe	200 (max 260 nm)	0.04

solution or as initial studies to define good conditions for protein structure studies by more refined methods such as NMR.

A typical CD spectrometer is shown in Figure 11.1. Plane polarized light is considered as composed of two circularly polarized components: left-handed and right-handed circularly polarized light. Together they build up plane polarized light. If only one component is selected, it interacts in its own way with the chiral peptide bond, and there is a somewhat different absorption associated with the left-handed and the right-handed component. If they are measured separately as A_L and A_R for the left-handed and right-handed components, respectively, then their difference, $A_L - A_R$, measured over the wavelength range 185–250 nm, will report on the overall chiral properties of the absorbing peptide bonds. The difference is called ellipticity and is usually normalized into a molar ellipticity $[\theta_M]$ spectrum if the protein concentration is known. The dimension of molar ellipticity is usually (for historical reasons) given as [deg × cm^2 × $dmol^{-1}$]. It turns out that different protein secondary structures will have different characteristic spectral shapes of their CD spectra, as illustrated in Figure 11.2. Each characterized secondary structure has well-defined maxima and minima associated with the CD spectrum, as well as well-defined intensities. The exact spectral shapes and parameters may vary somewhat between different sources of information.

As seen in Figure 11.2, the CD spectrum of an α-helix has a distinct shape with a double minimum at longer wavelengths. Its minimum at 222 nm occurs at a wavelength where most of the other secondary structures display weak CD signals. The maximum around 193 nm has about twice the intensity as its minima. For antiparallel

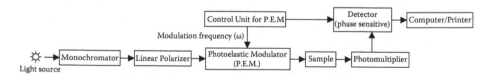

FIGURE 11.1   A schematic setup of a CD spectrometer. Linearly polarized light is passed through a photoelastic modulator unit operating at a certain frequency, producing alternately left-handed and right-handed circularly polarized light. The alternating absorption of the sample, A_L and A_R for the two types of circularly polarized light, respectively, is analyzed by a phase-sensitive detector. The difference, $A_L - A_R$, is typically shown as a function of the wavelength in a CD spectrum.

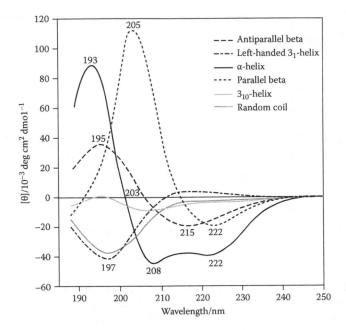

FIGURE 11.2 CD spectral signatures of basic secondary structure motifs of proteins or peptides. Typical maxima and minima are indicated by wavelengths in nanometers. (Figure courtesy of Jüri Jarvet.)

β-sheet one will see a minimum around 215 nm and an almost equally intense maximum at 195 nm. The parallel β-sheet has a minimum at 222 nm and a very intense maximum at 205 nm. The left-handed 3_1-helix is an interesting nonrandom structure that has been shown to appear in, for example, "unstructured" loops of protein structure and at low temperatures of peptides in aqueous solution. Its CD spectrum is similar but not identical to the spectrum reported for random coil. Whereas the CD of random coil is weakly negative around 220 nm, the left-handed 3_1-helix is weakly positive around this wavelength. Although the CD method has been available for decades, the theory is not developed to the stage where spectral shapes can be calculated *ab initio*. The interpretation has to be founded on empirical data.

For a short peptide in which only random coil and α-helix contribute to the secondary structures, it is easy to estimate the relative contribution of α-helix from the molar ellipticity at 222 nm. At this wavelength, an α-helix is characterized by a $[\theta_M]$ of $-38,000$ deg $\times$ cm^2 $\times$ dmol^{-1}, whereas the random coil has a small $[\theta_M]$, which can be neglected here. Figure 11.3 illustrates the CD spectra of a 22-residue peptide, the gastrointestinal hormone motilin, in aqueous solution and interacting with model membranes that induce α-helix to varying degree in the peptide. From its $[\theta_M]$ at 222 nm, one can estimate the amount of α-helix to vary between 15% and 40% under different conditions. Obviously, the results do not tell us if all peptide bonds are in 50% α-helix conformation, or if 50% of the sequence has 100% α-helix and the rest is completely unstructured. The reality is probably in-between. Other structural methods such as NMR are required to give atomic structure resolution, which is necessary for information on points such as this one.

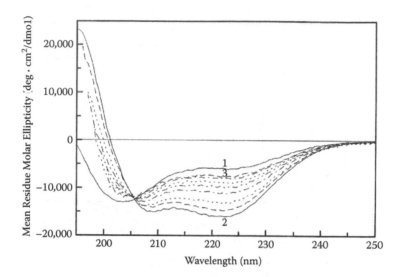

FIGURE 11.3 CD spectra of the 22-amino-acid hormone porcine motilin in pure water (curve 1) and in the presence of 0.1 mg/mL DOPG phospholipid vesicles, with added NaCl from 0 (curve 2) to 1000 mM (curve 3). The spectra show that the peptide in water is close to random coil secondary structure, which is converted to a dominating α-helix when the positively charged peptide binds to the negatively charged vesicles. Addition of increasing amounts of NaCl will force the peptide out into aqueous solution, thereby losing its helical structure. (Figure courtesy of Britt-Marie Backlund.)

For a real protein with many contributing secondary structures, the CD spectrum may include all secondary structure components with their relative quantitative weights. A simple evaluation as in the case of the short peptide is not possible. Instead, one must fit "the best" combination of secondary structure components to the recorded spectrum. This is done by computer algorithms where the spectral components of the different secondary structures are considered as known. Various fitting programs are available. Some are included in the CD spectrometer software from the instrument manufacturer, and others can be obtained from the Web.

CD spectra can be recorded almost as easily as light absorption spectra. One can follow, for example, the temperature dependence of the spectrum and study phenomena such as protein denaturation. In a native protein containing tryptophans, one can usually also see a relatively weak CD signal around 280 nm from the tryptophan side chains that are in a chiral environment in a structured protein. Overall, the sensitivity of CD is usually not as good as for fluorescence (see the following text), so it is not a preferred method for rapid kinetic studies of protein denaturation by stopped flow methods.

11.2   LINEAR DICHROISM

An optical method related to CD is linear dichroism spectroscopy, which can give very specific structural information. The method is based on the fact that the studied molecules are oriented (at least weakly), so that there is a preferred direction, for example, by the laminar flow of a solution, or by an electric field. Plane polarized light with a polarization direction parallel or perpendicular to this preferred direction will be absorbed differently by the studied molecules. Again, a difference spectrum, now called an LD spectrum, can be recorded and will give information about how the light-absorbing molecules are oriented relative to the preferred direction.

11.3 INFRARED SPECTROSCOPY

Infrared (IR) light involves longer wavelengths than visible light. Proteins and peptides absorb IR in various regions when making a transition between the vibrational levels of their chemical bonds. A useful region to study is that of the amide bonds in the protein or peptide backbone. Amide bonds have absorption bands around $1600–1700$ cm^{-1}, depending on the secondary structure of the peptide bond groups. The dimension describing the wavelength here is, by convention, given as wave number (cm^{-1}) describing the number of wavelengths contained in one centimeter. The higher the wave number, the shorter the wavelength and the higher the energy! In fact, it is possible to more or less quantitatively assign the contributions of the different IR absorption bands in this amide bond region to different secondary structures of the studied protein or peptide. A particular strength of IR spectroscopy is that it is independent of the state, whether solid or solution, of the sample, and similar or complementary results compared to CD results can be obtained. But CD spectroscopy requires an optically transparent sample, whereas IR spectra can typically be collected from the surface of a sample and are therefore independent of sample state.

11.4 FLUORESCENCE

Fluorescence spectroscopy is a sensitive technique based on emission of light after excitation of a fluorescent group (fluorophore). In a protein without prosthetic groups, the aromatic side chains, particularly of Trp, can be considered as intrinsic fluorophores. External fluorescent groups can also be attached to proteins *in vitro*, and modern cell biology in vivo often makes use of fusions of a target protein and another protein with an intrinsic fluorescent group formed by spontaneous posttranslational modification (such as green fluorescent protein, GFP see Section 11.6 and Chapter 16).

Figure 11.4 shows the principle of fluorescence emission after absorption of light. Figure 11.5 shows a sketch of a spectrofluorimeter. Fluorescence is based on absorption of a light quantum (excitation), which lifts the absorbing molecule to an excited state. After a short time (typically nanoseconds), the molecule returns to the ground state by light emission of another light quantum. There are also competing processes that return the system to the ground state without emission, leading to quenching of the fluorescence. The ratio between the number of emitted and absorbed light quanta is called the quantum yield. In the typical fluorimeter, the emitted light is measured at right angle relative to the exciting light. Fluorescence spectra are measured either as excitation spectra (when the emission wavelength is fixed and the excitation is

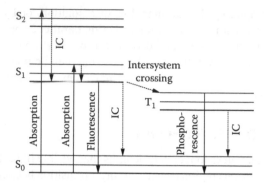

FIGURE 11.4 Energy level (Jablonski) diagram representing light absorption, fluorescence, and phosphorescence processes. Solid lines show absorption and radiative processes. Dashed lines show nonradiative internal conversion (IC) or intersystem crossing (ISC) processes. S_0 is the ground state, and S_1 and S_2 are the first and second excited states, respectively.

varied) or as an emission spectrum (when the excitation wavelength is fixed and the emission is varied).

Among intrinsic protein fluorophores, the side chain of Trp is by far the best due to its relatively high absorbance index and a quantum yield of 0.2 (Table 11.1). The emission spectrum of Trp free in aqueous solution has a maximum around 355 nm. In a more hydrophobic environment, such as in the interior of a protein, the emission spectrum can have its maximum as low as 320 nm and, at the same time, its quantum yield may be different. The side chain of Tyr has a lower absorbance index than Trp, and its quantum yield is only about 0.14. Its emission maximum is around 305 nm. The wavelength maximum of Tyr emission does not vary significantly depending on the polarity of the environment, but the intensity does. The Phe side chain is not really useful as a fluorophore. In general, the fluorescence properties of a molecule are related to the size of any planar conjugated system inside the molecule, and

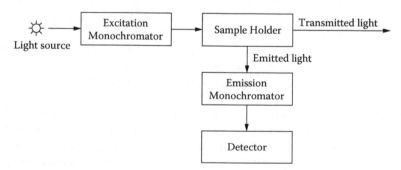

FIGURE 11.5 A schematic setup of a fluorimeter. A detector is placed at 90° angle to the exciting light. There are monochromators on both entrance and exit sides of the sample. A fluorescence emission spectrum is obtained when the excitation wavelength is kept constant and the emission monochromator sweeps through a wavelength region. A fluorescence excitation spectrum is obtained when the emission wavelength is kept constant and the excitation monochromator is swept.

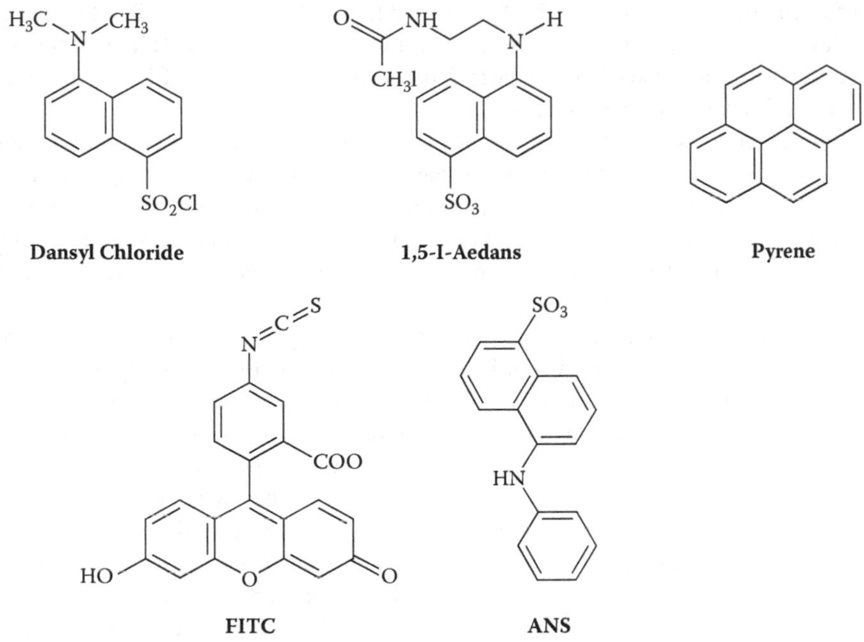

FIGURE 11.6 Chemical formulae of some common fluorescence dyes used to label biological macromolecules. The relatively large aromatic ring systems make them absorb relatively long wavelength light in the visible region and accordingly also have fluorescence emission in the visible region (or, for pyrene, long UV, close to the visible region).

hence the preference for the Trp side chain as an intrinsic fluorophore among normal amino acids.

Figure 11.6 shows some fluorescent markers commonly used for proteins and peptides. The fluorophore attachment can be effectuated directly during the chemical synthesis of a peptide, or by reaction of the protein or peptide after synthesis, making use of surface-exposed SH groups. In contrast to the heterocyclic side chains of a protein, the heterocyclic nucleic acid bases have extremely low quantum yields of fluorescence. It seems that evolution has selected nonfluorescent nucleic acid bases to yield high stability against destruction by light, whereas this is not an issue for the amino acid side chains in a protein.

11.5 FLUORESCENCE APPLICATIONS IN BIOCHEMISTRY AND BIOLOGY

An important application of fluorescence is that of FRET, fluorescence resonance energy transfer. This method makes use of two different fluorophores located at different positions in a molecule, and by exciting one and measuring the emission of the other, one can estimate the distance between the two. The energy transfer is dependent on the spectral (excitation and emission spectra) overlap of the two fluorophores as well as their orientation and distance relative to one another. Long-range distance

measurements on the order of 50 Å are possible by FRET. It is possible to buy particular pairs of synthetic fluorophores that are well suited for FRET experiments.
One can also make use of the intrinsic Trp fluorophore to make a FRET pair with a
synthetic fluorophore such as dansyl.

Fluorescence is useful in studies of molecular interactions, for example, those
between two proteins or between an enzyme and its substrate or an analog. Sensitivity
is high and, with a reasonably good fluorophore, it is easy to study micromolar solutions (typically a few hundred microliters). Fluorescence is often also used to follow
kinetic biochemical processes *in vitro*, such as protein folding or denaturation, in
combination with stopped-flow mixing.

Protein dynamics on the nanosecond time scale may be studied by time-resolved
fluorescence. The basic experiment is based on excitation by a short (typically picosecond) laser flash, then recording polarized fluorescence emission, and observing
how polarization is lost during the nanosecond process of fluorescence emission. The
loss of polarization is due to motion involving the fluorophore, and a rotational correlation time describing the motion can be determined. The overall rotational motion
of a typical protein in aqueous solution occurs on the nanosecond time scale, which
happens to be the lifetime of fluorescence for many fluorophores. Generally, one can
distinguish an overall motion on this time scale and a more rapid independent local
motion, typically associated with the fluorophore itself relative to the overall molecular
framework. An order parameter S^2 describing the angular space available for the local
motion can be evaluated, and its value varies from zero to one. An order parameter of
zero describes the state of completely free local motion, whereas an order parameter
of 1 means a completely rigid system where the local motion is not allowed.

Because of the high intrinsic sensitivity of fluorescence, it has in recent years
become possible to observe fluorescence emitted by single molecules, using advanced
optical methods such as confocal fluorescence microscopy. A recent technology is
fluorescence correlation spectroscopy (FCS), where a small number of fluorescent
molecules are observed inside the focal volume of a laser excitation source. When
molecules diffuse in and out of the focal volume, it is seen as fluctuations of the fluorescence signal. After statistical evaluation, a translational diffusion coefficient can
be determined for the diffusing molecules.

11.6   RECENTLY DEVELOPED FLUORESCENCE PROBES

Green fluorescent protein (GFP) originates from a species of jellyfish (Chapter 16).
Spontaneous rearrangement and oxidation of a triad of residues (Ser-Tyr-Gly) at the
center of the protein gives rise to a fluorescent group. The fluorescent group is also
formed when GFP is expressed as a fusion with another protein in a cellular environment. The phenomenon has turned out to be highly useful for visualization of cellular protein expression by fluorescence microscopy, fusing GFP to the protein under
investigation. The technology has been developed so that there are now possibilities
to express variant GFPs with different fluorescent colors. Of course, this also makes
it possible to study FRET between potentially interacting protein partners in a cellular environment if they are fused to differently colored GFPs that exhibit FRET
when they are close enough.

Quantum dots are stable, very bright fluorophores made from inorganic semiconductor materials (e.g., ZnS/CdSe). They have dimensions on the 10-nm scale, and they can be used to label biological material. Because of the parallel development of microscopy that makes it possible to detect objects with spatial resolution much better than expected from the wavelength of light (actually down to 10–20 nm), it is possible to use such fluorophores to study single molecules such as proteins on the surface of a living cell.

11.7 PHOSPHORESCENCE

Phosphorescence is a phenomenon related to fluorescence. As shown in Figure 11.4, there is a small but distinct probability that the excited molecule crosses over to a long-lived triplet state before deexcitation. Then, after a very long lifetime, sometimes on the order of seconds, deexcitation will occur by the emission of a phosphorescence light quantum, at which time the system returns to the singlet ground state. The long lifetime occurs because the singlet-to-triplet transition is essentially forbidden, and similarly the return triplet-to-singlet transition.

FURTHER READING

1. Campbell, I. D. and Dwek, R. A. 1984. Biological spectroscopy. The Benjamin Cummings Publishing Co., Inc., Menlo Park, CA, USA.
2. Serdyuk, I. N., Zaccai, N. R. and Zaccai, J. 2007. Methods in molecular biophysics. Cambridge University Press, Cambridge, UK.
3. Greenfield, N. J. 2006. Using circular dichroism spectra to estimate protein secondary structure. Nature Protocols 1, 2876–2890.
4. Zimmer, M. 2005. Glowing genes. Prometheus Books, Amherst, NY, USA.
5. Ballou, B., Lagerholm, C., Ernst, L., Bruchez, M. and Waggoner, A. 2004. Noninvasive imaging of quantum dots in mice. Bioconjugate Chem. 15, 79–86.

12 Nuclear Magnetic Resonance (NMR)

Astrid Gräslund

CONTENTS

The phenomenon of nuclear magnetic resonance (NMR) is based on the property of nuclear spin, which shows up as follows: certain nuclei behave as small magnets when they are exposed to a magnetic field. Physicists discovered the Zeeman effect, which is a quantum mechanical phenomenon describing the interaction between magnetic nuclei and an external magnetic field. The nuclei will be able to exist in well-defined states associated with different, discrete energies. In analogy with other types of spectroscopy, transitions can be induced between the different states if energy quanta that precisely match the energy difference between two states are fed into the system.

NMR theory was ahead of experiments, but in the late 1940s successful NMR experiments were first performed. It took 10 years more before the method was adopted in biochemistry. Technology development during the 1960s and onward has given us Fourier transform spectroscopy as well as multidimensional NMR. These modern methods can now be routinely applied to relatively large biomolecules up to 50 kDa or, in special cases, up to 1 MDa. For a relatively small protein of less than 30 kDa, it is possible to obtain information about its three-dimensional solution structure as well as descriptions of molecular dynamics and molecular interactions, all with atomic resolution. In fact, NMR is "the other" methodology for 3-D protein structures at atomic resolution, in addition to x-ray crystallography. In the database of protein structures (PPDB) at the end of 2008 there were about 7500? NMR structures reported, to be compared with 47000? x-ray structures. The comparative weakness of NMR is the molecular weight limitation and the somewhat lower resolution compared to x-ray crystallography. The comparative strength is that the structure is determined in solution under controlled conditions in terms of pH and solvent ions, and that additional information about dynamics and interactions may be obtained.

12.1 NMR THEORY AND TECHNIQUES

The simplest magnetic nucleus is a proton with spin quantum number ½, which gives rise to two energy levels in a magnetic field. Table 12.1 shows the most common nuclei in biomolecules, the isotopes for which they have ½ spin number, and their natural abundance. While ^1H with spin ½ is abundant in peptides and proteins and therefore immediately useful, the most abundant isotopes of carbon, ^{12}C, and nitrogen, ^{14}N, have spin quantum numbers 0 and 1, respectively. ^{12}C is therefore not NMR-active at all, and ^{14}N gives rise to complex spectra. For this reason, the isotopes ^{15}N and ^{13}C are often introduced into molecules to be studied by NMR, and one obtains spectra that can be relatively easily interpreted.

For simplicity, the following description will deal with a proton spectrum. For a proton in a magnetic field, the quantum theory description shows that transitions can be induced between the two energy levels if energy quanta fed into the system (typically of radiofrequency) exactly match the energy level difference. Figure 12.1 illustrates the energy levels and the transition. One great advantage of NMR as a spectroscopic method is the extreme resolution, so that in principle each proton within a molecule will have a distinguishable transition (at a distinguishable energy) that can be recorded. We therefore can consider each proton in a molecule to be a probe giving rise to a specific signal that gives information about the molecule with atomic resolution. Among the challenges that have been overcome over the years is the question of assignment, that is, which signal belongs to which proton in the molecule? A related question is: how can I manipulate the NMR signals to give additional information, regarding (a) which protons are neighbors in the covalent structure, (b) which protons are close in space, and (c) how rapidly does the molecule move and vibrate? Precise answers to such questions are now available from sophisticated NMR experiments.

TABLE 12.1
The Most Common Nuclei in Biomolecules—Spin Quantum Number and Natural Abundance

Nucleus	Spin Quantum Number	Resonance Frequency (MHz) at 11.74 Tesla	Natural Abundance (%)
^1H	½	500	100
^{13}C	½	125.7	1.1
^{15}N	–½	50.7	0.4
^{31}P	½	202.4	100

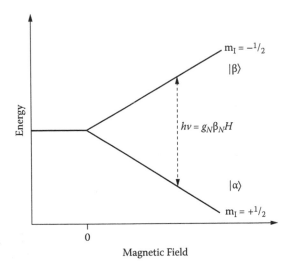

FIGURE 12.1 Energy-level diagram describing the Zeeman effect for a nucleus with spin ½ (typically a proton). The external magnetic field H splits the energy levels into two, corresponding to the magnetic quantum numbers $m_I = +1/2$ and $-1/2$. The energy separation between the two levels increases linearly as a function of H. Transitions between the levels can be stimulated by feeding electromagnetic radiation with a quantum energy $h\nu$ into the system, when $h\nu$ exactly matches the energy level separation $g_N\beta_N H$. The Planck's constant is h; ν is the frequency of the electromagnetic radiation; β_N is the Bohr magneton, a measure of the magnetic moment of the nucleus; and g_N is a constant.

A modern NMR spectrometer is built around a superconducting magnet with a very high magnetic field, and technology based on radiofrequency pulses that are absorbed by the sample and give rise to complex but interpretable NMR spectra. NMR spectrometers are characterized as 400 MHz, 500 MHz spectrometers, and so on, according to the typical resonance frequency of protons at the operating magnetic field. As already mentioned, one major strength of the NMR technique lies in the resolving power that gives rise to distinguishable resonances that differ only in extremely small values of energy. The NMR signals are in fact recorded as functions of time after very short excitation pulses by radio waves. The response signal is called free induction decay (FID), sometimes thought of as the ringing of a tuning fork after being hit. As in x-ray crystallography (phase space vs. 3-D space, see Chapter 14), the NMR signal can be considered to carry equivalent information when it is recorded in time space as a decaying sine wave or in frequency space as a frequency spectrum of energy absorption. After recording FID, this time-dependent function can be treated by mathematical procedures based on Fourier transformation, thereby transforming the function of time to a function of frequency (see Figure 12.2). Unlike crystallography, there is no serious phase problem in this case—we expect positive energy absorption signals and can adjust the spectrum numerically to give the right phase, at least in simple cases.

Figure 12.2 also illustrates one-dimensional and two-dimensional proton NMR spectra of a small peptide. In the 1-D spectrum, which shows energy

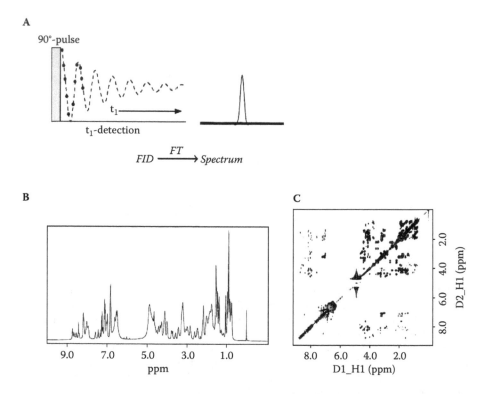

FIGURE 12.2 (A) Schematic representation of a pulsed NMR experiment. A short radio-frequency pulse (here called 90° pulse) excites the sample placed in the magnetic field. The signal emitted by the sample as its magnetization returns to equilibrium is recorded as a function of time (called *FID*, free induction decay). The time-dependent signal is treated mathematically by Fourier transformation (*FT*) and gives a frequency spectrum as a result. (B) Example of a one-dimensional proton spectrum of a short peptide, showing the different frequency regions where one can observe the different types of proton resonances: amide protons H^N, aromatic protons, H^α protons in the peptide backbone, and aliphatic protons. The zero point of frequency is defined by a standard substance. The frequency scale is given as parts per million (ppm) of the carrier frequency. (C) Example of a two-dimensional proton spectrum in which crosspeaks are seen as islands coming out of the baseplane. The diagonal represents the one-dimensional spectrum, and the crosspeaks represent correlations between pairs of proton resonances. (Figure courtesy of Lena Mäler.)

absorption as a function of frequency (or energy), the resonance of each proton in the molecule is separate and observable. The direct relation between frequency and energy arises because the energy E of an energy quantum can be expressed as $h\nu$, where h is Planck's constant and ν is the frequency of the electromagnetic radiation that is absorbed (see Figure 12.1). The NMR signals are typically grouped together as arising from protons in amides, aromatic groups, aliphatic or methyl groups, and so on.

The 2-D NMR experiment is performed by introducing a second timescale into the experiment—instead of one pulse excitation we use two or more short pulses to excite the sample, and we can now vary the delay between the application of the two pulses. Small incremental changes in the delay time followed by acquisition of a number of different FIDs, one after using each particular delay time, gives rise to a 2-D set of FIDs. This can be subjected to two rounds of Fourier transformation, and will then result in a 2-D frequency spectrum.

For the 2-D spectrum shown in Figure 12.2 we have chosen to do an experiment called correlation spectroscopy (COSY). The result is a "map"-like spectrum with frequency along the two axes. The 1-D resonances are seen as a mountain range in the diagonal. COSY crosspeaks appear as islands in the spectral plane between resonances belonging to protons that are attached to neighboring carbon atoms ("correlated"). It was 2-D experiments such as COSY and the related NOESY (nuclear Overhauser spectroscopy, which gives crosspeaks for resonances belonging to protons that are close in space) that first gave clues about how to assign resonances to protons in a sequence-specific way in relatively large biomolecules. COSY spectra could connect the resonances within a single amino-acid residue, and NOESY spectra could connect one residue to the next.

As mentioned in the introduction, studies of recombinant proteins nowadays often make use of ^{15}N- and ^{13}C-substituted molecules. The isotope substitutions are typically accomplished by producing the recombinant proteins in minimal media and adding ^{15}N-ammonium chloride and/or ^{13}C glucose to the medium. The NMR spectrometer can combine radio-frequency pulses in all three ranges needed for absorption by ^{1}H, ^{15}N, and ^{13}C nuclei, and 2-D or 3-D or even higher-dimensional spectra with a very high information content can be recorded. Figure 12.3 illustrates a section of a very common and useful type of 2-D spectrum, the HSQC (heteronuclear single quantum correlation) spectrum of a small protein. The two axes have ^{15}N and ^{1}H frequency scales, and each crosspeak corresponds to an amide group, that is, the ^{15}N–^{1}H group of each amino acid residue. This can be considered as a very useful "fingerprint" of a protein that after assignment has a specific and distinguishable probe of every single residue. These resonance crosspeaks are often used as probes of molecular interactions, such as between small molecules and a protein or between two proteins. In the case of two proteins, one could be labeled by ^{15}N and the other not, and in an ideal situation, one can directly see which resonances are disturbed by mixing the two proteins, that is, which residues are involved in the interaction. The amide crosspeaks are often used as probes also in studies of molecular dynamics, as will be described later.

12.2 NMR PARAMETERS

There are a number of basic parameters that are important for the interpretation of NMR spectra.

a. **The chemical shift,** defining the position of a particular resonance on the frequency scale. Often, the chemical shift is given in parts per million (ppm) scale. This scale describes the difference in frequency between actual resonance and that of a common reference, reporting this difference as parts per million of the carrier frequency (e.g., a chemical shift of 0.5 ppm for a proton with a 400 MHz spectrometer would mean that this resonance is

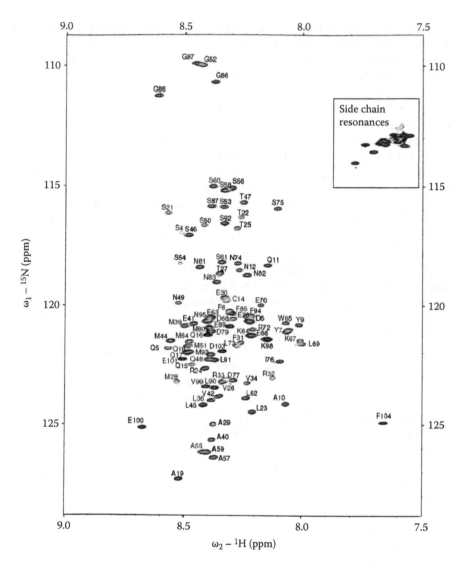

FIGURE 12.3 Example of a two-dimensional HSQC (heteronuclear single quantum correlation) spectrum, correlating amide ^{15}N–H resonances in a peptide. Each crosspeak represents one amide group. The horizontal scale represents the proton frequencies, and the vertical scale the ^{15}N frequencies. (Figure courtesy of Jens Danielsson.)

$0.5*10^{-6} * 400*10^{6} = 200$ Hz away from a reference signal). By reporting resonance positions using this scale, one is independent of the spectrometer frequency.

b. **The intensity** of the signal, measured as the area underneath the resonance, which is proportional to the number of nuclei involved. Typically, a methyl group would give rise to a resonance with three times the intensity of a single proton.

c. **Relaxation times T1 and T2,** which describe how the nuclear spin in question behaves within its environment after being excited by a radiofrequency pulse. The so-called transverse relaxation time T2 is directly related to, and inversely proportional to, the linewidth in Hz. Longitudinal relaxation time T1 describes the time course for how the excited system returns to equilibrium. An interesting aspect of NMR is that these two parameters, which can be directly measured for each resonance, are very directly related to motion in the system, that is, molecular dynamics. The theory is clear-cut and precise because the physical processes giving rise to relaxation by magnetic interaction between two neighboring spins are well known. It is interesting to note that it is the relaxation properties of a molecule that sets the size limit for biomolecules to be studied by NMR. The relatively slow motion of a large protein makes its NMR relaxation very rapid (short T2), and this results in a larger resonance linewidth. In the large molecule limit, the resonance vanishes in the noise. Typically, one would use relatively high temperature to study a protein in solution in order to enhance its mobility (but of course there is an upper limit when the native molecular properties are lost). There are also special NMR techniques developed to prolong the relaxation process for larger molecules, thereby sharpening the resonances. These techniques are based on the original TROSY spectroscopy (transverse relaxation optimized spectroscopy).

d. **Multiplet structure** of resonances, caused by strong interactions between nuclear spins. In a typical case, one proton resonance is split into two components separated by a few Hertz (the so-called J coupling) because of coupling to another proton with a different chemical shift bound to a neighboring carbon. This effect is in fact what was described in the previously mentioned COSY spectrum; in that experiment, two protons give rise to resonance crosspeaks if they are J-coupled.

These parameters are basic to any NMR experiment and are used in the interpretation of molecular properties that can be linked to spectral observations. Next, we take a close look at the common problem areas where NMR is used in studies of peptides and proteins.

12.3   NMR STRUCTURE DETERMINATION

NMR can be used to determine solution structures of peptides and proteins. A structure study is based on a variety of observations that first have to give the correct assignment of the spectrum. Originally, only proton spectra were used. COSY spectra give patterns that can be used to identify individual residues. NOESY spectra carry information that can be used to link one residue to the next along the backbone. The combined use of COSY and NOESY spectra makes it possible to achieve sequential assignment along the polypeptide sequence. Next, NOESY spectra are evaluated for "nontrivial" crosspeaks, that is, giving information about proton pairs that are not close in the chemical structure but still close in space. It is

Introduction to Peptides and Proteins

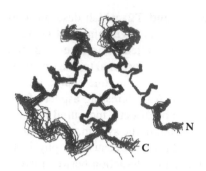

FIGURE 12.4 An overlay (ensemble) of the backbone structure of a small protein, determined by NMR. Each structure is calculated independently using NMR structure constraints but starting from different initial structures. (Reprinted from Mäler, L., Blankenship, J., Rance, M., and Chazin, W. J. Site-site communication in the EF-hand Ca^{2+}- binding protein calbindin D$_{9k}$. Nature Structural Biology, 7(2000), 245–250. With permission from Macmillan Publishers.)

known that the nuclear Overhauser effect, which produce crosspeaks in a NOESY spectrum (and is in fact based on the relaxation effects the two partner nuclei have on one another), only operates between nuclei that are closer than about 6 Å. Such observations give rise to a number of conditions for the molecule: nucleus a must be within 6 Å of nucleus b. Giving relatively loose boundaries, the distance conditions are evaluated based on the strength of the interactions (obtained as integrated areas under the crosspeaks) divided into weak, medium, and strong effects, with distance (r) constraints typically $r < 4$ Å, $r < 5$ Å, or $r < 6$ Å for the intense, medium and weak crosspeaks, respectively. A large number of distance constraints are collected and, toward the end of the spectral evaluation, the molecular system should be heavily "overdetermined" with them. Each observation carries only limited information since local mobility also will influence the size of the crosspeak. To the distance constraints can be added other types of geometrical information, for example, from J couplings, which carry information about dihedral angles. Then a mathematical method called distance geometry can be applied to the problem of calculating molecular 3-D structures that do not violate structural constraints. It turns out that many structures, usually deviating only in small details, will fulfill the structural constraint conditions. The results are usually presented as overlays of 10 or 20 structures (Figure 12.4).

If the protein has a well-defined 3-D structure, the different structure calculations will show very similar results. Weakly structured parts will show up as widely variable loop or coil structures with very little overlap. It is also interesting to note that the solution structure of a protein determined by NMR matches in close details that of the same structure determined by x-ray crystallography. This similarity is particularly obvious for the interior of the protein, whereas the surface properties vary for natural reasons: in NMR they are determined by interactions with the solvent (aqueous in most cases), and in x-ray crystallography by interactions with neighboring molecules in the crystal packing. Figure 12.5 gives

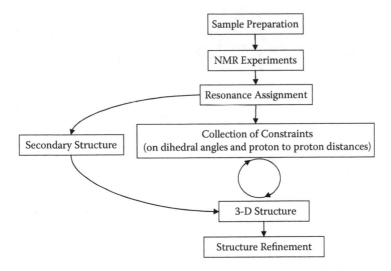

FIGURE 12.5 Overview of the different stages involved in the determination of an NMR solution structure of a protein. A protein sample of sufficient concentration, preferably labeled with ^{15}N and/or ^{13}C by recombinant production methods, is subject to NMR spectroscopy. The crucial sequence-specific assignment of resonances is already rich enough in information to evidence the presence of regular secondary structures along the protein backbone. This stage is followed by collection of various structural constraints, typically short proton–proton distances determined from observations of NOESY crosspeaks, and dihedral angles from COSY-type spectra. This information is used for computer calculations to obtain co-ordinates for 3-D structures, using programs such as DIANA based on distance geometry, or the more recent version CYANA. Finally, the determined structures may be refined by energy minimization and are presented as ensembles of structures, each fulfilling the conformational constraints nearly equally well. An initial crude structure can be iteratively used to obtain more resonance assignments and more distance constraints, and this can yield an iteratively improved final structure.

an overview of the procedures involved in a typical NMR structure determination of a protein.

12.4 NMR STUDIES OF MOLECULAR DYNAMICS

The weakly structured parts of a polypeptide chain that do not converge to a well-defined structure in the calculations are in most cases very highly dynamic. The dynamics of the polypeptide chain can be studied in a straightforward way in a ^{15}N-labeled protein sample by observing relaxation properties of the amide groups. A mathematical treatment will use the relaxation data to give dynamic parameters. In a common treatment, using the so-called model-free approach, one obtains an overall rotational correlation time (nanosecond timescale) for the whole molecule, a residue-specific order parameter varying between 0 (complete disorder) and 1 (complete order), and a residue-specific local rotational correlation time (picosecond to nanosecond timescale). The motional parameters can therefore be followed along

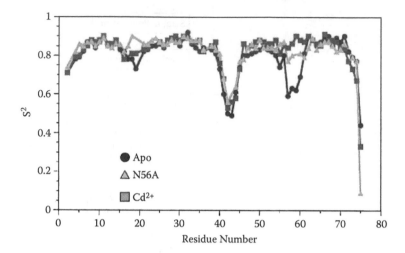

FIGURE 12.6 NMR relaxation studies yield the generalized order parameter S^2 along the polypeptide chain of a small protein calbindin (three different forms). S^2 approaches 1 for a rigid structure and drops to 0 for residues with complete motional freedom. Calbindin is composed of two relatively rigid domains linked by a flexible linker around residue 42, and has a calcium-binding site around residue 57. When this site is empty, as in the apo form of the protein, residues close to 57 also have increased motional freedom. (Reprinted from Mäler, L., Blankenship, J., Rance, M., and Chazin, W. J. Site-site communication in the EF-hand Ca^{2+}- binding protein calbindin D_{9k}. Nature Structural Biology, 7(2000), 245–250. With permission from Macmillan Publishers.)

the polypeptide chain. This is illustrated for the generalized order parameter S^2 in Figure 12.6.

12.5   NMR VERSUS CRYSTALLOGRAPHY AS A 3-D STRUCTURE DETERMINATION METHOD

What are the weaknesses of NMR as a tool in studies of proteins or peptides? First of all, NMR is a relatively insensitive method, and relatively large amounts of sample are needed. For a 3-D structure study, a sample of typically 500 microliters of 1 mM protein, preferably with isotope labeling, is needed. As a method for structural biology, NMR must be compared with x-ray crystallography. Here the main disadvantage is the size limitation of NMR, which does not apply to x-ray crystallography. NMR studies are also much less streamlined than modern x-ray crystallography and still require "human intervention" along the study to a much larger extent even if many attempts at automation of assignment procedures are under way. For these reasons, in structural genomics projects in which automation and speed are high priorities, x-ray crystallography is the method of preference. Only if crystallography is not successful will NMR structure studies be carried out. However, NMR serves a direct role also in structural biology laboratories that mainly focus on crystallography. NMR provides a very reliable screening method to determine if there is a well-developed 3-D structure in a particular protein construct in a particular solvent; even a simple 1-D proton spectrum will directly give evidence of the presence or absence of ordered tertiary structure in the molecule.

When the next step in understanding the human proteome is considered, namely, that of interactions between different proteins or between proteins and nucleic acids (sometimes called the interactome), then NMR is one of the major tools with great promise. These studies should make use of the possibility of differential isotope labeling and easy visualization of interaction areas of the molecules. Also, for the important question of experimental description of the detailed molecular dynamics of a macromolecule, NMR is probably the most promising method.

12.6   NEW NMR DEVELOPMENTS

New developments in NMR techniques involve the use of multiple isotope labeling, as mentioned earlier. Also, deuterium labeling of a macromolecule is sometimes used in studies of large molecules since deuterium instead of protons on most places in a molecule will affect the relaxation of the remaining protons, which will give slower relaxation and sharper resonances. This effect is often more than enough to compensate for the direct loss of signals from diluting the protons with deuterons.

New spectroscopic methods have been developed that allow much larger structures to be studied than before. The previously mentioned two-dimensional TROSY experiment can yield relatively well-resolved spectra for proteins up to several hundred Dalton in molecular weight, that is, almost an order of magnitude larger molecules than were possible to study before.

Other new developments involve the observation of residual dipolar couplings when the protein is weakly aligned in the sample. Normally, in high-resolution NMR, the molecules tumble freely and rapidly to give highly resolved spectra. By introducing agents such as filamentous bacteriophages or polyacrylamide gels that can orient themselves and give rise to weak orientation of the protein molecules, additional geometric constraints can be determined and used in the structure calculation to give refined structures.

The further use of automated procedures is another development in NMR methodology. The aim here is to make resonance assignments and structure calculations with a minimum of interventions by the spectroscopist.

12.7   SOLID-STATE NMR IN BIOLOGY

Besides high-resolution NMR, solid-state NMR can also be used to study biological macromolecules. Because of the lack of rapid motion in the sample, the resonance linewidths are often large. Special techniques involving rapid spinning of the sample in the magnet will enhance the spectral resolution. Solid-state NMR has been used for studies of small isotope-labeled cofactors to large molecular assemblies, such as the visual pigment inside the large membrane protein rhodopsin.

Modern solid-state NMR techniques allow full assignment also of uniformly isotope-labeled molecules of moderate size. There are promising applications in structure studies of samples composed of material in a so-called amyloid state, for example, for the amyloid beta peptide involved in Alzheimer's disease (see Chapter 29).

12.8   MAGNETIC RESONANCE IMAGING

The dominating use of NMR in the world is nowadays found in hospitals, in the MRI (magnetic resonance imaging) equipment used as an alternative or complement to x-ray tomography for diagnostic purposes. The basic principle is the same as in NMR spectroscopy. However, for imaging, one uses varying magnetic fields, the so-called field gradients, over the object. This makes it possible to relate the absorption of certain energies to a 3-D image of the object, in fact its proton density, which, of course, is high inside and very low outside the patient. Further refinements make it possible to determine relaxation properties of the signals and therefore distinguish between different kinds of tissue, giving contrast to the overall image.

FURTHER READING

1. Wüthrich, K. 1986. NMR of proteins and nucleic acids. John Wiley & Sons, Inc., USA.
2. Cavanagh, J., Fairbrother, W. J., Palmer III, A. G. and Skelton, N. J. 2007. Protein NMR spectroscopy: principles and practice. Elsevier/Academic Press, San Diego, USA.
3. Ferentz, A. E. and Wagner, G. 2000. NMR spectroscopy: a multifaceted approach to macromolecular structure. Quart. Rev. Biophys. 33, 29–65.
4. Palmer III, A. G. 2001. NMR probes of molecular dynamics: overview and comparison with other techniques. Annu. Rev. Biophys. Biomol. Struct. 30, 129–155.

13 Methods to Follow Protein Folding

Astrid Gräslund

CONTENTS

The folding of a protein is a spontaneous process that should take place once the polypeptide chain leaves (or has left) the ribosome. The final 3-D structure needed for a particular biological activity is coded for by the sequence of amino acids as postulated by Anfinsen, but this does not tell us how this process occurs in a particular environment.

The problem of protein-folding processes has been the subject of intense research over several decades, and the emerging picture is that the processes are certainly not simple. Small globular proteins seem to fold on their own in an aqueous environment as well as in a cellular environment. In contrast, larger proteins or proteins made up of several components often in the cell require assistance from molecular chaperones, proteins that "help" other proteins to fold. Membrane proteins are another special class that require a special cellular machinery to fold properly. Then one should not forget that an individual protein certainly can misfold and that the cell has the protein degradation functions, for example, the ubiquitin–proteasome pathway, to take care of such situations.

In the laboratory, we can study protein folding experimentally. Most studies have dealt with small proteins that fold spontaneously in an aqueous solution. This chapter will describe some of the methods that are available for this type of studies.

13.1 PROTEIN DENATURATION STUDIES

Proteins keep their native 3-D structure only under specific conditions in terms of temperature and solvent. There are particular thermophilic organisms that keep their proteins in a native state also at highly elevated temperatures, even above 100°C, but normally proteins should be kept well below this temperature. Protein structures are held together by a combination of mainly "hydrophobic" forces, hydrogen bonding, ionic forces, disulfide bonds, and so on. By slowly raising the temperature of an aqueous solution of a small protein in a functional and folded state, one can follow a spectroscopy signal arising from, for example, tryptophan fluorescence or circular dichroism (CD), and easily characterize the folded/unfolded transition as a function of temperature. The transition can be described by a two-state model with the folded and unfolded states in rapid equilibrium in each molecule. Therefore, if the protein is an enzyme, its enzymatic activity will typically also follow the same transition curve.

Another classical method to induce protein denaturation is by adding agents that break up the hydrogen bonds necessary for the native structure. Guanidinium hydrochloride or urea at molar concentrations are typical denaturants used in such studies. Similar to temperature studies, unfolding can be followed by suitable spectroscopic methods and be described as a two-state process. Figure 13.1 illustrates such a denaturation curve for apomyoglobin (myoglobin without the heme group) using urea as denaturant and the CD signal at 222 nm as readout. Extreme pHs can also be used to induce unfolding of proteins. In simple cases the unfolding/folding transition is reversible, both with temperature-induced, denaturant-induced, and pH-induced transitions. In this kind of studies, when the unfolding processes are followed by nonkinetic techniques, the results only show the equilibrium parameters, such as the temperature, denaturant concentration, or pH at which 50% of the native structure remains (or rather when there is 50% probability that the molecule is in its native state). In Figure 13.1, the urea concentration at the midpoint of the urea-induced denaturation corresponds to 3.2 M urea for the wild-type protein and lower midpoint concentrations for the mutant proteins.

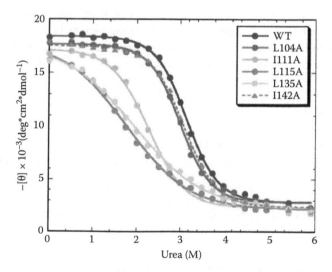

FIGURE 13.1 Protein unfolding as a function of urea concentration, exemplified by apomyoglobin. The CD signal is followed from a highly α-helical state towards a random coil. Selected single residue mutants show lower stability and resistance to urea unfolding, compared to the wild-type protein. (Reprinted from Nishimura, C., Dyson, H. J., and Wright, P. E. (2005). Identification of native and non-native structure in kinetic folding intermediates of apomyoglobin. J. Mol. Biol., 355: 139–156. With permission from Elsevier.)

13.2 PROPERTIES OF THE "DENATURED" STATE OF A PROTEIN

In recent years it has become apparent that a "denatured" state of a protein is not really a 100% random structure. Depending on the method of denaturation, it has been found that even if the tertiary structure of a protein is mostly lost together with the biological activity, secondary structures may remain to some extent, and there may be an overall globular shape of the molecule. It is in fact rather difficult to prepare a totally random coil form of a protein. Several studies have shown that a so-called "molten globule" structure can be stabilized and studied after mild acid denaturation (reaching typically pH around 2) of a protein. Interestingly, the molten globule form of a protein is similar to what has been described as a folding intermediate in the normal folding pathway for a protein.

13.3 KINETIC STUDIES OF PROTEIN FOLDING

Rapid kinetic methods can also be used to study protein folding. In this type of study, one must use a rapid mixing device, whereby a native protein solution is mixed with a denaturant or vice versa, a denatured protein solution is diluted into an aqueous solution. The process is followed by rapid spectroscopy in real time, based on fluorescence or CD. The two solutions are entered into a mixing chamber by two syringes and an outlet is passed into the spectroscopic equipment. There the flow is "stopped" by another syringe and the mixture is allowed to react inside the spectrometer while a spectral parameter is recorded. The result of a stopped-flow kinetic study of protein folding by optical spectroscopy gives an overall picture of the folding process with a time constant, depending on the conditions. The time resolution is usually limited by the sample mixing time in the stopped-flow mixing device, which is typcially on the order of 50 ms. Figure 13.2 shows kinetic traces monitored by

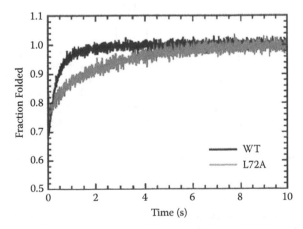

FIGURE 13.2 Time-resolved protein folding, followed by CD for apomyoglobin in a stopped-flow kinetic experiment, after dilution of a urea solution. The single-residue mutant folds slower than the wild-type protein. (Reprinted from Nishimura, C., Dyson, H. J., and Wright, P. E. (2005). Identification of native and non-native structure in kinetic folding intermediates of apomyoglobin. *J. Mol. Biol.*, 355: 139–156. With permission from Elsevier.)

the CD signal at 225 nm when apomyoglobin (and one more slowly folding mutant protein), originally in 5.5 M urea, was allowed to refold in an aqueous buffer where the final urea concentration became below 1 M.

Another experimental approach that in principle can give spatial resolution on the atomic scale coupled with stopped-flow time resolution is based on hydrogen exchange studies of amide protons with NMR. For the NMR study, the protein should be labeled with ^{15}N so that the individual amide groups can be followed by HSQC spectroscopy (see Chapter 12 on NMR). A typical experiment could be performed as follows: A fully protonated and unfolded ^{15}N-labeled protein in 6 M urea H_2O solution at pH 6 is refolded by rapid dilution into the deuterated buffer, pH 6, without urea. The refolding is allowed to proceed for a certain time (say, 30 ms). Because of the moderate pH, the exchange rate of any amide proton is slow during refolding. After refolding has reached a certain stage, a pH "pulse" to pH 10 of short duration (20 ms) is applied. Unprotected amide protons, that is, those that are not yet part of a folded structure, exchange rapidly for deuterium during this pH pulse. The experiment is repeated with different samples exposed to different refolding times. The results can be analyzed using two-dimensional HSQC NMR spectroscopy. The time scale of hydrogen-to-deuterium exchange can be followed for each individual amide group since only amide groups with protons bound contribute to the intensity of their particular crosspeak. High proton occupancy at short folding times show regions of the protein that are folded at early times. Regions of the protein that remain protonated only after long refolding times are late to participate in the folding process. Figure 13.3 shows typical HSQC results of pulse-labeled apomyoglobin samples after a short and a long refolding

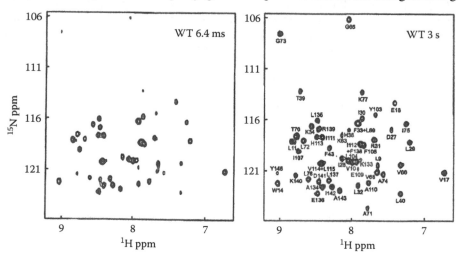

FIGURE 13.3 Kinetic appearance of NMR spectra during the protein refolding process in an aqueous solution after being completely unfolded at a high concentration of urea. After 6.4 ms of refolding, the structure is only well defined in certain parts of the protein, and these give rise to well-defined resonances in an NMR spectrum. After 3 s, the whole protein is refolded. (Reprinted from Nishimura, C., Dyson, H. J., Wright, P. E. (2005). Identification of native and non-native structure in kinetic folding intermediates of apomyoglobin. J. Mol. Biol., 355: 139–156. With permission from Elsevier.)

time. It should be pointed out that the pH pulse labeling strategy resolves at least partly the classical problem when studying proton exchange: the question whether the proton exchange is limited by geometry or by chemistry. At pH 10 the chemistry is immediate and favorable for exchange, whereas the short exposure to high pH does not interfere too much with the refolding process.

Also without atomic resolution by NMR methods, hydrogen–deuterium exchange can be used to quantitatively follow protein-folding processes using similar quench-flow pulse-labeling strategies. Using mass spectrometry, one can quantitatively determine the increased mass of the protein resulting from deuterium exchange of the unprotected exchangeable protons after a certain time of refolding.

Experimental techniques to study kinetic protein folding in real time vary in time resolution. Methods in which folding is initiated by mixing two solvents in a stopped-flow apparatus usually work on a millisecond timescale. If folding can be triggered optically or by a temperature jump, much higher time resolution can be achieved, down to nanoseconds.

13.4   ENERGY LANDSCAPES AND PROTEIN-FOLDING PATHWAYS

Theoretically, the protein folding process may be visualized by using a schematic diagram of an energy landscape in which the configuration should be considered to have many more dimensions than can be depicted on paper (Figure 13.4). The native state has the lowest energy but can be reached through a number of pathways. Local

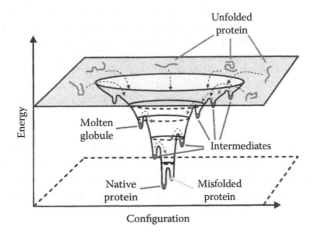

FIGURE 13.4   Energy landscape representation of the protein-folding process ("funnel model"). The unfolded protein, at a high energy level, takes up a large configuration space. As the energy moves toward a global minimum, the available configuration space becomes smaller and smaller. Intermediate false minima represent intermediate states that can be populated during the folding. In favorable cases, the global minimum is reached and the protein achieves a native configuration. In other cases, the protein is captured in a false minimum, too deep to be exited under the normal folding process, which results in a misfolded protein. (Reprinted from Schultz, C. P. (2000). Illuminating folding intermediates, Nat. Struct. Biol., 7: 7–10. With permission from Macmillan Publishers.)

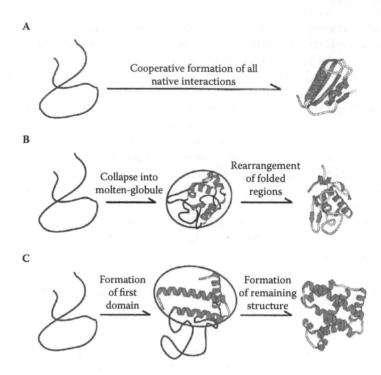

FIGURE 13.5 Protein folding can follow different pathways. (A) A small protein can often fold and spontaneously form all native interactions without any populated intermediates. (B) Another protein first collapses into a so-called molten globule state with secondary-structure interactions but not yet the tertiary, long-range interactions. These are formed during a second stage of folding rearrangements. (C) A third possibility is that a multidomain protein first folds one domain, which then helps the other domains to fold properly. (Reprinted from Hamada, D. et al. (2000). Evidence concerning rate-limiting steps in protein folding from the effects of trifluoroethanol. Nat. Struct. Biol., 7: 58–61. With permission from Macmillan Publishers.)

minima are seen close to the globular minimum and may represent misfolded structures. Small proteins usually do not have folding intermediates. For larger proteins (>100 residues), intermediates are common. A molten globule state may represent such an intermediate. Figure 13.5 illustrates folding pathways for small and large proteins.

FURTHER READING

Lindorff-Larsen, K., Rogen, P., Paci, E., Vendruscolo, M. and Dobson, C. 2005. Protein folding and the organization of the protein topology universe. Trends Biochem. Sci. 30, 13–19.

Nishimura, C., Dyson, H. J., and Wright, P. E. 2005. Identification of native and non-native structure in kinetic folding intermediates of apomyoglobin. J. Mol. Biol. 355, 139–156.

14 Mass Spectrometry

Astrid Gräslund

CONTENTS

Mass spectrometry methods separate molecules based on size and charge. The molecules to be studied must be converted into gas-phase ions before analysis. That is, they have to be converted from solution or solid phase into the gas phase, and ionized in the process. After ionization, the ions are separated in the mass analyzer based on their mass-to-charge ratio and then detected. Figure 14.1 shows the principal scheme of a mass spectrometer with the three modules of ionization, mass analysis, and detection.

The principle of separation based on molecular size and charge goes back more than 100 years, but it took until the 1970–80s before the principles could be used to study large molecules such as proteins. For large biomolecules, the difficult problem was to perform their chemical preparation in the proper way, that is, to produce the correct molecular ions in the gas phase. Early attempts typically used heat to provide the energy to volatilize the molecules, and this was not successful for proteins. Instead, the utility of modern mass spectroscopy methods to study biomolecules such as peptides and proteins are mainly based on the development of two ionization methods: matrix-assisted laser desorption/ionization (MALDI) and Electrospray ionization (ESI). These methods have been shown to be able to convert the large polar, zwitterionic molecules into ions in the gas phase without degrading them in the process.

Mass spectrometry has become one of the most powerful tools for the analysis of proteins. The sample requirements are on the order of 1 pmol to investigate a single protein. The power of the method is due to the fact that it gives fast and accurate information on the molecular weight of the investigated molecule, which can lead to instant identification. In the emerging fields of proteomics and systems biology, mass spectrometry is one of the cornerstone methods. In a typical application, the whole proteome of a cell, as the cell responds to changes in the environment, can be characterized. Modern methods that couple efficient separation techniques with mass spectrometry allow thousands of proteins to be identified in

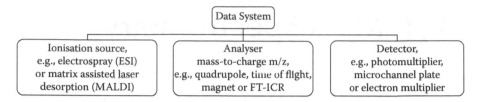

FIGURE 14.1 The different modules of a mass spectrometer. The molecules to be analyzed are first ionized using typically the principle of electrospray ionization (ESI) or matrix-assisted laser desorption ionization (MALDI). Then, in the analyzer module, the mass-to-charge ratio is determined by measurement of time-of-flight and finally detected.

a single automated experiment from minute samples. It becomes possible to follow the changing proteome qualitatively and, to some extent, also quantitatively. Mass spectrometry also gives important information in the field of protein–protein interactions.

14.1 ESI

The ESI technique for ionization is based on spraying a fine mist of the liquid solution of the protein to be analyzed in the mass spectrometer. Figure 14.2 illustrates the

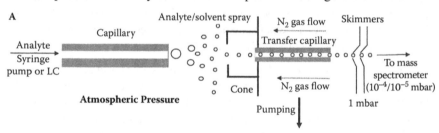

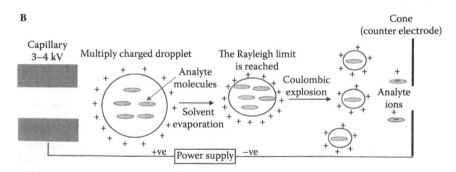

FIGURE 14.2 (A) The principle of ESI. Droplets of the solution to be analyzed are forced out of a thin capillary. When the solution passes through the thin opening kept at a high voltage, the analyte molecules become ionized. In the drying chamber, the solvent in the droplets evaporates, and the droplets become smaller and smaller. Finally, the small droplets explode (coulombic explosion) because of the high concentration of the charge inside. Naked ionized protein molecules are collected by a counter electrode and led into the analysis chamber and detected. (B) A more detailed illustration of solvent evaporation and coulombic explosion.

electrospray ionization process. A voltage difference between the capillary carrying the liquid and the inlet to the spectrometer creates small, charged droplets of the analyte. In the spectrometer, these droplets evaporate until the charge on their surfaces becomes too high for surface tension to hold them together. They reach the so-called Rayleigh limit, whereupon they explode and give rise to bare ions of the originally dissolved molecules. The ions have various charges, depending on variable states of protonation at proton-accessible sites. They are analyzed in vacuum in the mass spectrometer according to the mass-to-charge ratio. The process can result in ions with varying charges in the range from +2 to +40 or higher. High accuracy in molecular weight determination is obtained by averaging the signals from ions with multiple charges. The signals from ions with different charge states may be considered as independent measurements of the molecular weight, and therefore the averaging gives significant improvement in accuracy.

14.2 MALDI

In MALDI technology, a laser beam is used to ionize the sample, which is dissolved in a "matrix" with suitable properties to absorb the energy of the laser light and transfer it to the dissolved molecules. Figure 14.3 illustrates the MALDI process.

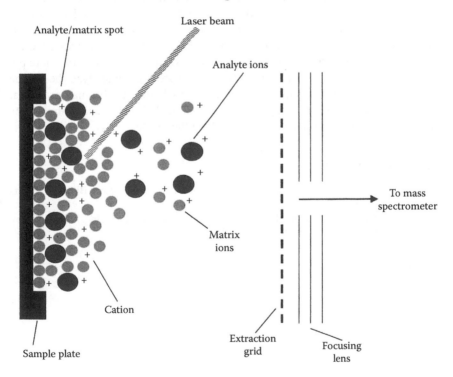

FIGURE 14.3 The principle of MALDI. The sample to be analyzed is mixed with a matrix material and deposited on a plate that is subjected to a high voltage. Laser light of a suitable wavelength is applied to the deposit, which excites the matrix as well as the analyte molecules. The analyte molecules leave the deposit plate and go away as intact molecular ions, which are led into an analysis chamber and detected.

With a suitable combination of laser wavelength and matrix composition, the analyte is thermally desorbed after exposure to pulsed laser light, and ions enter into the gas phase for analysis. In common MALDI spectrometers, a pulsed nitrogen laser with a wavelength of 337 nm is used for excitation. The molecular weight of a protein, up to nearly 40 kDa, may typically be determined with an accuracy of about 0.01%. Generally, the ESI method is considered to provide an even higher accuracy in the determination of molecular weight. It is also the least damaging of the two ionization methods, so it can, in principle, also be used to study molecular complexes held together by relatively weak noncovalent interactions.

14.3   TANDEM MASS SPECTROMETRY (MS–MS)

Both ESI and MALDI ionization techniques are comparatively mild, and the molecular ions undergo little fragmentation in the process. The molecular masses of the investigated proteins and peptides can be obtained with high accuracy. The covalent structure of the ion is not immediately available. However, by modern advanced techniques, additional valuable information in protein studies can be obtained by MS, namely, on the amino acid sequence of the analyte. This requires the use of a tandem mass spectrometer (MS–MS), in which the different produced ions are isolated ("trapped" by different ion trap techniques), fragmented by collisions with an inert gas, and then analyzed separately. Recently, new MS methods have been developed employing very sophisticated technology, and MS is one of the more powerful methods used in proteomics research. These are used to analyze the outcome of advanced separation techniques involving electrophoresis or chromatography, and high throughput coupled with high sensitivity is needed. Advanced computer programs are employed to deconvolute and analyze the results.

FURTHER READING

Yates III, J. R. 2004. Mass spectral analysis in proteomics. Annu. Rev. Biophys. Biomol. Struct. 33, 297–316.

Steen, H. and Mann, M. 2004. The ABC's (and XYZ's) of peptide sequencing. Nature Reviews, Molecular Cell Biology 5, 699–721.

15 Chemical Synthesis of Peptides and Proteins

Ülo Langel

CONTENTS

The technique of chemical peptide synthesis goes back more than 100 years. Ever since the natural peptide structure became available, scientists have been eager to copy Nature and make synthetic peptides in their laboratories. Milestones of early peptide chemistry are the first synthesis of a protected peptide by the azide method (Curtius, 1881); first dipeptide synthesis by chloranhydride method (Fischer, 1901) and synthesis of an 18-amino-acid-long peptide with sequence $LG_3LG_3LG_9$ (Fischer, 1907); introduction of Cbz and tert-butyloxycarbonyl (*t*-Boc) protective groups (Bergmann and Zervas, 1932; Carpino, 1957); synthesis of first bioactive peptides glutathione and carnosine (1935) and oxytocin (1953, Nobel prize to du Vigneaud in 1955).

The modern era in peptide synthesis started with Bruce Merrifield's (1921–2006) work in 1959 introducing the idea of solid phase peptide synthesis

(SPPS), for which he was awarded the Nobel Prize in 1984. Synthetic peptides have been used for numerous applications in biochemical research, which has stimulated the development of new methods for the synthesis of these molecules. Here we present SPPS, with a short introduction to its methods.

15.1 SOLID PHASE PEPTIDE SYNTHESIS

The principle of SPPS is that the peptide is anchored to an insoluble support, a resin, which is stable in the chemical reactions carried out during synthesis (Figure 15.1). SPPS is superior to other methods due to its speed, versatility, possibility for automatization, etc. Protected peptides are often poorly soluble in organic solvents, and an inherent advantage of SPPS is that attachment to the crosslinked polymer resin suppresses aggregation and promotes solvation of the peptide chain. Chemical reactants can then freely diffuse into the interior of the resin beads, and the reaction rates are close to those in solution. In contrast to solution synthesis, a large excess of reagents can be used to rapidly drive the reactions to completion since excess reagents are easily washed away after each step. The intermediates in the synthesis are not purified

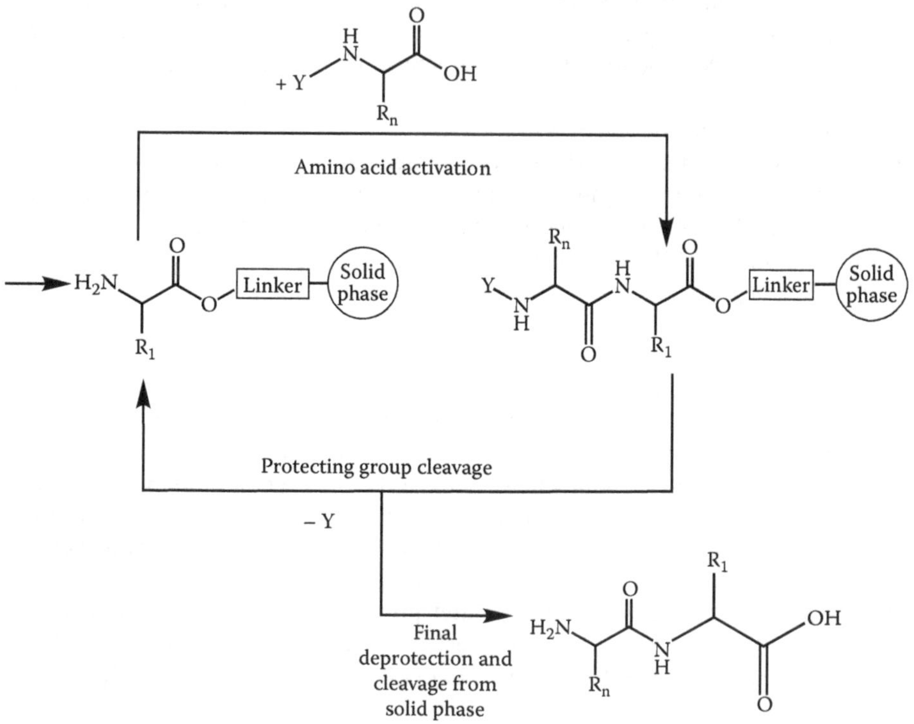

FIGURE 15.1 General scheme of solid phase peptide synthesis (SPPS) as iterative process of peptide chain assembly where Y is a temporary protective group and the side chains of groups R_n are protected in accordance with their functionality.

and isolated, and this makes SPPS easier to perform and decreases the time required for the synthesis of a peptide.

The peptide is assembled in a stepwise manner by repeated cycles of formation of a peptide bond between the amino group and an N^α-protected amino acid, followed by deprotection of the temporary N^α-protecting group (see Figure 15.1). At the end of the synthesis, the assembled peptide is cleaved from the solid support. The functional groups of the amino-acid side chains are protected with permanent protecting groups, which are also cleaved after the synthesis is completed. Synthesis is typically carried out from the C-terminus to the N-terminus of the peptide, but solid phase synthesis of peptides carried out in the opposite direction has also been reported, although this method is not applied often.

The main disadvantage of SPPS is that every step in the synthesis essentially has to go to completion to obtain the correct peptide in a satisfactory yield. Side products formed during each synthetic step accumulate during synthesis, and it is usually difficult to purify the correct product from a mixture of peptides with similar physicochemical properties. With improved methods for synthesis and purification of peptides, many of the initial problems have been overcome, and SPPS has become a routine method for the synthesis of peptides.

15.2 COUPLING AND ACTIVATION: RAPID IN SITU PROTOCOLS

$$\text{R-COOH} \longrightarrow \text{R-COX} \xrightarrow{\quad H_2\overset{|}{N}-R_1 \quad} \text{R-CO-NH-R}_1 \tag{15.1}$$

The activation of the carboxyl group (C-activation) remains the underlying principle of all coupling methods in use (see Equation 15.1), where X is an electron-withdrawing atom such as chlorine, or a group (such as the azide group), which renders the carbon atom of the carboxyl group sufficiently electrophilic to facilitate nucleophilic attack by the amino group. The tetrahedral intermediate thus formed is stabilized by the elimination of X, which is usually a good leaving group. The choice of the electron-withdrawing group X in the reaction of peptide bond formation is crucial in yielding the desired, highly pure peptide. Many strategies have been used, the most popular of which are the activation of the α-carboxylic groups of amino acids as carboxylic acid halides (chlorides and fluorides), azides, anhydrides, and active esters (substituted phenyl esters, esters prepared by carbodiimides, and alcohols). Some examples are shown in Figure 15.2.

15.3 PROTECTION GROUP STRATEGY

An activated carboxyl component, $H_2N\text{-CHR-CO-X}$, can acylate all reactive groups in addition to the desired α-amino group if the latter is not protected. Thus, for a defined course of the coupling reaction, it is necessary that only a single nucleophile (α-amino group) should be available for acylation. This requires the masking of the amino group of the carboxyl compound. Similarly, the reactive groups in the side

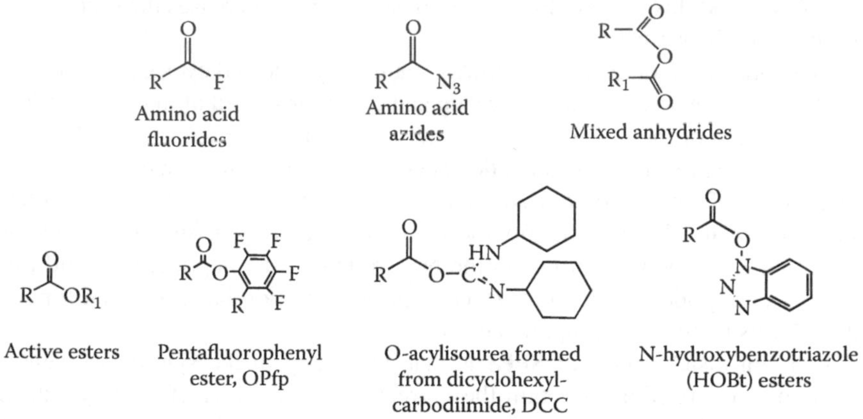

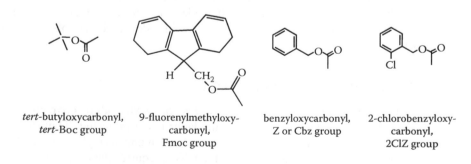

FIGURE 15.2 Some examples of C-terminally activated amino acids in SPPS. Group R is a protected amino acid.

chains of trifunctional amino acids, such as Lys residues, must be blocked; otherwise, branching of the chain will occur. Some examples of protective groups are represented in Figure 15.3.

In order to synthesize peptides using the methods of SPPS, one has to choose the protective group on the α-amino groups. In SPPS, in practice, the α-amino groups are protected with two temporary protecting groups, Fmoc or t-Boc, which are cleaved after each coupling reaction with base or acid, respectively. The choice of Fmoc or t-Boc strategy defines the choice of side-chain protective groups, cleavage method, attachment of the C-terminal amino acid and, sometimes, even solid phase matrix. Alternative methods to Fmoc or t-Boc strategies have been suggested, such as Alloc-strategy, but they are limited due to the poor commercial availability of such amino acids. Presently, both the t-Boc/benzyl-based strategy and Fmoc/t-butyl-based strategy are used, and it has not been clearly established which synthetic strategy gives the best results. For practical reasons, mostly the Fmoc strategy is used routinely today since it does not require the application of "dangerous" liquid HF cleavage used in t-Boc strategy.

FIGURE 15.3 Commonly used protections for α-amino group.

The functional groups of the amino-acid side chains are protected with permanent protecting groups, which are cleaved only after synthesis is completed (see Figure 15.3). Two different protection strategies can be used—one in which differential protection relies on a different sensitivity to the same deprotection agent, or a strategy termed orthogonal protection in which the protecting groups are removed by different chemical mechanisms.

The t-Boc group, which is cleaved by acids such as trifluoroacetic acid (TFA) in a solvent (e.g., dichloromethane), is used as a temporary protecting group for the α-amino function of the amino acid. Most of the side-chain functional groups are protected with benzylic protecting groups, which are stable in TFA but are cleaved along with the anchor to the solid support by a strong acid such as hydrobromic acid (HBr) in TFA, liquid HF, or trifluoromethylsulfonic acid (TFMSA). The t-Boc group was first applied in the 1950s and is the most frequently applied Nα-protecting group. This group is stable for treatment with nucleophiles, bases, and catalytic hydrogenation, but is cleaved with acids. The main problem with this strategy is that, in liquid HF, several side reactions occur, particularly aspartimide and carbocation formation (see the following text). These side reactions can often be overcome with the careful choice of reaction conditions but can cause problems in the synthesis of certain "difficult" sequences.

A new protection strategy was developed in 1972 by Carpino and Han based on the **9**-fluorenylmethyl-oxycarbonyl (Fmoc) group as the Nα-protecting group. The Fmoc group is stable to acid but is cleaved by a base, for example, piperidine in N,N-dimethylformamide (DMF). With this strategy, most of the side-chain functional groups are protected with t-butyl groups that are stable in bases but are cleaved by a moderately strong acid such as TFA, which is relatively easy to handle compared to HF. The cleavage with bases using the Fmoc strategy can cause aspartimide formation, depending on the sequence, such as the occurrence of sensitive Asp-Gly. Base-cleavage of the Fmoc group is also very sensitive to the aggregation of the peptide chain in SPPS, and often, with this strategy, deletion peptides are observed with SPPS.

The benzyloxycarbonyl group (Z or Cbz) and its analogs (2ClZ) are also widely used in SPPS and are cleaved by HF treatment. Other groups for α-amino protection are based on allyl derivatives (cleaved under nearly neutral conditions by Pd(0)-catalyzed allyl transfer), including allyl esters and allyloxycarbonyl groups.

15.4 ORTHOGONAL SCHEMES FOR PROTECTION

In the synthesis of complex peptides, it is sometimes desirable to selectively cleave a protecting group in the presence of other protected functional groups, and, for this purpose, a variety of orthogonally cleavable protecting groups and resin linkers are available. An orthogonal system was introduced in 1977 by Barany and Merrifield using a set of completely independent classes of protecting groups. The members of each class can be selectively removed in any order and in the presence of members of all other classes.

Ideally, the remaining protection should be inert to the reaction conditions used for cleavage of the orthogonal protecting group that are, for example, cleaved by

acid, base, nucleophiles, catalytic hydrogenation, photolysis, and palladium catalysis. Presently, one of the most commonly used types of orthogonal protecting groups
and linkers are allyl-based structures that are cleaved by palladium catalysis. The
allyl-based structures are reported to be stable to the conditions of both *t*-Boc synthesis and Fmoc synthesis and are cleaved under mild conditions.

In the synthesis of peptides containing multiple inter- or intramolecular disulfide
bonds, the selective cleavage of orthogonal protecting groups is often a prerequisite
for achieving the correct pairing of disulfides, and for this purpose, several different
protecting groups for Cys have been developed. Orthogonal protection is also useful in the synthesis of cyclic peptides, which can be synthesized on solid phase by
attachment of the peptide to the resin by the functional group of an amino side chain.
Other examples of complex syntheses requiring orthogonally cleavable protecting
groups include the synthesis of multiple antigenic peptide systems (MAP) consisting
of a branched, nonimmunogenic lysine core on which a peptide antigen is synthesized, which is used for the production of antibodies, and the synthesis of template-
assembled synthetic proteins (TASPs), which are artificial proteins produced by the
synthesis of peptide chains on a small, organic template.

15.5 SIDE REACTIONS IN SPPS

The chemical synthesis of proteins on solid phase is limited by side reactions that
decrease the yield and make the purification of the correct product more difficult.
Although there has been continuing development of new methods for the synthesis
of peptides, and many side reactions can now be avoided, a number of side reactions
remain common.

15.5.1 INCOMPLETE ACYLATION/DEPROTECTION

Incomplete acylation is a serious side reaction since deletion peptides lacking one
or more amino-acid residues are produced. Deletion peptides usually have physicochemical properties similar to the correct product and are therefore difficult to separate using chromatographic methods. A convenient way to prevent the formation of
deletion peptides is to cap the remaining free amino groups after acylation. The terminated peptides thus produced are more easily separated from the correct peptide.
The capping reagents used, such as acetic anhydride or acetyl imidazole, are often
more efficient acylation reagents since they are less sterically hindered than activated
amino acids and can be used in larger volumes, but in many cases, the amino groups
that are resistant to coupling will also react slowly with the capping reagent.

There are two main reasons for incomplete acylation that are fundamentally
different. The most common reason for incomplete acylation in SPPS is aggregation of the growing peptide chains to β-sheet, which makes the amino groups
inaccessible to acylation. Interchain aggregation is sequence dependent but difficult to predict. Among the methods that are most frequently used to disrupt
the hydrogen bonds between the aggregating peptide chains are the use of polar
solvents, low resin loadings, elevated temperatures, or the addition of chaotropic
salts or detergents. For exceptionally difficult sequences, the concept of protecting

the peptide bonds to eliminate the possibility of hydrogen bond formation may be useful, such as by using N-(2-hydroxy-4-methoxybenzyl) (Hmb) protection, N-(2-hydroxybenzyl) (Hbz) protection, or pseudoprolines. The incomplete deprotection of the α-amino group due to intermolecular aggregation of the peptide chains in Fmoc chemistry remains one of the principal problems in Fmoc synthesis of peptides. In *t*-Boc chemistry, deprotection is generally quantitative since TFA, which is used for cleavage of the Boc group, is an excellent solvent for the peptide resin.

The second major reason for incomplete acylation in SPPS is that slow couplings may be encountered when bulky, sterically hindered amino acids are coupled (or coupled to). There is a significant difference in the acylation rate of the β-branched amino acids (i.e., Thr, Ile, and Val) compared to the other natural amino acids. Many nonproteinogenic amino acids are considerably more sterically hindered, and much effort has been made in developing coupling reagents that are particularly well suited for coupling sterically hindered amino acids, for example, the bromophosphonium and uronium salts such as PyBrOP (bromotripyrrolidinophosphonium hexafluorophosphate) and TBTU (2-(1H-benzotriazol-1-yl)-1,1,3,3-tetramethyluronium tetrafluoroborate) reagents, respectively.

15.5.2 RACEMIZATION

Racemization is a well-known side reaction occurring in the activation process of amino acids and peptides in SPPS yielding the mixture of optical isomers of amino-acid residues. Many factors, such as activation methods and coupling reagents; substituents at Nα of amino acids; presence or absence of bases, temperature, and solvents can affect the racemization process but can usually be neglected. There are two particular cases where racemization can be significant: the first in the case of the anchoring of the first amino acid to the solid phase with the application of a strong base such as DMAP, and the second in the case of the coupling of His.

The best-studied mechanism of racemization is the formation of azlactones. The explanation for the tendency to racemization of these compounds lies in the abstraction of proton at C^α of amino acid (the chiral center) by bases due to resonance stabilization of the carbon ion generated in the process (Equation 15.2).

$$(15.2)$$

15.5.3 UNDESIRED CYCLIZATION

Dipeptide esters readily cyclize forming diketopiperazines. Ring closure can take place spontaneously because the thermodynamic stability of the six-membered ring overcomes the energy barrier in the formation of a cis-peptide bond, and the reaction

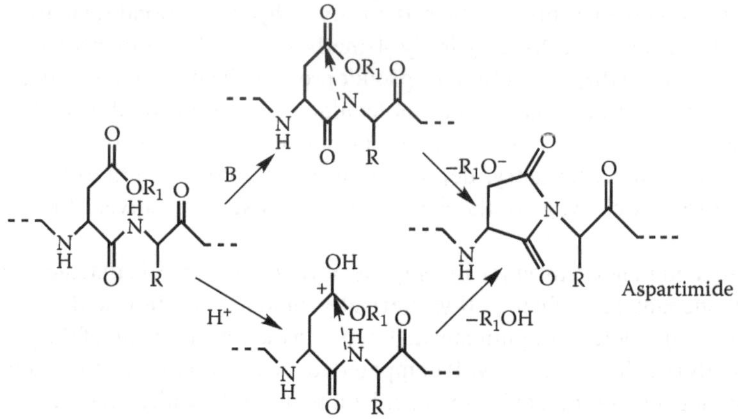

FIGURE 15.4 Base- and acid-catalyzed aspartimide formation and possible equilibriums.

is accelerated by bases. In most cases, the losses suffered by diketopiperazine formation are minor, but certain residues such as Gly, Pro, N-methyl amino acids, Val, and
Ile enhance the tendency for cyclization.

Acid-catalyzed aspartimide formation is a common side reaction in SPPS when a
strong acid such as HF is used for cleavage of the peptide from the resin (Figure 15.4).
A proposed mechanism for acid-catalyzed aspartimide formation is that the carbonyl
oxygen of protected Asp is protonated, which enhances the electrophilicity of the
carbonyl carbon and facilitates attack from the peptide bond nitrogen. The rate of
aspartimide formation in the Asp-X model is most rapid when Gly, Ser, or Thr are in
the $n + 1$ position; moderately rapid with Ala, Leu, or Phe; slow with Met, Asp, and
Glu; and extremely slow with Val or Ile.

Glu may form glutarimide by an analogous reaction, but this side reaction
does not occur as frequently as aspartimide formation of aspartic acid. Instead,
other side reactions, such as γ-acylation under acidic conditions, dominate for
glutamic acid.

15.5.4 ALKYLATION

One of the objections against the Boc/benzyl synthetic strategy is the need for a
strong acid, such as HF or TFMSA, for the final deprotection, which can cause
various side reactions such as alkylation of sensitive amino-acid residues, N →
O-acyl shift, aspartimide formation, and acylation of the γ-carboxyl group of glutamic acid.

Cleavage of the benzylic protecting groups used in Boc chemistry by anhydrous HF or TFMSA proceeds by an S1 mechanism, and benzylic cations are
generated. The carbocations can alkylate the nucleophilic side chains of Cys,
Tyr, Met, and Trp (Figure 15.5). The addition of a scavenger such as *p*-cresol,
which traps the carbocations, suppresses this side reaction but does not always
give full protection.

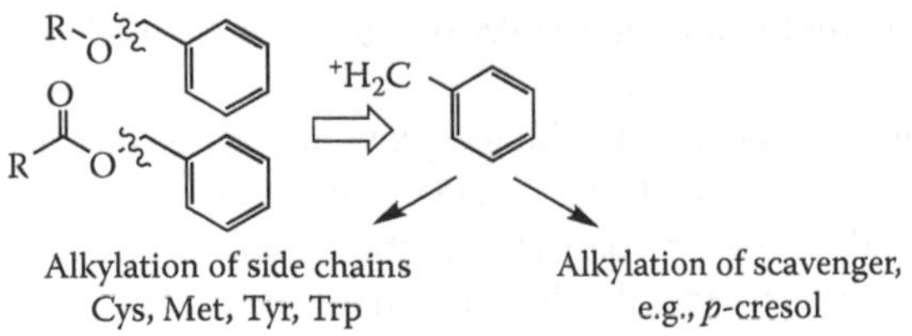

Alkylation of side chains   Alkylation of scavenger,
      Cys, Met, Tyr, Trp   e.g., *p*-cresol

FIGURE 15.5 Acidolytic cleavage of benzylic protecting ether and ester groups, respectively, in anhydrous acid.

15.5.5 SIDE REACTIONS DUE TO OVERACTIVATION

The term "overactivation" points to the ambiguity created in acylation reactions in which the activated derivative of the carboxyl component is too powerful to be selective and causes acylation not only of the amino group, which is expected to form a peptide bond, but also of less good nucleophiles, such as hydroxyl groups. Anhydrides are such overactivated derivatives and also the intermediates generated in the activation of carboxylic acids with carbodiimides. Carbodiimides react with the unprotected -SH group of Cys to form isothioureas, which in turn yield, by β-elimination, dehydroalanine derivatives. The imidazole moiety of His can add carbodiimides to produce substituted guanidines that are easily converted back to His.

15.6 SOLID SUPPORTS

The novel feature of solid-phase peptide synthesis was the application of solid supports in chemical synthesis. In the simplest form, SPPS involves a heterogeneous reaction mixture composed of an insoluble resin-bound peptide chain, a soluble activated amino-acid derivative, and a solvent. Many supports have been examined, but only a few have met the requirements and found wide use. Lightly crosslinked polystyrene and polyacrylamide resins have been the most successful, but some novel resins have recently become more popular.

Copoly(stryrene-divinylbenzene) is a copolymer of styrene and divinylbenzene (DVB) (Equation 15.3).

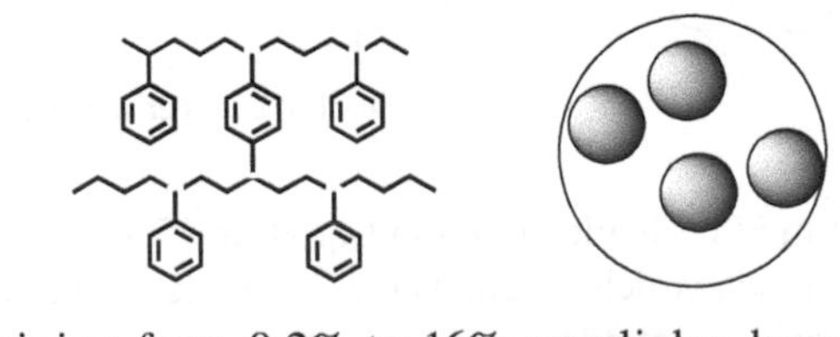

A   B

(15.3)

Preparations containing from 0.2% to 16% crosslinker have been studied, but those containing 1% DVB have the best overall properties (A). These polymers are physically stable and completely insoluble in commonly used solvents. They form beaded gels (B) that become highly and reversibly solvated and swollen in organic solvents of intermediate polarity such as dichloromethane (DCM), toluene, and dimethylformamide (DMF), but do not swell in water, methanol, and hexane.

It is important to consider that, with the growing peptide chain, the swelling properties of the resin change, and this change is different in different solvents. Both polymer and peptide solvation contribute to the swelling.

Styrene resins can be readily derivatized by nearly all methods of aromatic chemistry to provide functional groups for the attachment of amino acids for peptide synthesis. The peptide chains are not restricted to the surface of the beads but are uniformly distributed throughout the resin matrix. Since swollen beads are rapidly permeated by the reagents, diffusion is at least ten-fold faster than coupling reaction and is not a rate-limiting step.

In search of solid supports for application in different specific studies, a variety of materials has been used, such as the monoester of 1,4-(dihydroxymethyl)phenyl-silica, controlled-pore glass, phenol-formaldehyde polymer, cellulose powder and paper, Sephadex LH-20, cotton sheets, threads or beads of cellulose (Perloza), and even proteins where the amino groups are used.

15.7   HANDLES, LINKERS, AND SPACERS

It is possible to prepare solid supports with altered polarity and solvation properties by introducing functional groups or by grafting one polymer to another. It is also possible to fine-tune the resin derivatives so that the anchoring bond holding the peptide chain to the support can have any desired sensitivity to cleavage conditions. The resin derivatives can be designed for cleavage by strong, medium, or weak acids, by bases and other nucleophiles, and by photolysis, hydrogenolysis, or other catalytic reagents. They can also be designed to yield peptides with C-terminal free carboxylic groups, esters, amides, or hydrazides.

Some linkers have been designed to serve as spacers, with the idea that the peptide could be assembled better if it were farther removed from the resin support. The design of linkers is closely associated with protecting group strategies that were covered earlier.

$$(15.4)$$

The handle concept in SPPS is illustrated in Equation 15.4. Shown are the protective group Y (t-Boc or Fmoc), which is removed at each step, and X (O or NH, to yield C-terminal peptide acids or amides on cleavage).

Many additional possibilities exist to achieve C-terminally modified peptide after cleavage from the solid phase, such as C-terminal hydrazides, N-alkylamides, p-nitroanilides, esters, aldehydes, and others, by the choice of appropriate linkers and handles.

15.8 CHEMICAL SYNTHESIS OF PROTEINS AND FRAGMENT CONDENSATION

The longer or more complex peptides are difficult to synthesize due to side reactions during synthesis that reduce the yield of the correct product and make the purification more difficult. The prospect of synthesizing more complex peptides would also have implications on de novo design and synthesis of artificial proteins, where a few structural motifs have been successfully designed, synthesized, and shown to fold into the predicted tertiary structure. So far, the most promising method for the chemical synthesis of proteins has been to synthesize peptide segments, which are ligated to form the full-length protein. Several strategies have been reported; the peptide segments can be condensed in solution or on solid phase, and the amino-acid side chains can be protected or unprotected.

15.8.1 CHEMICAL LIGATION OF UNPROTECTED PEPTIDES IN AQUEOUS SOLUTION

Recently, there has been much interest in the development of new methods for the ligation of unprotected peptides since protected peptides often have poor solubility, which complicates the ligation. For the ligation of unprotected peptides, very selective chemical reactions have to be employed in the ligation of the peptide segments in order to avoid interference by the functional groups of the amino-acid side chains. Several different chemistries have been developed for the "highly chemoselective reaction," all of which give rise to an unnatural covalent structure at the ligation site. The selectivity of these ligation reactions means that unprotected peptide segments can be used in chemical synthesis. Furthermore, the ligation reactions can be carried out in aqueous solution, and they are rapid and quantitative.

Native chemical ligation (NCL; Dawson and Kent, 1984) is a technique for constructing a large peptide from two smaller peptides: a C-terminal thioester peptide and an N-terminal cysteine peptide. A second-generation ligation chemistry was introduced in 1994 allowing the straightforward preparation of proteins with native backbone structures from fully unprotected peptide-building blocks.

The first step is the chemoselective reaction of an unprotected synthetic peptide-thioester with the thiol side chain of another unprotected peptide segment containing an amino terminal Cys (Figure 15.6). This gives a thioester-linked intermediate as the initial covalent product. Without change in the reaction conditions, this intermediate undergoes spontaneous, rapid intramolecular reaction to form a native peptide bond at the ligation site. The target full-length polypeptide is obtained in the desired final form without further manipulation. Even in the presence of Cys-residue-free SH groups in one or both segments, the ligation reaction is completely regioselective.

15.8.2 ENZYME-CATALYZED FORMATION OF PEPTIDE BOND

The principle of microscopic reversibility defines that proteases can catalyze both proteolysis and ligation. Under physiological conditions, though, the equilibrium lies

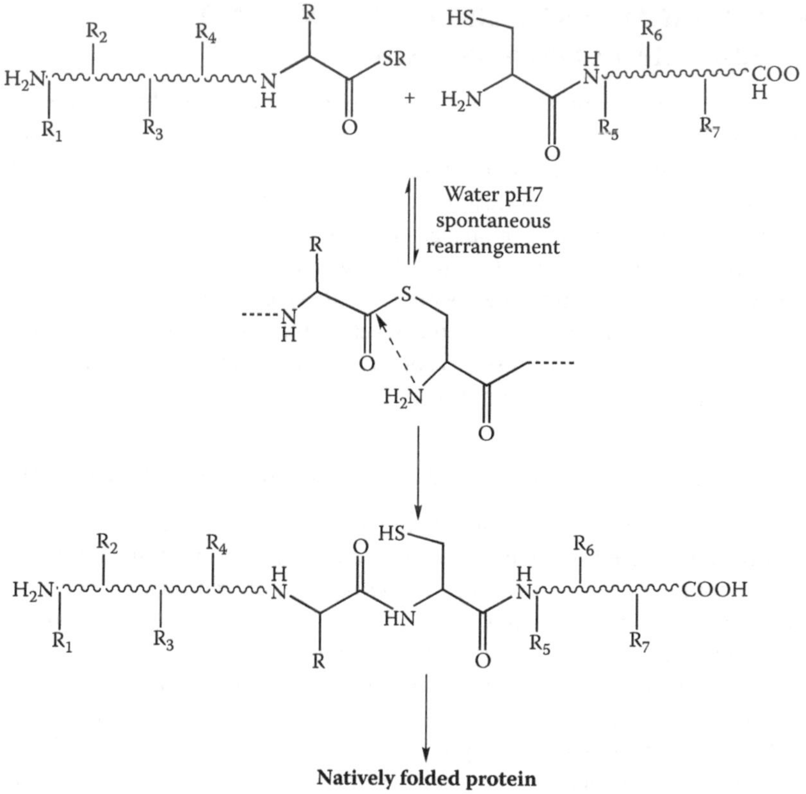

FIGURE 15.6 Native chemical ligation from two unprotected peptide fragments, one with a C-terminal thioester group and the second with an N-terminal Cys residue. (From Dawson, P. E., Muir, T. W., Clark-Lewis, I., and Kent, S. B., 1994, Science, 266(5186), 776–779. Reprinted with permission from AAAS.)

strongly in favor of proteolysis. Van't Hoff first proposed in 1898 that, by shifting this equilibrium, one might convert a hydrolase to a ligase, thus providing a catalytic approach to segment condensation reactions.

$$R\text{-}COOH + H_2N\text{-}R' \rightleftharpoons R\text{-}CO\text{-}NH\text{-}R' + H_2O \qquad (15.5)$$

The equilibrium of the reaction (Equation 15.5) lies far to the left, and thus, the hydrolysis of the peptide bond but not its formation is a spontaneous process. In the absence of catalysts, the rate of the reaction is extremely low. This reaction can be accelerated by intervention of proteolytic enzymes. Such enzymes, as true catalysts, do not affect the equilibrium of the reaction. The equilibrium can be shifted by an increase in the concentration of one of the reactants or by removal of a component from the reaction mixture. Hence, the same enzymes that catalyze the hydrolysis of the peptide bond can also be used for the synthesis of peptides. Hydrolytic enzymes papain, trypsin, and chymotrypsin have been used for the synthesis of selected target peptides. Altering reaction conditions such as solvent polarity, temperature, and pH can shift the equilibrium point. Varieties of experiments have induced proteases to work backward and function as peptide ligases. In practice, however, this approach has significant limitations, that is, the proteases tend to be insoluble in the solvents applied for reverse product ratio.

Protease engineering (see Chapter 16) to favor ligation relative to hydrolysis offers many advantages. Emil T. Kaiser and colleagues demonstrated that a subtilisin variant, thiolsubtilisin, where the active site Ser was chemically converted to Cys (S221C), was efficient for synthesis of amide bonds. The ratio of aminolysis to hydrolysis is 600-fold greater for thiolsubtilisin relative to subtilisin. Subtilisins with even higher activity to synthesize amide bonds have been introduced. To acetylate these enzymes, due to steric hindrances, highly activated esters must be used. A double mutant S221C/P225A-subtilisin (also called subtiligase) has been introduced. The latter has proved to be most successful in fragment condensation methodology.

15.9   COMBINATORIAL PEPTIDE CHEMISTRY

Combinatorial approach is today a major approach to the discovery of new drugs in major pharmaceutical companies and university laboratories leading to novel drug leads. Combinatorial peptide chemistry, introduced in the late 1980s, creates physical peptide libraries that are mixtures of peptides consisting of hundreds to millions or more of single components that have defined characteristics: equal length, defined amino-acid positions, mix positions, and such. Peptides may consist of common l-amino acids, d-enantiomers, unusual residues, backbone-modifying peptide mimetics, sugar- or lipid-containing building blocks, and others. In addition, the peptides may be cyclized via disulfide, lactone, or other linkages. The method of portioning-mixing (PM) makes it possible to synthesize millions of peptides easily in the form of multicomponent mixtures or peptide libraries. The method is based on Merrifield's SPPS method, but the coupling cycle is replaced by three simple operations (Figure 15.7).

The resin for the synthesis of the library is divided into portions for the coupling of 20 (or more) amino acids. Uniform couplings can therefore be achieved since the competition between slow- and fast-coupling amino-acid derivatives no longer exists. After washing steps, the polymer is mixed again, washed, and the N-terminal protection group removed simultaneously for all peptides in one vessel. For the next coupling, the distribution of the polymer in each vessel is purely statistical.

In the linear, soluble peptide libraries, the active sequence is found by an iterative process going from unspecific sublibraries to better-characterized sublibraries and ending with single peptides. Alternatively, positional scanning of sublibraries is performed, and the final identification of the peptides is revealed by comparing the results.

New encoding systems translate the stepwise buildup of a desired polymer either to an oligonucleotide sequence or to a peptide sequence. After the coupling of each building block by a conventional strategy, either a di- or tripeptide or a trinucleotide is coupled to the same bead by an orthogonal strategy. After the identification of the bead that carries the active compound, either the DNA is amplified by PCR and

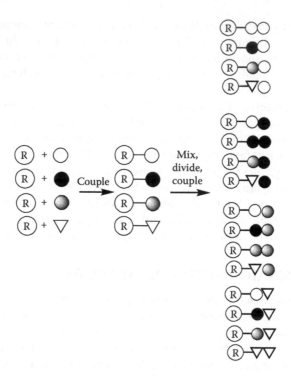

FIGURE 15.7 Scheme for the "divide-couple-recombine" method for the two first steps of a small tetracomponent combinatorial library where R denotes the solid-phase particle, and the rings and triangle are four different amino acids.

sequenced, or the bead is subjected to Edman degradation and the coding peptide identified. A translation protocol leads to the active compound. If the 20 common l-amino acids are varied in all positions, the enormous diversity of combinatorial chemistry is synthesized. Thus, the total number of peptides is 400 (20^2) for dipeptides, 8000 (20^3) for tripeptides, 160 000 (20^4) for tetrapeptides, etc.

FURTHER READING

1. Goodman, M., Ed. 2004. Synthesis of Peptides and Peptidomimetics. New York: George Thieme.
2. Benoiton, N. L., Ed. 2005. Chemistry of Peptide Synthesis. Boca Raton, FL: CRC Press.

Introduction to Part III

Tiit Land

Understanding biochemical functions and the cellular roles of proteins is one of the ultimate goals in biochemistry, and has a very high importance for biotechnology and medicine, particularly for drug design. Experimental studies aimed at giving insights into protein functions include methods for the determination of expression, localization, structure, biochemical activity, protein–protein interactions, and protein–ligand interactions. In addition, there are several computer-based techniques and approaches available that can be used to predict protein binding sites with ligands, protein structures, and functions. Development of new computational methods and algorithms has become more important than ever as we have entered a postgenomic era where many large-scale genome sequencing projects have provided us with a tremendous amount of data, including new putative protein sequences, but without clues to their functions. In Part III, some of the mentioned methods are introduced and the relevant issues are discussed.

In Chapter 16, protein engineering and gene silencing techniques are described. These methods are very important for studying biochemical functions and the mechanisms of proteins, for producing novel bioactive molecules with altered functions, and for developing research tools and possibly therapeutics agents based on controlling gene expression. In Chapter 17, the biochemical basis of protein–protein interactions and protein–ligand interactions are described, and methods for identifying these interactions are introduced. Studies of protein–protein interactions are extremely important in a functional context because individual proteins are very often members of multiprotein complexes that carry out biochemical functions. In addition, one protein can transiently associate with another target protein to modify, regulate, or translocate it to different cellular compartments. In Chapter 18, principles and methods of a rapidly developing field of science, bioinformatics, are described. As mentioned above, the genomics era has provided us with a large amount of data, which has to be stored and then can be used for analysis and identification of relationships between datasets. In some cases, the function to a new gene can be assigned by using sequence comparison methods.

In Chapter 19, we describe additional computational methods that can be used for the prediction of three-dimensional structures of proteins from their amino acid sequences. Knowing structures of proteins can, in turn, be helpful for function prediction, as the tertiary structure of a protein carries out its biochemical function. Finally, some relevant aspects and methods of another rapidly developing field, proteomics, are described in Chapter 20. The importance of proteomics is impossible to underestimate when having as a goal to completely understand biological processes. Its importance can be also illustrated by comparing proteomics with less dynamic genomics and transcriptomics, which do not provide any information about posttranslational modifications and protein–protein, protein–DNA, or protein–ligand interactions.

16 Protein Engineering and Gene Silencing

Ülo Langel

CONTENTS

The diverse fields of protein and nucleic acid engineering, particularly by their ability to modulate gene expression, are very central to both the understanding of secrets of life and creation of novel drugs, materials, and technologies. This chapter introduces protein engineering and gene silencing with the aim of demonstrating the high impact of these technologies in understanding biochemical mechanisms as well as in applications in drug discovery.

16.1 PROTEIN ENGINEERING

In its broadest sense, protein engineering is the process of creation of proteins with novel, valuable properties by the application of techniques from molecular biology, biochemistry, cell biology, and bioinformatics. Two main goals can be defined for protein engineering. The first is the efficient understanding of biochemical mechanisms and their functioning, and the second the production of molecules with novel, designed, and improved properties and useful functions. Protein engineering was made possible by developments in recombinant DNA technologies and site-directed mutagenesis (recognized in part by the Nobel Prize to M. Smith in 1993) in the 1970s and 1980s, enabling specifically altered genetic information coding for a protein.

The two general strategies for protein engineering are rational design and directed evolution. In rational design, the detailed knowledge of the structure and function of the protein is applied in order to introduce the required changes, such as mutations in the novel proteins. In directed evolution, random mutagenesis is applied to a protein, yielding a selection of variants with required properties, followed by further mutations or selections for proteins with improved properties. This strategy mimics natural evolution and complements rational design, not requiring the structural information of a protein. The drawback of rational design in protein engineering is that the effects of various mutations are difficult to predict and that the detailed information of the structure is often unavailable. The drawback of directed evolution is the requirement for high-throughput screenings and, often, robotic equipment for that purpose. Often, rational design and directed evolution experiments are carried out in parallel, yielding synergism from both strategies.

16.1.1 SITE-DIRECTED MUTAGENESIS: THE CASE OF FLUORESCENT PROTEINS

In order to confirm the hypothesis that part of a gene codes for an amino-acid sequence that is responsible for the function of this protein, it is sometimes necessary to delete or alter this part of the gene sequence without altering the rest of the expressed protein. Site-directed in vitro mutagenesis has been used to introduce these alterations and create proteins with improved properties (i.e., to carry out protein engineering).

The site-directed mutagenesis approach in protein engineering is well exemplified by optimization and development of Aequorea victoria jellyfish wild-type green fluorescent protein, GFP, to achieve fine-tuned photophysical properties (Nobel Prize to Osamu Shimomura, Martin Chalfie, and Roger Tsien in 2008). This yielded wavelength shifting in emitted fluorescence, and hence, color change from yellow to deep red. The wild-type GFP was quickly modified to produce fluorescent proteins (FP) emitting in the blue (BFP), cyan (CFP), and yellow (YFP) regions of spectra in the 1990s (Figure 16.1). The orange and red spectral regions became first available when the first red FP was discovered in a reef coral, leading to multiple red-emitting FPs. Today, this enables the use of the suitable fluorescent protein with altered excitation and emission wavelengths, enhanced brightness and photostability, reduced oligomerization, and improved pH resistance relative to the original GFP.

These alterations of emission color have been achieved by manipulations of local environmental variables around the chromophore, including the position of charged amino acids, hydrogen bonding networks, and hydrophobic interactions within the protein matrix. Spectral shifts are attributed to differences in the covalent structure and the extent of π-orbital conjugation of the chromophore (Figure 16.1 [1–4]). The structure–function relationship by further site-directed mutagenesis has yielded

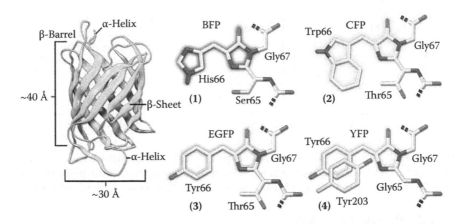

FIGURE 16.1 FP β-barrel architecture and approximate dimensions, and chromophore structures of common Aequorea FP derivatives. (1) BFP, (2) CFP, (3) EGFP, (4) YFP. Portions of the chromophores that are conjugated and give rise to fluorescence are shaded. (Reproduced with permission from the Company of Biologists: Shaner, R. C., Patterson, G. H., Davidson, M. W. 2007. J. Cell Sci. 120: 4247–4260.)

proteins with engineered properties such as finely tuned color variants of fluorescent proteins.

All fluorescent proteins known today adopt a similar three-dimensional cylindrical structure where a polypeptide backbone is organized into 11 strands of an extensively hydrogen-bonded β-sheet that surround a central α-helix containing the chromophore (Figure 16.1). The β-sheets are linked together with by Pro-rich loops, and the amino-acid side chains in each sheet project into the protein interior or toward the surface. The interior of the protein is tightly packed, and there is little room for diffusion of ions or other intruding small molecules. The rigid structure of the FP interior is responsible for the unique chemical environment that initiates autocatalytic chromophore formation by three of the amino acids in the central α-helix. Changes in this environment yield variations in spectral and other physical properties. For example, a common tripeptide, Met-Tyr-Gly (MYG), form different chromophores spanning huge, 177-nm emission space (from emission maximum of 486 nm in cyan FP to 663 nm in far-red FP), depending on the nature of the chemical and physical environment inside the β-barrel.

The first new, engineered FPs involved the mutated amino-acid residues in the chromophore structure or its immediate vicinity where the emission wavelengths were shifted by tens of nanometers. Chromophores such as SHG (in blue FP), TWG (in cyan FP), TYG (in enhanced GFP), and GYG (in yellow FP) are illustrated in Figure 16.1.

Rational design of FPs with even wider range of affected spectral properties by protein engineering became available after the establishment of the crystal structure of the Aequorea FPs and the discovery of red-shifted FPs in corals, yielding the new FPs with higher spectral shifts and other properties. However, often such protein engineering is limited, and the strategies of random mutagenesis are applied to further optimize the design.

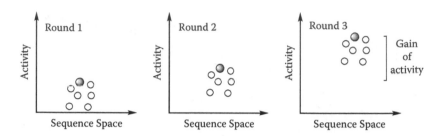

FIGURE 16.2 In an evolutionary strategy of protein engineering, each successive library is based on the highest-activity mutant of the previous library. (Reprinted with permission from MacMillan Publishers: Yuen, C. M. and Liu, D. R., 2007, Nat. Methods, 4(12), 995–997.)

16.1.2 PROTEIN EVOLUTION

It is often necessary to modify and thus optimize the characteristics of proteins either for diagnostic properties or to create novel drugs. As mentioned earlier, traditional gene technology requires extensive knowledge about protein structure and dynamics and is limited by available computer algorithms. This makes the methods slow and the results far from optimal. Optimizing proteins by in vitro evolution seeks to introduce diversity into proteins randomly without the need for knowledge about the structure of the protein (Figure 16.2). However, a drawback is that, in addition to variants with beneficial mutations, harmful and superfluous mutations will also be produced. Subsequent DNA recombination using the best variants identified can help solve these problems. Fragmenting and then recombining genes with different point mutations will produce new genes with mutations in every possible combination. This allows for the screening out of harmful and superfluous mutations.

A promising new approach to further improving the properties of FPs, called evolutionary optimization, applies the structure-based random library where specific residues are targeted. In this way, the optimized CFP and YFP (discussed earlier) variants were achieved and called YPet and CyPet, respectively, demonstrating improved pH stability. In YPet and CyPet, the beneficial amino-acid substitutions are distributed throughout the proteins, near and far from the chromophore.

16.1.3 SITE-SPECIFIC CHEMICAL LABELING OF PROTEINS

Site-specific chemical modification is a process of stoichiometric altering of protein with the quantitative derivatization of a (single) unique amino-acid residue without modifying other amino-acid residues or changing the protein conformation. The site-specific introduction of radioisotopes, fluorophores, and other labels/probes into proteins without interfering with its function is a valuable tool in biochemistry and drug development. Several strategies are available for site-specific modifications such as incorporation of noncoded amino acids and enzymatic posttranslational modifications.

Site-directed mutagenesis is often used today to alter proteins at genetic level where the noncoded amino acids are introduced into the protein sequence during biosynthesis. Introduction of unnatural amino acids with defined chemical, steric, and

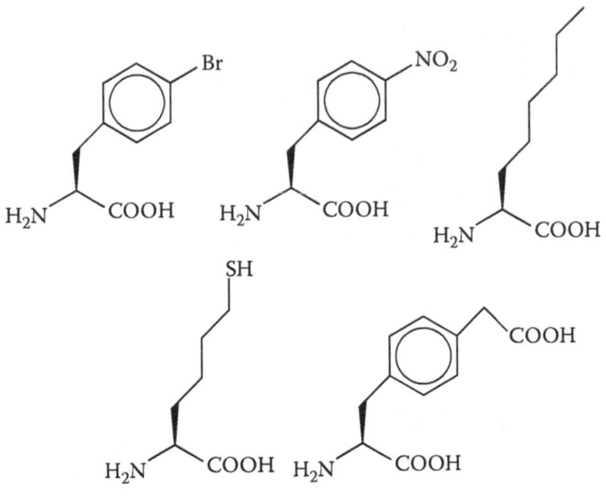

FIGURE 16.3 Examples of noncoded amino acids that have been incorporated into proteins by site-specific modification.

electronic properties at determined sites in proteins provides powerful tools to study protein structure, function, and biochemical properties. Over 30 noncoded amino acids have been selectively incorporated into proteins by methods of site-directed mutagenesis; among them, amino acids with posttranslational modifications, photoaffinity labels, and other chemical moieties have been used. See Figure 16.3, in which the selection of such amino acids with variable reactivity is presented.

16.1.4 PHAGE DISPLAY

A bacteriophage (phage) is a virus that infects only bacteria, causing its DNA expression by the cells and replication. The two major components of a phage are the genetic material and the protein coat. The purpose of the protein coat is to protect the genetic material from being damaged as the phages reproduce and go from cell to cell. Phage display uses bacteriophages such as the M13 coliphage and the strategies of genetic engineering to insert foreign DNA fragments into a suitable phage coat protein gene, followed by the expression of the modified gene as a fusion protein and displayed on the surface of the phage (Figure 16.4). The main purpose of phage display technology is to determine the proteins and peptides interacting with molecules of interest. Phages used in phage displays are normal ones that have been genetically modified to express only one extra protein on their surface.

Phage display technologies have been applied for several purposes in protein engineering. For example, phage display is a powerful complementary method to random mutagenesis for selection of desired variants of protein mutants as well as a method for construction of antibodies with desired properties. Phage display has become a powerful method to identify peptides and proteins that interact with a given protein.

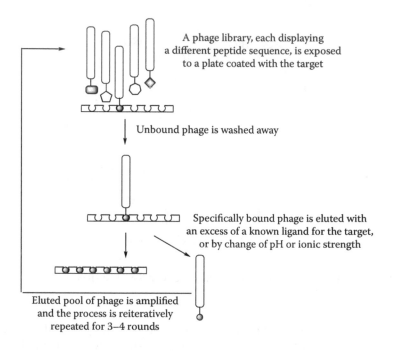

FIGURE 16.4   Scheme of phage display.

16.2   GENE SILENCING

Controlling gene expression has in recent years become an attractive research tool and potential source of therapeutic agents. By affecting DNA and RNA, one can modulate the expression of genes that have deviated from the normal levels, and hence, intervene with human disease processes. Most of the important molecular work in the body is carried out by proteins. Consequently, many disease processes are characterized by faulty protein production or inappropriate protein activity.

DNA/RNA therapeutics seek to prevent production of abnormal proteins or boost production of normal proteins. Conventional therapies typically act by interfering with mature proteins, usually enzymes or cell-surface receptors. DNA/RNA therapeutics, however, are designed to prevent deleterious proteins from being made, either by blocking their production or repairing the mutated genes. Additionally, controlling gene expression with short oligonucleotides is a powerful tool in studying gene–protein functions, enabling clarification of the role of newly discovered genes and proteins, which is especially valuable in the era of genomics and proteomics. Here, we briefly describe the possible applications of short oligonucleotides in gene regulation, with special focus on nonviral oligonucleotide delivery methods.

Recent success in several whole-genome sequence projects clearly suggests the need for rapid transformation of genomic information into functional data. Techniques for the knockdown of specific genes (i.e., gene silencing) is important in the study of protein functions, as well as a means for future therapeutics, as summarized in Figure 16.5. It is noteworthy that gene silencing has been found to be an effective technique in a

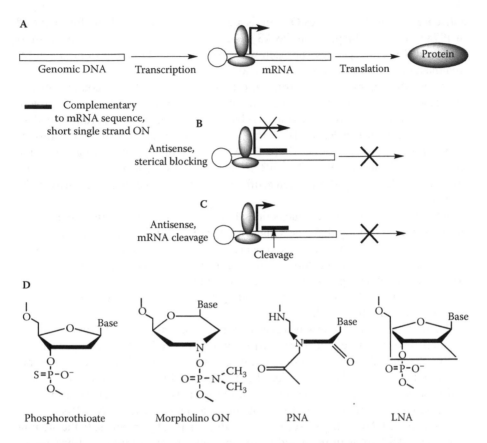

FIGURE 16.5 Scheme of antisense gene silencing strategies: A—Noninfluenced gene expression. B—Short antisense ON causes sterical hindrance to translation. C—Short antisense ON recruits mRNA-degrading enzymes (e.g., Rnase H), yielding mRNA fragmentation. D—Some modified nucleotides used in ON-silencing strategies.

wide range of organisms, including plants, fungi, and mammals, suggesting a huge potential for gene silencing in research and development of therapeutics.

16.2.1 mRNA as Target

Targeting of RNA with exogenous drugs (especially oligonucleotides) is a relatively new approach. The U.S. Food and Drug Administration (FDA) approved the first RNA targeting drug in 1998, and several products are now in phase III clinical trials. In the following text, several approaches in which RNA is employed either as a tool or a target are reviewed where gene expression is inhibited via RNA.

16.2.1.1 Antisense Technology

The first report showing that gene expression can be modified by exogenous ssDNA-inhibiting protein expression in a cell-free system was published in 1977 (Patterson, Roberts, Kuff, PNAS USA, 74, 4370). Inhibition of Rous sarcoma virus formation by antisense oligonucleotides (AS ON) in tissue culture was shortly thereafter reported in 1978 (Zamecnik, Stephenson, PNAS USA, 75, 280). Today, AS is by far the most advanced RNA therapeutics approach with one approved ON drug (fomivirsen for treatment of cytomegalovirus retinitis) and several in clinical trials.

In the AS strategy, a single-stranded oligonucleotide binds to a complementary messenger RNA, forming a double-stranded molecule. It "knocks down" protein biosynthesis by translational arrest or selective degradation (by recruiting endogenous nucleases such as RNase H) of the mRNA in the cell. This strategy abolishes synthesis of the protein, for which the RNA codes simply by eliminating the source message (see Figure 16.5B,C). The mRNA sequence is known as the "sense" strand, and the complementary short (often artificial) ON recognized by it is defined as the "antisense" strand.

AS ON are usually 15–21 nucleotides long. The specificity of this approach is based on the probability that any sequence longer than a minimal number of nucleotides (13 for RNA and 17 for DNA) occurs only once within the human genome. The target region should be chosen with regard to the mechanism of the antisense approach. Targeting regions around the translational start codon may lead to success by preventing the ribosome from binding to RNA, but other regions have been targeted successfully as well.

There are several reasons why a gene may not be accessible to antisense. First, the targeted gene may have redundant partners within the cell that can compensate for its loss. Second, the antisense effect produces only the loss-of-function phenotype, which will not always be detectable. The third possibility is that the mechanism of antisense precludes access to certain genes. These kinds of problems have stimulated the search for improvements of AS technologies and more efficient AS ONs.

Unmodified ONs are highly susceptible to degradation by enzymes. In the attempt to produce AS ONs with increased resistance, several structurally modified oligonucleotides have been introduced. In Figure 16.5D, some modified ON, frequently used in AS technology, are presented. Besides improved nuclease stability, improved target affinity for some artificial ON (peptide nucleic acids or PNA, locked nucleic acids or LNA) has been demonstrated, making them good drug candidates.

16.2.1.2   RNAi and siRNA, miRNA and shRNA

RNAi (RNA interference) refers to the endogenous mechanism of gene silencing by small noncoding RNAs complementary to an mRNA target. The mechanism was first detected in plant and worm studies, and is now a well-characterized process of gene expression control in multiple species. The report on discovery of RNAi in 1998 resulted in Nobel Prize to C. Mello and A. Fire in 2006.

siRNAs (short interfering RNA) are small, double-stranded RNAs, typically 21–23 base pairs in length, that are involved in gene silencing through degradation of mRNA, thereby reducing translation (Figure 16.6). This mechanism is similar to the antisense-mediated degradation of mRNA, except that RNA interference (RNAi) is an endogenous process and therefore involves other enzymes. The RNAi pathway is initiated by long, double-stranded RNAs that trigger the enzyme Dicer, which processes these long RNA transcripts into siRNAs. These siRNAs are then

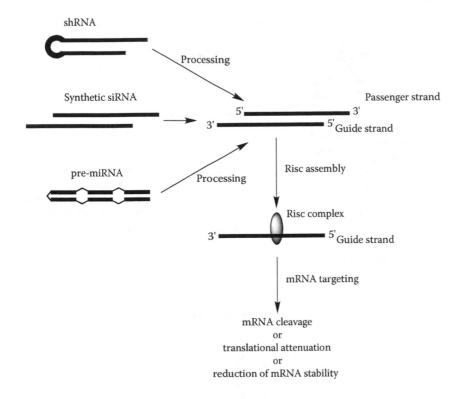

FIGURE 16.6 Gene silencing with shRNA, siRNA, and miRNA.

incorporated into the RNA-induced silencing complex (RISC) that targets and degrades complementary mRNAs. RNAi is an important and highly specific process, mainly active during embryonic development and as a defense against viruses utilizing dsRNA.

MicroRNAs (miRNAs) are short, noncoding RNAs that posttranslationally regulate gene expression (Figure 16.6). Over 300 miRNA genes have been identified in the human genome, of which many are cancer associated. miRNAs are endogenous triggers of the RNAi pathway. Short hairpin RNAs (shRNAs) are another expressible RNAi pathway recruiting oligonucleotides (Figure 16.6). shRNA libraries today are often used to identify genes, for example, in tumor suppression pathways.

16.2.1.3   Ribozymes

Ribozymes are RNA molecules with catalytic properties, or RNAs that are a part of larger enzyme complexes such as ribonuclease P. Ribozymes were discovered 25 years ago. They mediate phosphodiester bond cleavage and formation as well as peptide bond formation. Artificial ribozymes have been shown to catalyze a broad array of other chemical reactions, and their number is continuously growing. The Nobel Prize in Chemistry was awarded to Sidney Altman and Thomas R. Cech in 1989 for their discovery of the catalytic properties of RNA, which are briefly introduced in Figure 16.7A and B.

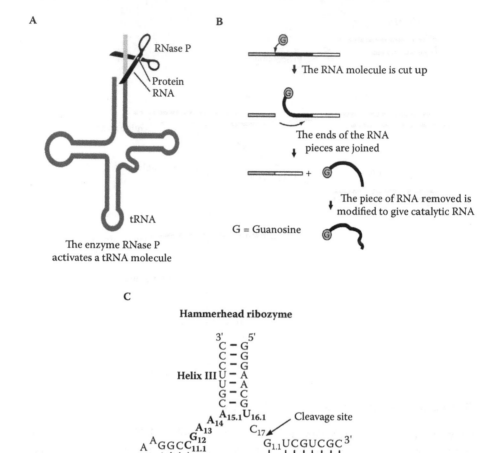

FIGURE 16.7 Ribozyme mechanisms. A—Sidney Altman studied the enzyme RNase P, which is found in intestinal bacteria. RNase P activates a special kind of RNA molecule called tRNA (transfer RNA) by removing a portion that is unnecessary for its function. The enzyme RNase P has the unusual property of containing not only a protein molecule but an RNA molecule as well. B—Thomas Cech studied an RNA molecule from the primitive uni-cellular animal Tetrahymena. He discovered that an unnecessary piece of RNA is removed from the middle of this molecule, the loose ends thus formed are then joined together, and that the RNA molecule itself catalyzes this reaction functioning as an RNA-synthesizing enzyme. This means that catalytic RNA can also make new RNA. C—Secondary structure of a ham-merhead (top) and a hairpin (bottom) ribozyme with bound substrates; the cleavage site in the oligonucleotide substrate is shown with an arrow. (With permission from Annual Reviews: Kraut, D. A., Carroll, K. S., Herschlag, D., 2003, Annu. Rev. Biochem., 72: 517–571.)

Natural ribozymes involve binding sequences along with a catalytic core capable of cleaving a specific RNA molecule, including mRNA. By cutting the target RNA into two ineffective strands, they prevent protein biosynthesis and are important gene silencers. Their potential as molecular therapeutics is demonstrated with cases where Phase I and II clinical trials are under way. DNAzymes (or deoxyribozymes) are artificial RNA-cleaving DNA analogs of ribozymes. As two examples, the conserved sequences of the cleaving RNAs, hammerhead and hairpin ribozymes, are presented in Figure 16.7C. Both ribozymes catalyze strand scission to yield 5-hydroxyl and 2′,3′-cyclic phosphate. Each self-cleaving RNA has been converted into a multiple turnover ribozyme by separating a catalytic core (outlined nucleotides).

16.2.1.4   RNA Decoys and Ligands: Aptamers, Decoy ON, Spiegelmers

Aptamers (derived from Latin aptus—"to fit") are artificial ON (RNA or DNA) ligands with high affinity and specificity generated against certain targets, such as amino acids, drugs, proteins, or other molecules. In nature they exist as nucleic-acid-based genetic regulatory elements called riboswitches. Artificial ligands are isolated from combinatorial libraries of synthetic ONs first described in 1990 as SELEX technology. Aptamers have shown affinity for their targets, comparable or better than monoclonal antibodies, sometimes with K_D values in the picomolar range. Spiegelmers (from German Spiegel—"mirror") are synthetic mirror-image RNA or DNA ON ligands based on unnatural enantiomeric forms of l-ribose or l-2′-deoxyribose, characterized by nuclease resistance.

16.2.2   DNA as Target

Molecules that interact with DNA can serve as regulators of gene expression, and these molecules are therefore potential therapeutic agents for genetic diseases and cancers. A summary of the possibilities to use short ONs in DNA regulation is presented in Figure 16.8.

One interesting biomedical tool based on ONs and used to interfere with protein expression is the decoy strategy. This approach is designed to operate on transcription factors and to regulate their activity. Transcription factors are generally nuclear proteins that play a critical role in gene regulation, exerting either a positive or negative effect on gene expression. These regulatory proteins bind specific consensus sequences found in promoter regions of target genes. The consensus-binding sequences are generally 6–10 base pairs in length and are occasionally found in multiple iterations upstream or downstream of transcription initiation sites. The binding of transcription factors and subsequent interactions of these proteins with each other as well as with RNA polymerases or their cofactors yield a complex set of factors that determine the relative transcriptional activity. The decoy strategy was first used as a tool for investigating transcription factor activity in cell culture systems. Similar decoys can also be devised as therapeutic agents either to inhibit the expression of genes that are transactivated by the factor in question or to upregulate genes that are transcriptionally suppressed by the binding of a factor.

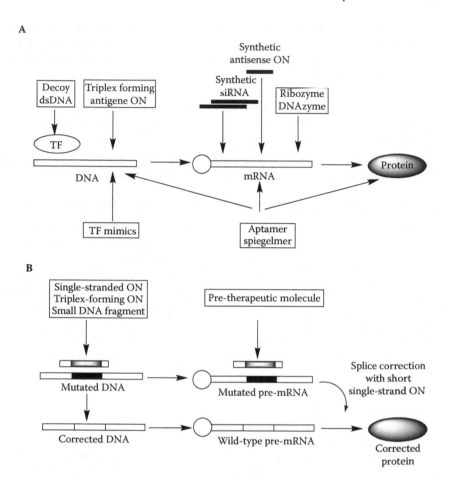

FIGURE 16.8 Target sites for nucleotide-targeting molecules. A—Targets for gene silencing; TF-transcription factor. B—Gene correction strategies. (Modified with permission from Elsevier: Fichou, Y. and Ferec, C., 2006. Trends Biotech. 24(12): 563–570.)

Triplex-forming oligonucleotides (TFO) are single-stranded ONs that contain purine-rich genomic sequences to bind to the major groove of DNA duplexes to form specific and stable bonds, yielding prevention of gene transcription.

Transcription factors (TF) have become attractive targets for therapeutic intervention. TF mimics are artificial polyamino acid sequences usually comprising the DNA-binding domain (e.g., fragment of zinc-finger) and effector domain mediating the desired function. TF mimics could be used as either transcriptional activators or repressors, depending on the effector domain.

16.3 RESTORATION OF GENE EXPRESSION

Correction of specific mutations is based on targeting of the locus to be converted by homologous DNA-derived sequences, and rely on endogenous cellular machinery to recognize and mediate the desired base change. These short DNA strategies are potential alternatives to viral gene therapy. DNA-based molecules are not immunogenic, and maintain a tissue-specific expression pattern that is crucial for therapeutics. In Figure 16.8B, single-stranded ON, triplex-forming ON, and small DNA fragments are presented as examples of possibilities to restore gene expression by short ONs.

Expression of alternatively spliced mRNA variants at specific stages of development or in specific cells and tissues contributes to the functional diversity of the human genome. Aberrations in alternative splicing have been found as a cause of or contributing factor to the development, progression, or maintenance of numerous diseases. The use of antisense ONs to modify aberrant expression patterns of alternatively spliced mRNAs is a novel means of potentially controlling such diseases (Figure 16.8B). Oligonucleotides can be designed to repair genetic mutations, modify genomic sequences in order to compensate for gene deletions, or modify RNA processing in order to improve the effects of the underlying gene mutation. Sterics block the ON approach, such as PNA (discussed earlier) has proved to be effective in the experimental model for various diseases.

16.4 DELIVERY OF OLIGONUCLEOTIDES

For the oligonucleotides to reach their target and perform their action, several barriers have to be overcome. First, the oligonucleotide must be internalized by the cell. Inside the cell the oligonucleotide has to escape degradative organelles and be resistant to the actions of degradative enzymes. Finally, the oligonucleotide must be able to bind specifically to its RNA/DNA/protein target. Much experimental effort is often required to overcome these hurdles.

Liposomes, cationic polymers and, a more recent approach, cell-penetrating peptides have been widely used for ON delivery at least in vitro. Liposomes are artificial lipid spheres with an aqueous core containing water-soluble molecules such as ONs. Liposomes are capable of passing through the membrane of the target cell, either by endocytotic or membrane fusion mechanisms, to deliver the cargo. Cationic lipids have been used to synthesize liposomes.

Cationic polymers, based on the DNA/RNA-polymer complex, can interact with negatively charged cell surfaces and deliver the cargo. Polyethyleneimine, poly-Lys, and their derivatives are most often used cationic polymers for ON delivery. Cell-penetrating peptide technology is currently becoming a promising solution for the delivery of ON therapeutics and will be discussed in Chapter 27.

FURTHER READING

1. Shaner, N. C., Patterson, G. H., and Davidson, M. W. 2007, Advances in fluorescent protein technology. J. Cell Sci. 120: 4247–4260.
2. Yuen, C. M. and Liu, D. R. 2007. Dissecting protein structure and function using directed evolution. Nat. Methods 4(12): 995–997.
3. Wang, H. and Chen, X. 2007. Site-specifically modified fusion proteins for molecular imaging. Front. Biosci. 13: 1716–1732.
4. Sidhu, S. S. and Koide, S. 2007. Phage display for engineering and analyzing protein interaction interfaces. Curr. Op. Str. Biol. 17: 481–487.

5. Fichou, Y. and Ferec, C. 2006. The potential of oligonucleotides for therapeutic applications. Trends Biotech. 24(12): 563–570.
6. Lu, Y. and Liu, J. 2006. Functional DNA nanotechnology: emerging applications of DNAzymes and aptamers. Curr. Op. Biotech. 17: 580–588.
7. Masiero, M. Nardo, G., Indraccolo, S., and Favaro, E. 2007. RNA interference: implications for cancer treatment. Mol. Aspects Med. 28(1): 143–166.
8. Kim, D. H. and Rossi, J. J. 2007. Strategies for silencing human disease using RNA interference. Nature Rev. Genetics 8(3): 173–184.

17 Protein–Ligand Interactions

Ülo Langel

CONTENTS

The interactions of proteins with their counterparts determine their function and biological importance. Cells and organisms efficiently and specifically interrelate with their environment, distinguishing among multiple molecules those with which they have a biological response. The interactions of proteins ("ligand binding") are often used to classify proteins such as enzymes, cell-surface receptors, immunoglobulins, etc.

A protein ligand (ligare is "to bind" in Latin) is a molecule that is able to bind to and form a complex with a protein, resulting in a biological function. Most often, ligands bind reversibly by intermolecular forces such as ionic bonds, hydrogen bonds, and van der Waals forces; the interaction equilibrium involves association and dissociation processes. The interaction between the ligand and receptor protein is characterized by the equilibrium-binding constant, K_D, which is a quantitative measure of the affinity of the interaction (see the following text). Depending on the character of the protein, ligands can be termed substrates, activators, inhibitors, neurotransmitters, etc. They can also be of very different character, from small molecules to macromolecules.

17.1 PROTEIN–PROTEIN INTERACTIONS

Proteins interact with each other in a highly specific manner, and these interactions are crucial for many biological functions. Their impairment may lead to the development of several disorders. Hence, the characterization of protein–protein interactions and, especially, their networks is of highest impact for defining cellular processes at the molecular level as well as for the understanding of diseases and for novel therapeutic approaches.

17.1.1 PROTEIN INTERFACES IN PROTEIN–PROTEIN INTERACTION NETWORKS

The understanding of protein–protein interactions at atomic/molecular level yields information about the physical basis of affinity and molecular recognition and is a prerequisite for exploring the biological functions of proteins. These interactions are very complex and can be characterized by their size, shape, and surface complementarity. The noncovalent hydrophobic and electrostatic interactions, as well as the flexibility of the molecules determine the nature of the interactions. Protein–protein interaction sites are formed by protein surfaces called interfaces. The standard-size (1200–2000 Å^2), small (<1200 Å^2), and large (2000–4000 Å^2) interfaces have been described. Small-size interfaces are generally found in short-lived and low-stability complexes, whereas large interfaces occur in G-proteins and other signal-transducing mechanisms, as well as in some protease-inhibitor interactions.

Protein–protein interfaces are often hydrophobic and, consequently, hydrophobic forces are prevalent in these interactions. As described in Chapter 2, nonpolar regions of the amino-acid residues contribute mainly to these interactions through van der Waals forces. Electrostatic forces are also often involved in protein–protein interactions, where they determine the lifetime of the complexes by, for example, increasing the association rate. Additionally, the average number of hydrogen bonds is proportional to the area of subunit surfaces. Detection of specific amino-acid residues contributing to the specificity and strength of protein interactions is therefore an important step in defining the interactions in protein–protein complexes.

L-Alanine-scanning mutagenesis is still a valuable tool for the analysis of protein–protein interfaces as well as for detection of hot spots in these interactions. l-Ala substitutions remove the side chains adjacent to the β-carbons without introducing additional conformational freedoms and allowing the role of side-chain functional groups to be characterized by these mutations. Using this l-Ala scan, it has been found that individual residues are unevenly distributed across the interfaces, and that only a few key residues or "hot spots" contribute significantly to the binding free energy of protein–protein complexes. Hot spots are defined by l-Ala scans as the sites where the mutation causes a significant increase in the binding free energy (2–4 kcal/mol or up to three orders of magnitude in a binding-affinity constant). Around 10% of the interfacial amino-acid residues are defined as hot spots and, often, they overlap with structurally conserved residues. The most often occurring hot spots are Trp (21%), Arg (13%), and Tyr (12%). The role of these residues in interfaces between the human growth hormone and the growth-hormone-binding

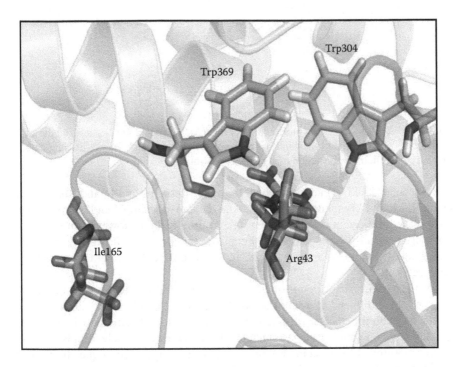

FIGURE 17.1   The human growth hormone complexed with its receptor. The four hot spot residues are highlighted in pink. (Reprinted from Moreira, I. S. et al. 2007. Proteins. Structure, Function and Bioinformatics. 68: 803–812. With permission from Wiley-Liss.)

protein is illustrated in Figure 17.1. The side chain of Trp has special function in the interfaces due to its large size, hydrophobic and aromatic nature, and its contributions to the π-interactions. The residues of Leu, Ser, Thr, and Val are disfavored as hot spots.

Proteins recognize each other along a large, millimolar-to-femtomolar range of affinities, yielding diverse functional requirements. The binding affinity is related to the shape and chemistry of the binding surface, and hence, is more or less specific to a given interaction. Specificity of binding is present even in weak protein–protein interactions, and the knowledge of these specificity-determining sites is of importance in drug design. Structural rearrangements of the proteins are often observed in the binding process. Most often, conformational changes in loop regions and amino-acid side chains occur, though even larger backbone rearrangements have been demonstrated as discussed in Chapter 23 for 7TM receptors.

17.1.2 DOCKING OF PROTEIN–LIGAND INTERACTIONS

The prediction of the possible ligand-binding sites of proteins and ligand–receptor-binding geometries are major goals of computational drug design and docking methods. The major challenges of these methods are the realistic prediction of ligand–receptor-binding energies and the understanding of ligand and receptor flexibilities.

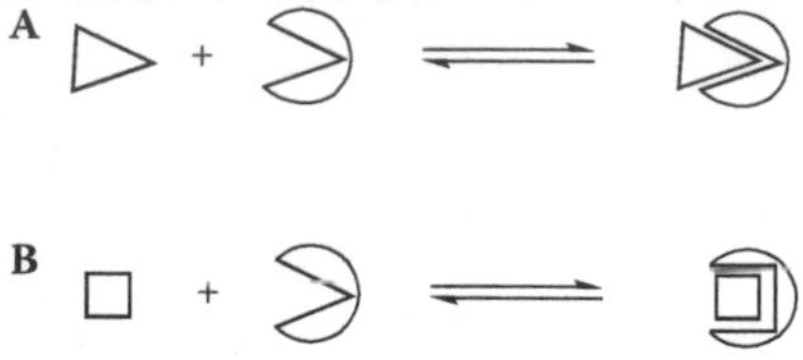

FIGURE 17.2 Two models of molecular docking. (A) "Lock-and-key," where, on the surface of the lock (protein), the key (ligand) is inserted. (B) "Induced fit," where the ligand and the protein adjust their conformation to achieve a "best-fit."

Molecular docking methods, which treat both proteins and ligands as rigid molecules (lock and key model, Figure 17.2A), are relatively simple and inexpensive and have played important roles in screening and drug design. However, in most cases, docking approaches more realistically consider the molecular flexibility for both the ligand and the protein, and hence, the induced fit model of interaction is in force (Figure 17.2B).

Fast and robust shape complementarity methods of docking describe the surfaces of a protein and ligand as a set of dockable features. Usually, these methods cannot model the dynamic changes in protein–ligand interactions, but multiple ligands can be quickly scanned with the shape complementarity. Other approaches apply the hydrophobic features of the protein or a Fourier shape descriptor technique. To perform a docking between protein and ligand, the knowledge of the structure of the protein is required, usually determined by x-ray crystallography or NMR spectroscopy.

Docking is a powerful tool in drug design where docking has been applied to screen large databases for hits (leads) of potential drugs in silico yielding the bioactive molecules that are likely to bind the protein target. Further, docking has been applied for drug lead optimization by prediction of the orientation of a ligand in the protein–ligand complex, possibly leading to design of more potent and selective ligands.

17.2   METHODS TO STUDY PROTEIN–LIGAND INTERACTIONS

17.2.1   Protein–Protein Interactions

To identify protein–protein interactions, multiple experimental methods are available, based on distinct physical principles characterized with their own strengths and weaknesses. These include the yeast two-hybrid (Y2H) method, affinity-purification mass-spectrometry (AP-MS), protein microarrays, coimmunoprecipitation, bimolecular fluorescence complementation (BiFC), fluorescence resonance energy transfer (FRET), surface plasmon resonance, and others.

The yeast two-hybrid (Y2H) screen has been used to generate global protein–protein networks for several lower organisms, and even in the human proteome. The Y2H is today a main method to identify and map physical protein–protein interactions, sometimes named interactomics. It is based, as shown in Figure 17.3, on the phenomena that, in most eukaryotic transcription factors (TF), the activating and binding can function in close proximity to each other without direct binding. The TF, split into two fragments (AD and BD), activates the transcription when the two fragments are

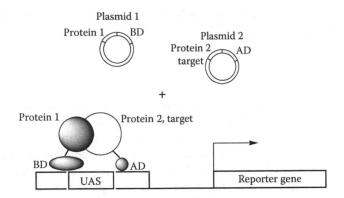

FIGURE 17.3   Scheme for the two-hybrid screen. The downstream reporter gene is activated upon the binding of a transcription factor onto an upstream activating sequence (UAS). The transcription factor is split into two separate fragments called the binding domain (BD) and activating domain (AD). The BD is the domain responsible for binding to the UAS, and the AD is the domain responsible for the activation of transcription. Two fusion proteins with interacting proteins 1 and 2 (target) yield reporter gene activation and confirm their interaction.

in close proximity caused by the interaction between artificial fusion proteins in yeast. This method can qualitatively identify the binding partners of a protein.

Another major method for detection of protein–protein interactions is the AP-MS, which is suitable for detection of multiprotein complexes under near-to-physiological conditions. AP-MS is based on immune-affinity purification methods such as tandem affinity purification (TAP) in conjunction with mass spectrometric (MS) protein identification strategies. Epitope-tagged proteins are transduced into cells and immune-purified with reagents specific to the tag. All co-purifying specific interactors and nonspecific ("false positives") proteins are identified by MS analysis. Quantitative proteomics methods are used today to characterize protein interactions where the relative abundance of the differentially labeled proteins with stable isotopes are compared in different samples.

Protein microarrays are today widely used to identify binary protein–protein interactions. Purified recombinant proteins are, in a high-density manner, immobilized onto a surface-coated glass slide that, in turn, is probed with fluorescently labeled target proteins to detect the interaction.

Coimmunoprecipitation is a complementary assay for protein–protein interactions, especially for unmodified, endogenous proteins. The protein of interest is isolated with a specific antibody, and the interaction partners that stick to this protein are subsequently identified by Western blotting.

17.2.2   Quantification of Receptor–Ligand Binding

17.2.2.1   Theoretical Aspects

Interaction of a receptor protein with a ligand is determined by the overall free energy of the interaction, which can be defined as the affinity of the two to each other.

When a receptor (R) interacts with the ligand (L), the receptor–ligand complex (RL) is formed reversibly (Equation 17.1).

$$R + L \underset{k_{-1}}{\overset{k_1}{\rightleftharpoons}} RL \tag{17.1}$$

However, often the biochemical situation is much more complicated, involving many components in binding and signal transduction events. The binding of several ligands to the same receptor, competition among several receptors for the same ligand, ability of different ligands to induce different active receptor conformations, and interaction of a receptor with one or several G proteins (promiscuous coupling) are only a few such examples. Several more complex schemes have been proposed and discussed in the literature, such as the cubic ternary complex model. All available models, however, usually describe specific cases of ligand–receptor–effector interactions, and the best-fitted model should be selected in each experiment for evaluation of the thermodynamics of the interactions.

The dissociation equilibrium constant K_D is defined as a ratio of product and reactant concentrations in equilibrium, shown in Equation 17.2, and often is used as an affinity measure

$$K_D = \frac{k_{-1}}{k_1} = \frac{[R] \cdot [L]}{[RL]} \tag{17.2}$$

for a protein–ligand interaction. The relationship between the equilibrium constant and the free energy change is presented in Equation 17.3:

$$\Delta G = \Delta G_O + RT \ln K_D \tag{17.3}$$

where G is free energy, ΔG is the change in free energy for the reaction, R is the gas constant, T is the absolute temperature, and ΔG_o is the free energy change for the reaction under standard conditions. Under equilibrium, there is no change in the concentration of the reactants and products, and thus, no change in free energy, $\Delta G = 0$, (Equation 17.4).

$$\Delta G_o = -RT \ln K_D \tag{17.4}$$

The concentration change of R and RL in time are illustrated in Figure 17.4A. When ligand binding is measured by making use of a radioactive ligand that is easily measured, then the amount of bound ligand is referred to as $[B]$, instead of $[RL]$.

A simple way to determine the binding affinity K_D is to represent the experimental data as a Langmuir binding isotherm, plotting $[RL]$ versus $[L]$ (Figure 17.4B). This is sometimes also called the "direct binding curve" and Equation 17.5 describes it.

$$B_L = \frac{B_{max} \cdot [L]}{K_D + [L]} \tag{17.5}$$

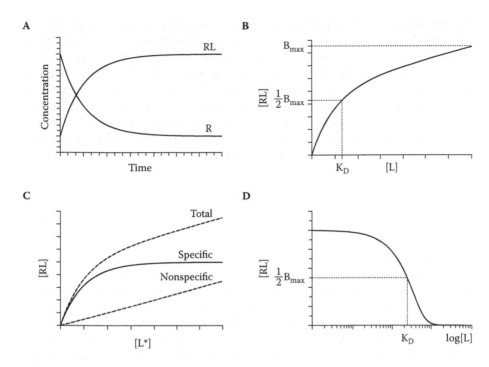

FIGURE 17.4 Experimental characterization of hypothetical ligand–receptor interaction. (A) Increase of RL complex and decrease of R upon time. (B) General Langmuir binding isotherm and schematic explanation of experimental meaning of K_D and B_{max}. (C) Presentations of total, nonspecific and specific binding. (D) Graphical presentation of "displacement method."

This equation is identical in form to the Michaelis–Menten equation for enzyme kinetics, where B_L is the amount of ligand bound to the receptors at concentration $[L]$, a value measured experimentally, and $[L]$ is a known value of the concentration of the radioactive ligand. The equation provides a means for evaluating both K_D and B_{max} (amount of ligand bound to the receptors at saturating concentration) values. Unfortunately, ligands do not bind exclusively to the receptor of interest; they also bind to other proteins and can partition into lipid membranes. This "unwanted" binding is referred to as "nonspecific binding," in contrast to the "specific binding" to the receptor. The amount of nonspecific binding depends on the physicochemical properties of the ligand employed. In practice, one has to subtract nonspecific binding (from some nondefined adsorption) from the total binding in order to obtain the specific binding and applicable binding curve (Figure 17.4C).

Another popular way to determine the ligand binding affinity of the receptor is to apply a constant concentration, usually at sub-K_D range, of a radioactive ligand and displace it with growing concentrations of a ligand of interest (with unknown affinity; Figure 17.4D). This method is known as a "displacement or competition method"; it enables one to characterize the binding affinity of many novel ligands without a need to label all of them (radioactively). In addition, low-affinity ligand binding can be characterized in these competition-binding experiments using a high-affinity radio tracer. The practical meaning of the affinity is that the K_D value equals the ligand concentration that occupies 50% of the available receptors, $B_L = B_{max}/2$ (Figure 17.4B). In fact, when the K_D value is determined by the displacement method, one obtains experimentally the value of IC_{50}, which is the concentration of the nonlabeled displacer that inhibits the binding by 50%. To calculate the real K_D (or K_i to distinguish between labeled ligand and the inhibitor of binding) value, one has to apply the Cheng and Prusoff correction (Equation 17.6), which compensates for occupancy of the receptor by a radio tracer of known concentration ($[L]$) and affinity (K_D).

$$K_i = \frac{IC_{50}}{1 + [L]/K_D}$$
(17.6)

depending upon the concentration of the labeled ligand in relation to the K_D value.

Often the receptor–ligand interaction is more complicated than described and concerns reaction stochiometry different from one to another. The Hill equation is commonly used to estimate the number of ligand molecules that are required to bind to the receptor, sometimes in order to produce a functional effect. A. V. Hill empirically applied it in 1910 to describe the binding of oxygen to hemoglobin. Now it is widely used to analyze the binding equilibrium in a receptor–ligand interaction. The Hill equation is derived from a binding reaction scheme in which n molecules of ligand bind to a receptor (Equation 17.7).

$$R + nL \underset{k_{-1}}{\overset{k_1}{\rightleftharpoons}} RL_n$$
(17.7)

The Hill equation for the oxygen-binding curves by hemoglobin was proposed as in Equation 17.8.

$$[RL] = \frac{K_D \cdot [L]_n}{1 + [L]_n/K_D}$$
(17.8)

After rearrangement, the equation changes to Equation 17.9, where Y = occupancy:

$$\log\left[\frac{Y}{(1-Y)}\right] = \log K_h + h \log[X]$$
(17.9)

and a plot of $\log[Y/(1-Y)]$ against $\log[X]$ should be a straight line of slope h (Hill plot). This plot provides a simple means of evaluating h and K_h. The constant h is now known as the Hill coefficient (n, n_H); it is widely used as an index of cooperativity. A simple mass-action binding curve would have $n_H = 1$. In contrast, $n_H > 1$ indicates positive cooperativity, whereas $n_H < 1$ indicates negative cooperativity or multiple populations of sites (such as R coupled and uncoupled to G-protein). In receptor binding, the constant K_h is K_D.

Efficacy is the ability of a ligand, after binding to the receptor, to activate the transduction mechanisms that lead to a cellular response. Receptor agonists are defined as the ligands that, upon binding to their specific receptors, cause a change in the receptor conformation that leads to the activation of the signal transduction cascade. In other words, agonists stabilize an active receptor conformation ($R*$). Inverse agonists bind preferentially to the inactive conformation of the receptor (R), yielding a lowering of basal signaling. Partial agonists are not able to produce maximal effects, even when saturating the receptor, due to their modest preference for the active receptor site. The competitive receptor antagonists have no preference for R or $R*$ but displace other ligands from the active or inactive binding site, thereby preventing signaling.

Peptide hormone-receptor interactions and protein–protein interactions are key biochemical processes that have great potential as specific targets for novel therapeutic agents. In many cases, it is desirable to identify nonpeptide ligands for the target receptors since they have potential for more suitable pharmaceutical properties such as stability to proteases and oral bioavailability.

17.2.2.2   Receptor–Ligand Labels and Binding Assays

The quantitative characterization of receptor–ligand interactions is of very high impact for the screening of novel, pharmacologically relevant molecules, and numerous assays are available for screening and quantification of receptor ligands. Both radioactive and nonradioactive assays are available for this.

Radioactive assays are fast and relatively easy to use, though they are hazardous to human health and produce radioactive waste. These assays also require the separation of radiolabeled ligand from the receptor-bound complex, and are therefore considered to be only of medium throughput. Radioisotope labels such as ^{3}H, ^{125}I, and ^{32}P are most common in labeling ligands because the introduction of these labels has relatively low effect on the affinity of the ligand when designed favorably.

An example of the radioactive labeling of peptides and proteins with ^{125}I (also applicable for ^{131}I-labeling) is presented in Figure 17.5A, where the chloramine T method is briefly presented for labeling of Tyr (sometimes even Cys and His) side chains of proteins. This labeling is a method of choice when mild oxidation during the labeling procedure is not a danger to the peptide/protein, such as when sensitive to oxidation amino acid residues (Met, Cys) are present in the sequences. Specific activity of the labeling is calculated after purification of the radiolabeled ligand. Several other iodination methods are available today, such as the Bolton–Hunter method (primary amino groups of target proteins are labeled), Iodo-Beads method (Figure 17.5B; labels Tyr side chains with an immobilized form of chloramines T), and the Iodo-Gen method (Figure 17.5C; gentle labeling of phenolic groups in proteins).

Labeling of proteins with tritium serves as an alternative to radioiodination. One popular and efficient method applies the reductive methylation procedure by sodium [^{3}H]-borohydride to reduce ketones and aldehydes to alcohols so as to obtain labeled proteins with specific activities approaching those obtained with radioiodine. Another, milder technique for random tritium labeling of proteins is to perform a catalytic ^{1}H-^{3}T exchange in solvents such as water, dimethylformamide, and

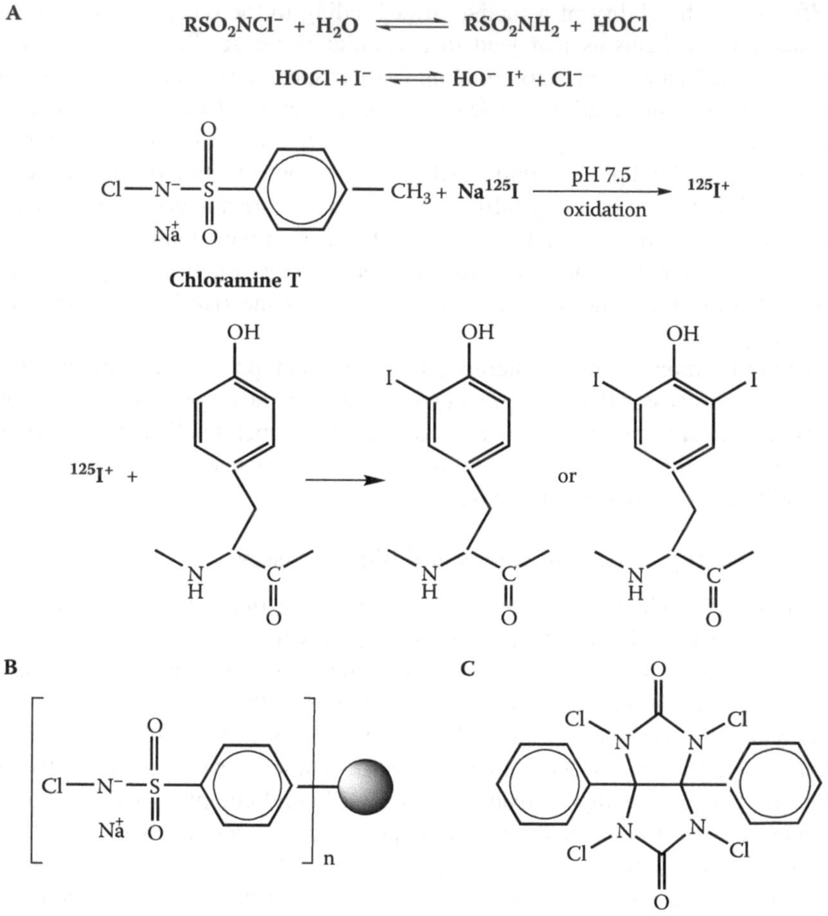

A

$$RSO_2NCl^- + H_2O \rightleftharpoons RSO_2NH_2 + HOCl$$

$$HOCl + I^- \rightleftharpoons HO^- I^+ + Cl^-$$

Chloramine T

B

C

FIGURE 17.5 Iodination of peptides and proteins. (A) Iodination of side chain of Tyr of a peptide/protein with the chloramine T method. Iodonium ion (I^+) is first obtained by mild oxidation of $Na^{125}I$ by chloramine T, followed by its interaction with nucleophilic centers of the Tyr side chain. (B) Structure of Iodo-Beads iodination reagent, N-chloro-benzenesulfonamide sodium salt, immobilized form of chloramine T on polystyrene beads. (C) Structure of Iodo-Gen, 1,3,4,6-tetrachloro-3c~, 6c~-diphenylglycouril. (D) IPy2BF4 reagent, bis(pyridine) iodonium (I) tetrafluoroborate.

other aprotic solutions. In these reactions, noble metals are the typical catalysts, and tritium-labeled water and tritium gas are used as sources of the isotope. In a simple approach to random labeling of protein molecules with tritium–tritium gas exposure, the compound is sealed in an ampoule containing carrier-free tritium gas and kept in this ionizing field for several weeks. However, considerable radiation decomposition of the protein occurs, causing the formation of contaminants.

Nonradioactive assay methods are becoming more popular and are based on optical methods such as colorimetry, fluorescence or luminescence detection, fluorescence polarization, fluorescence energy transfer, or surface plasmon resonance.

Today, there is an endless variety of fluorescent labels with enhanced brightness; great photostability and improved physical properties (water solubility, pH stability) can be used to produce efficiently labeled peptides and proteins. Some popular examples of fluorophores (and luciferin) are presented in Figure 17.6B. Labeling of peptides and proteins with these chemical entities follows the rules of chemical interactions, and all functional groups in targets can be used. For example, as in Figure 17.6A, primary amino groups of peptides/proteins can be modified with fluorophores containing reactive groups such as carboxylic acids, isothicyanate, sulfonyl chloride, or succinimidyl ester. The variety of these interactions is huge, and hence, the labeling possibilities as well. When these possibilities are considered, together with the information about the impact of the functional groups in peptides/proteins, the assays can be modified without changing the bioactivity of the biomolecule after labeling.

Receptor–ligand-binding assays are classified according to the need for separation of free ligand from the receptor-bound ligand. In the case illustrated in Figure 17.7A, the free and bound ligands are separated by a rapid separation method such as filtration through glass-fiber filter or gel filtration, centrifugation, or even dialysis. Such an assay is most often used in quantitative radioreceptor analysis. Different assays that do not require separation are presented in the following text.

Scintillation proximity assay or SPA (Figure 17.7B) applies the increase of light emission as a result of energy transfer from the radioactive decay of the radioligand to the bead containing the scintillant, such as polyvinyltoluene microspheres, with a polyhydroxysurface coating. The light emission only occurs if the radiolabeled ligand (L) and receptor are in close proximity (around 10 μm); otherwise the energy of radioactivity is absorbed by the surrounding buffer. The receptor is immobilized on a solid support (bead) with the prerequisite that the receptor is available for this immobilization. Immobilization of the receptor is based on the interaction of glycoproteins and glycolipids with wheat germ agglutinin by capturing antireceptor protein antibodies on an anti-IgG-coated surface. SPA is today available for several, but not all, receptors, and it is a preferred method for high-throughput screening of novel ligands as drug candidates.

Several strategies are available for nonradioactive assays, all depending on the application of some kind of (fluorescent) labels. Hence, in each case, the design of the labeling without interfering with the biological activity of the biomolecules is crucial. A few of these nonradioactive assays are briefly described in the following text, and the surface plasmon resonance assay is described in Chapter 25.

The scheme for fluorescence resonance energy transfer, FRET, is presented in Figure 17.7C. This assay is based on the transfer excitation from a donor to an acceptor molecule when both come into close proximity (10 nm). The receptor R is labeled with an acceptor (e.g., fluorescent antibody [Ab]), and the ligand is labeled with a donor fluorophore. The luminescence variant of FRET, called bioluminescence resonance energy transfer (BRET), is available where a luminescent donor and fluorescent acceptor are used. BRET is often used to detect protein–protein interactions, such as between receptor dimers or between signaling molecules.

Fluorescence polarization assay, shown in Figure 17.7D, measures the change in polarization of light emitted from a fluorescently labeled ligand L as a

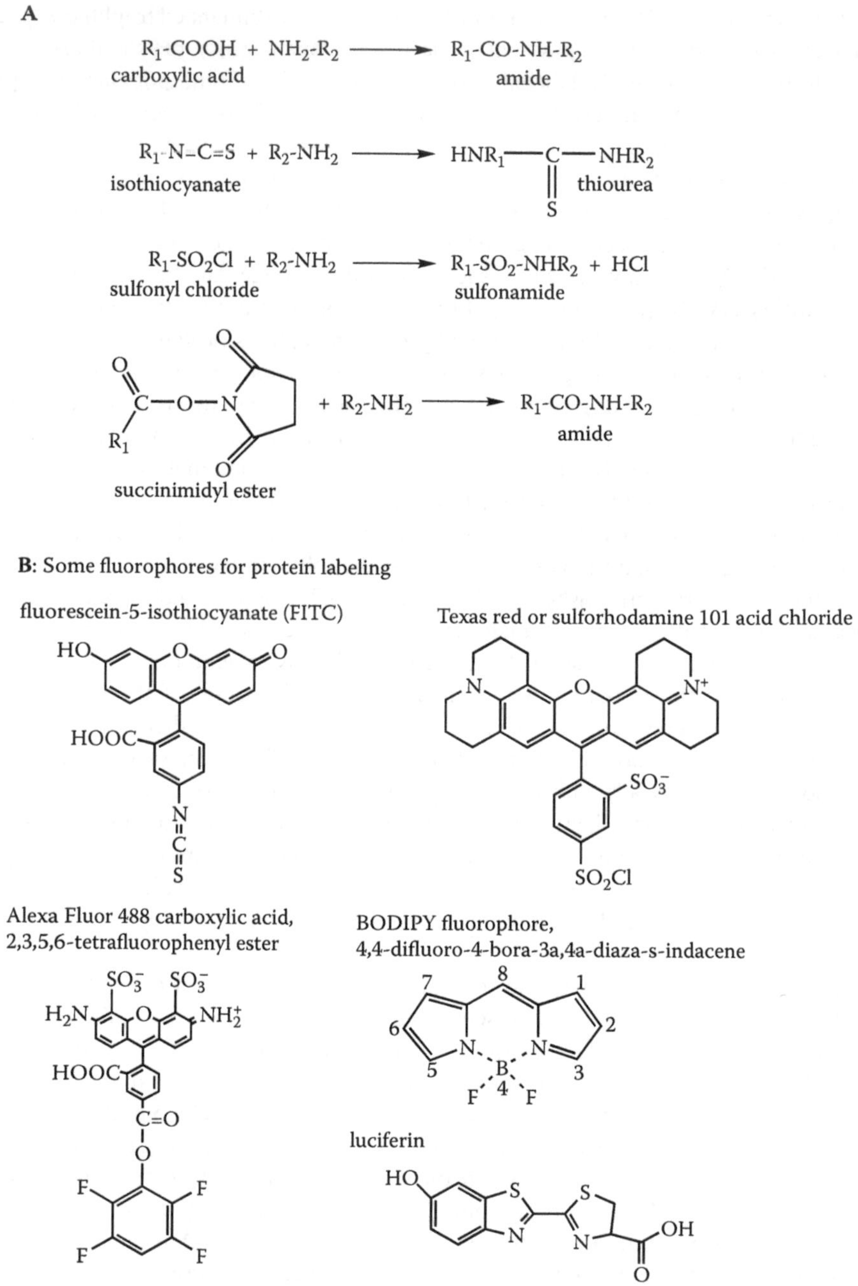

A

$$R_1\text{-COOH} + NH_2\text{-}R_2 \longrightarrow R_1\text{-CO-NH-}R_2$$
carboxylic acid   amide

$$R_1\text{-N-C=S} + R_2\text{-NH}_2 \longrightarrow HNR_1\text{---}\underset{\underset{S}{\|}}{C}\text{---}NHR_2$$
isothiocyanate   thiourea

$$R_1\text{-SO}_2\text{Cl} + R_2\text{-NH}_2 \longrightarrow R_1\text{-SO}_2\text{-NHR}_2 + HCl$$
sulfonyl chloride   sulfonamide

succinimidyl ester $\quad + R_2\text{-NH}_2 \longrightarrow R_1\text{-CO-NH-}R_2$
                                                            amide

B: Some fluorophores for protein labeling

fluorescein-5-isothiocyanate (FITC)   Texas red or sulforhodamine 101 acid chloride

Alexa Fluor 488 carboxylic acid,   BODIPY fluorophore,
2,3,5,6-tetrafluorophenyl ester   4,4-difluoro-4-bora-3a,4a-diaza-s-indacene

luciferin

FIGURE 17.6 Labeling of peptides/proteins with nonradioactive labels. (A) Introduction of amino modifications by interaction of primary amino groups of a protein (R_2) with carboxylic acid, isothiocyanate, sulfonyl chloride, and succinimidyl ester derivatives of fluorophores, R_1. (B) Some examples of popular fluorophores, and luciferin yielding bioluminescence in enzymatic reaction with luciferase for labeling.

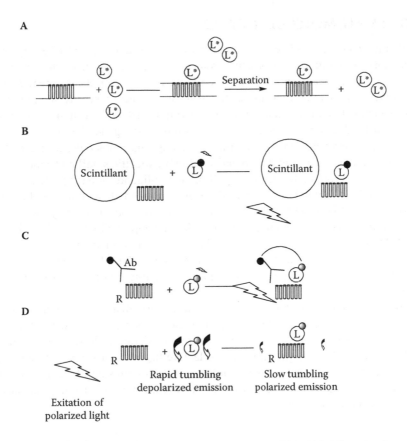

FIGURE 17.7 Schemes for receptor–ligand-binding assays. (A) Labeled ligand and receptor (here GPCR) form equilibrium complex after incubation, followed by rapid separation (filtration or centrifugation) of free (L*)- and receptor bound (RL*)-labeled ligand. (B) Scintillation proximity assay. (C) Fluorescence resonance energy transfer (FRET). (D) Fluorescence polarization assay.

consequence of a mobility change of the labeled ligand. Excitation of a fluorescent low-MW ligand by polarized light results in depolarized emitted light due to rapid rotation of the ligand. Upon binding of the fluorescent ligand to a high-MW receptor (R), the rotational speed decreases and the emitted light stays partially polarized.

In high-throughput screening (HTS) of ligand–protein interactions, a large number of diverse chemical structures is tested (in parallel) to find the interactors for a given protein. HTS is often applied for drug screening where the possible novel drug development targets are called "hits." Being simple, rapid, efficient, and even cost efficient, the HTS is the method of choice everywhere when the large number of chemical entities has to be screened efficiently. DNA microarrays are often used in HTS to study the expression of biological targets associated with human disease.

17.3   SMALL-MOLECULE LIGANDS

Small molecules can influence protein–protein interactions, and offer significant advantages over influencing protein–protein interactions by macromolecules such as other proteins. Many small molecules diffuse readily over plasma membrane and the blood–brain barrier; they are relatively stable and act reversibly and rapidly in a dose-dependent manner to interact with the protein partner. This makes it extremely important for drug development to find small molecule ligands for proteins.

There are many challenges in finding and using small molecules to interfere with protein functions such as to design and synthesize specific small molecule effectors for each target protein. To identify small-molecule ligand for each protein function, high throughput methods for ligand discovery are needed, such as small-molecule microarrays (SMM), where the small molecules on the microarrays are derived from natural products, combinatorial libraries, or approved drugs. The scheme of an SMM is illustrated in Figure 17.8, where the protein of interest is probed with the SMM and binding is detected using a fluorescence-based scanner. Several small molecule attachment chemistries, as well as screening and profiling methods, are available today.

The SMM screening has found small molecules that interact with transcriptional regulators, enzyme inhibitors, cell surface receptors, etc., with affinities in the low-micromolar range. These ligands with moderate affinities can be and are further developed into high-affinity and selective ligands by the combination of synthetic chemistry and microarray technologies. Often such small ligands will inhibit protein–protein interactions, and serve as powerful drug leads as they can successfully block these interactions.

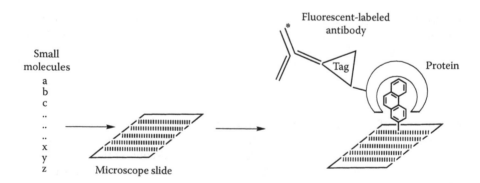

FIGURE 17.8   Small-molecule microarrays (SMM). Small molecules are arrayed onto functionalized microscope slides (left) and screened by, for example, incubation with a protein (purified or complex mixture), followed by incubation with fluorescently labeled antibody (against the protein or an epitope tag) and detection (right).

FURTHER READING

1. Moreira, I. S., Fernandes, P. A., and Ramos, M. J., 2007. Hot spots—a review of the protein–protein interface determinant amino-acid residues. Protein, 68(4): 803–812.
2. May, A. and Zacharias, M. 2005. Accounting for global protein deformability during protein–protein and protein–ligand docking. Biochim. Biophys. Acta, 1754(1–2): 225–231.
3. Ruffner, H., Bauer, A., and Bouwmeester, T. 2007. Human protein–protein interaction networks and the value for drug discovery. Drug Disc. Today, 12(17/18): 709–716.
4. Lalonde, S., Ehrhardt, D. W., Loqué, D., Chen, J., Rhee, S. Y., and Frommer, W. B. 2008. Molecular and cellular approaches for the detection of protein–protein interactions: latest techniques and current limitations. Plant J., 53(4): 610–35.
5. Duffner, J. L., Clemons, P. A., and Koehler, A. N. 2007. A pipeline for ligand discovery using small–molecule microarrays. Curr. Op. Chem. Biol., 11(1): 74–82.
6. Davenport, A. P., Ed. 2005. Receptor binding techniques. Methods in Molecular Biology, 2nd edition. Totowa, NJ: Humana Press.

18 Sequence Analysis and Function Prediction

Tiit Land

CONTENTS

In this chapter we briefly describe the principles and methods used by bioinformatics, which is a rapidly growing field of science dealing with the collection, maintenance, analysis, and interpretation of large amounts of experimental data. At the beginning of the genomics era, the main goal of bioinformatics was the creation and maintenance of databases in order to store and retrieve biological information such as nucleotide and protein sequences. After 1994, when the number of nucleotide sequences began to grow exponentially, the necessity to create computer-based algorithms to analyze them became more evident. Therefore, one of the most important tasks of bioinformatics today is to extract, analyze, and identify relationships between data sets that often comprise nucleotide and protein sequences, protein domains, and protein structures. To serve that purpose, the field of bioinformatics includes the development of algorithms and statistical tools for the analysis and interpretation of relationships between data sets, such as prediction of protein structure and function.

A whole book would be required to describe all the databases, algorithms, and statistics used in the field of bioinformatics. In this chapter we describe briefly some important protein databases and focus on the use of algorithms for the analysis and interpretation of sequence data sets, including prediction of protein function. In Chapter 19 we consider how bioinformatics can assist in the prediction of protein structures and interactions.

This chapter is not intended for a computer scientist, and a reader who is interested in detailed analysis of algorithms and statistical tools is recommended to read specialized literature on this topic.

18.1 PROTEIN DATABASES

The first complete protein sequence determined was bovine insulin, sequenced by Frederick Sanger and colleagues in the early 1950s. About 10 years later, there were more than 100 protein sequences published, and the first protein sequence database, Atlas of Protein Sequence and Structure, was created by Margaret Dayhoff, who is also credited as a founder of the field of bioinformatics. However, the atlas contained very few uncharacterized proteins and was mainly used to investigate sequence diversity between homologous proteins (such as globins) from diverse organisms. After the introduction of rapid DNA-sequencing methods in the mid 1970s, more and more protein sequences were predicted by translating sequenced DNA (or cDNA), and thus, the number of uncharacterized protein sequences began to increase. As several large-scale genome-sequencing projects have been completed, a large amount of data concerning the number and distribution of proteins has become available. However, there are several issues that should be considered when predicting the protein-coding regions in DNA sequences. First, predicting correct start and stop codons of a gene and the splicing pattern may be exceedingly difficult. Second, after predicting a putative open reading frame (ORF) and translating it into a protein sequence, how do we know that this particular ORF is expressed as there is no experimental evidence? By searching databases, it may be possible to identify other proteins with similar sequences that have been demonstrated to be expressed, which is an indication that the particular DNA codes for a protein since a conserved ORF is likely to be expressed. Third, mRNA from some genes can be edited, resulting in the generation of splice variants and leading to the biosynthesis of different polypeptides from a single gene. Finally, sequence databases often contain raw data derived directly from experiments and various sequencing projects, making it possible that deduced sequences are not correct due to frameshifted fragments, sequences from pseudogenes, and sequencing errors. Some databases are highly curated, that is, entries in the database are analyzed and verified by human experts.

Biological databases can be classified in several ways. In this chapter, we describe protein databases as sequence or structure databases. A comprehensive list of protein databases is available at the ExPaSy Web site (http://www.expasy.org).

18.1.1 Selected Sequence Databases

Currently, the principal source of protein sequences are translations of DNA sequences deposited in primary nucleotide sequence databases. These are GenBank, maintained by the U.S. National Center for Biotechnology Information (NCBI); the DNA Databank of Japan (DDBJ); and the European Molecular Biology Laboratory (EMBL) Nucleotide Sequence Database, maintained by the European Bioinformatics Institute (EBI). New DNA sequence data can be deposited with any of these three databases, which form an International Nucleotide Sequence Database Collaboration, and the data between the databases is shared and updated on a daily basis so that DNA sequences kept in each database are essentially identical. These DNA sequence databases are repositories for raw sequence data submitted directly from experiments and DNA-sequencing projects. However, the entries are heavily annotated, containing not only the DNA sequence itself but also literature references, putative start and stop codons for translation, and deduced protein sequence when appropriate.

18.1.1.1 Entrez Proteins

The NCBI protein database, called Entrez Proteins (http://www.ncbi.nlm.nih.gov/sites/entrez?db=protein) offers a complete set of all protein sequences deduced through DNA translations. Entrez, a data retrieval tool developed by the NCBI, is an integrated search-and-retrieval system that can be used to search all databases at once with a single query string. In addition to translated protein sequences from GenBank/DDBJ/EMBL, the proteins in the Entrez system include sequences from highly curated protein sequence databases Swiss-Prot, Protein Information Resource (PIR), Protein Data Bank (PDB), and Protein Research Foundation (PRF) (see the following text). While this makes the Entrez Proteins nearly complete, ensuring that almost any protein sequence information can be retrieved, the database is excessively large and redundant. For practical purposes, NCBI has also created a nonredundant database, in which identical protein sequences from the same organism appear as a single entry. In addition to the sequence information, each Entrez Protein has links to all databases maintained by the NCBI, such as GenBank where the corresponding nucleotide sequence can be directly retrieved, PubMed, Taxonomy database, Molecular Modeling Database (MMDB) that contains three-dimensional structures of proteins and polynucleotides, and Online Mendelian Inheritance in Man (OMIM) for proteins associated with human diseases.

When searching these sequence databases and retrieving sequence information, it is important to bear in mind that the sequences in the databases are in fact repositories for raw data and that the databases are not heavily curated. The submitter is responsible for the correctness of the sequence and its annotation, and further updates or corrections or both come from the submitter. Despite this, Entrez Proteins are currently the ultimate resource for almost any protein sequence.

18.1.1.2 Swiss-Prot and TrEMBL

In contrast to Entrez Proteins, Swiss-Prot (http://www.expasy.org/sprot/) is not just a repository for protein sequences but is a collection of confirmed protein sequences that is extensively annotated with information about protein structure and function, it domains structure, posttranslational modifications, variants, and bibliographic references. Swiss-Prot entries are linked to various external databases such as PubMed, DNA sequence databases GenBank/DDBJ/EMBL, and protein sequence motifs and domain databases PROSITE, Pfam and BLOCKS (see the following text). The Swiss-Prot was started in 1986 by Amos Bairoch at the University of Geneva and has been developed by the Swiss Institute of Bioinformatics (SIB) and the EBI. The quality of the data in Swiss-Prot is very high because it is curated by human experts. New sequences are added into the database after analysis and verification by experts, which, however, makes the Swiss-Prot relatively poorly covered and somewhat out of date. The latest Swiss-Prot release 56.0 of July 22, 2008, contained 392,667 sequence entries, 20,069 of that human protein sequences. In order to speed up the entry of protein sequences into the Swiss-Prot, a supplemental database, TrEMBL, has been developed. The sequences in the TrEMBL are translations of all coding sequences in the EMBL nucleotide sequence database, and each entry is in the Swiss-Prot format style, waiting for curation and entry into the Swiss-Prot. Since TrEMBL entries are generated automatically without curation by human experts, some proteins included in the database can be hypothetical, unlike the Swiss-Prot that contains only confirmed protein sequences. The latest TrEMBL release 39.0 of July 22, 2008, contained 6,070,084 sequence entries. TrEMBL is considered to be less redundant than Entrez Proteins.

18.1.1.3   Sequence Motifs and Domain Databases

Classification of proteins into families with common patterns, such as sequence motifs and domains, can be helpful in assigning functions to uncharacterized proteins. The protein sequence motif can be defined as a conserved amino-acid sequence pattern that has functional significance. A protein domain is a part of a protein sequence that forms a stable three-dimensional structure and can be independently folded. Domains are often evolutionarily conserved and may contain one or more sequence motifs. Proteins, in turn, can consist of a single protein domain or several domains.

Several sequence motif and domain databases have been constructed using different algorithms that identify motifs, fingerprints of collection of motifs, domain profiles, and hidden Markov models (HMMs), which are built from multiple sequence alignments of the template family. These databases include more information than the standard profile, such as the positions of common insertions and deletions of amino acids in the protein family. As will be described later in this chapter, identifying sequence motifs and domains can be helpful in providing clues to the functions of newly discovered proteins.

PROSITE (http://www.expasy.org/prosite/) is a database of protein families, domains, and functional sites. It is maintained at the SIB and is tightly linked to Swiss-Prot. PROSITE provides a rapid tool for determining a possible function for uncharacterized proteins, which sequence is deduced by translating genomic or cDNA sequences. The reader who is interested in details of PROSITE is recommended to read the PROSITE User Manual (http://www.expasy.org/prosite/prosuser.html).

The BLOCKS database (http://blocks.fhcrc.org/) consists of short ungapped multiple alignments of highly conserved regions of proteins. The database can be used to search a given protein or DNA sequence that will be translated in all six reading frames against the blocks. The PRINTS database (http://www.bioinf .manchester.ac.uk/dbbrowser/PRINTS/) is a collection of protein motif fingerprints that are groups of conserved motifs of a protein family. Multiple alignments kept in BLOCKS and PRINTS databases can be useful for the identification of patterns and motifs that are conserved over great evolutionary distances (see the following text).

The Pfam database (http://pfam.sanger.ac.uk/) contains multiple protein sequence alignments and hidden Markov models covering many common protein domains and families. The database was jointly developed by three groups in the United Kingdom, United States, and Sweden. Each entry in the Pfam is an alignment of the most conserved domains of related proteins from Swiss-Prot and TrEMBL databases.

18.1.2 SELECTED STRUCTURE DATABASES

Determination of the three-dimensional structure of a protein is very difficult when compared to primary sequence analysis. However, structures are much more useful when identifying very distantly related homologs, predicting protein functions, and using protein structure as a template for predicting a three-dimensional structure for a protein sequences with unknown structure (which will be discussed in Chapter 19).

18.1.2.1 Protein Data Bank

While Swiss-Prot is a major database of confirmed protein sequences, the PDB is a repository for three-dimensional structures of proteins and nucleic acids. The PDB was founded in 1971 by Edgar Meyer and Walter Hamilton at the Brookhaven National Laboratory and is currently maintained by the Research Collaboratory for Structural Bioinformatics (RCSB) at Rutgers University (http://www.rcsb.org/pdb/ home/home.do). As in Swiss-Prot, data in PDB are extensively curated and are of very high quality. As of July 29, 2008, the PDB contained 52,103 structures, 48,092 of those proteins.

18.1.2.2 SCOP and CATH

Structural Classification of Proteins (SCOP) is a manually curated hierarchical database of protein structures developed and maintained at the MRC Laboratory of Molecular Biology in Cambridge, United Kingdom (http://scop.mrc-lmb.cam.ac.uk/ scop/). SCOP classifies protein structures into three levels: families, superfamilies, and folds. First, two hierarchical levels—family and superfamily describe near and distant evolutionary relationships, and proteins are defined having common folds if they have the same major secondary structures in the same arrangement and with the same topological connections. All known structures are grouped into different classes: all-α proteins, all β-proteins, α- and β-proteins with α-helices and β-strands (α/β), α- and β-proteins with segregated α-helices and β-strands ($\alpha + \beta$), multi-domain proteins (α and β), membrane- and cell-surface proteins and peptides, and small proteins.

The CATH database (http://www.cathdb.info/) is also a hierarchical database that classifies protein folds into four major levels: Class (C), Architecture (A), Topology (T), and Homologous superfamily (H). The CATH classification is based on secondary structure content (C-level), orientations of secondary structures (A-level), connections between secondary structures (T-level), and sequence and structure comparisons (H-level).

18.2   SEQUENCE ALIGNMENT

Sequence comparison methods have provided modern biosciences with an extremely powerful tool. Initial characterization of any new DNA or protein sequence starts with searching biological databases to find homologs of the gene or the protein. This information can often give insights into the functions and mechanisms of the newly sequenced DNA or protein molecules. Furthermore, comparing a newly sequenced DNA or protein with all previously characterized sequences can give a tremendous insight into its evolution. Thus, the two main objectives of DNA or protein sequence analysis are to identify homologous sequences and to use this information for prediction of biological functions, and to explore evolution.

18.2.1   PAIRWISE SEQUENCE ALIGNMENT

The details of computational methods developed for sequence alignments are beyond the scope of this chapter. Briefly, when two sequences are aligned, the computer algorithm looks for identical characters or a series of character patterns that appear in the same order in both sequences. Introduction of gaps into one of the sequences is often needed to capture most of the identical characters in both alignments. Inserted gaps compensate for the insertions or deletions of nucleotides that may have taken place during evolutionary time. Finally, the number of shared residues is determined and the alignment score is calculated, which includes penalties for gaps that are needed for preventing the insertions of an unreasonably high number of gaps into the aligned sequences. The calculated alignment score shows the degree of similarity between the two sequences and the closeness of their evolutionary relationship.

A significant sequence similarity between two molecules suggests that they have a common evolutionary origin. Although both DNA and protein sequences can be searched in databases to find homologs, protein comparisons are often more effective. First, proteins are built from 20 different amino acids, whereas DNA molecules are built from 4 different nucleotides. Second, one amino acid can be coded by more than one codon, and some amino acids are specified by six codons. Third, DNA sequences are much larger than protein sequences. Finally, amino-acid residues can be compared based on physicochemical characteristics, not only as the sequence identity or nonidentity. For example, when two physicochemically similar amino-acid residues, such as leucine and isoleucine, are aligned with one another in a pairwise alignment between two protein sequences, the functional properties between the two proteins are likely unchanged. Further, evolutionary changes in protein sequences tend to involve substitutions between chemically or physically similar amino acids since such changes are less likely to affect the function or structure of the protein. Substitution of one amino-acid residue with another one with similar physicochemical properties is called conservative substitution.

18.2.2 MULTIPLE SEQUENCE ALIGNMENTS

Although pairwise alignments are important for finding related proteins, multiple alignments are critical to the analysis of protein families and remote homologs. Multiple sequence alignments can also be helpful in finding amino-acid residues that are important for biological activities of protein families. For example, the conservation of a particular amino-acid residue between two functionally related protein sequences could occur by chance, but if the same amino acid is conserved in more sequences within the same protein family, it may have functional importance.

The most commonly used method for multiple alignment is Clustal. This program first performs pairwise alignments to assess the degree of similarity between each sequence, and the sequences are clustered by score to produce a guide tree that is similar to a phylogenetic tree. Sequences are then aligned step by step so that the two most similar sequences are aligned first and other sequences are added in order of similarity.

18.2.3 SUBSTITUTION MATRICES

As discussed, conservative substitutions should be taken into account in the alignment score when aligning protein sequences. In general, a conservative substitution should be less penalized than a replacement of one amino-acid residue with another one that has completely different chemical or physical properties. However, not all amino-acid substitutions that are not conservative result in structural and functional changes of proteins: they depend on which location within the amino-acid sequence of the protein a nonconservative substitution has occurred. For example, when such replacements occur in less structured regions of a polypeptide chain and not in the active site, then it can have relatively little effect on protein function. Unfortunately, there is often no prior information about the location of substitutions within a particular protein sequence. Thus, the question is: how can we assign weightings to particular substitutions? To solve this problem, many substitutions that have occurred in evolutionarily related proteins have been examined and the so-called substitution matrices have been deduced. These are 20×20 matrices where the numbers define probability values of the transformation of one amino-acid to another one of all possible amino-acids pairs during certain evolutionary time. The simplest possible matrix would be the identity matrix in which the score for the exact match of two amino acids is 1, and that for the mismatch is 0. This identity matrix does not work in reality when comparing evolutionarily closed proteins since it does not take into account amino-acid substitutions. More sophisticated scoring schemes have been developed that take into account conservative substitutions. For example, the score for the substitution of a leucine for an isoleucine can be +2, but that of an arginine for an isoleucine can be −4. Thus, a large positive score in substitution matrices means that a particular substitution is likely to occur relatively frequently, and a large negative score corresponds to a rare substitution.

The first substitution matrix, called PAM (Percent Accepted Mutation), was created by Margareth Dayhoff in the 1970s. PAM matrices were derived by first aligning a small number of closely related protein sequences and counting amino-acid substitutions within the families. The PAM1 matrix was calculated from sequence comparisons having not more than 1% divergence. The observed substitutions were then extrapolated to more distant evolutionary relationships. For example, PAM70 matrix is supposed to apply for protein sequences that differ by 70%, or a 0.7 change per aligned amino-acid residue. Thus, the PAM1 matrix should be used for constructing alignments of closely related proteins, and PAM70 matrix for more divergent sequences. For even more distantly related proteins, the PAM250 matrix has been deduced, which should reflect evolutionary changes with an average 2.5 substitutions per amino-acid residue.

Although PAM matrices have been extensively used in protein sequence alignments, their main limitation is that they are derived from alignments of closely related protein families and then extrapolated to more distantly related sequences. In 1992, Steven and Jorja Henikoff constructed a series of substitution matrices called BLOSUM. As in the case of PAM matrices, they are derived from an analysis of related proteins, but unlike PAM, the alignments used to construct BLOSUM matrices include more diverse protein families. Further, all BLOSUM matrices are based on observed alignments and are not derived by extrapolating data from alignments of closely related sequences. Finally, BLOSUM matrices are constructed using conserved ungapped multiple alignments of related proteins from the BLOCKS database (BLOSUM = BLOcks SUbstitution Matrix). Since the BLOCKS database includes alignments of only highly conserved regions of distantly related proteins, BLOSUM matrices are based on local alignments in contrast to PAM matrices that are based on global alignments. For these reasons, BLOSUM matrices are now most often used in protein sequence alignments. In the BLOSUM62 that is currently the default matrix in popular sequence alignment algorithms such as BLAST (see the following text), the substitution scores are derived from sequence alignments with no more than 62% identity. This means that the BLOSUM matrix with a higher number should be used for aligning more closely related sequences, whereas a matrix with a lower number should be employed for alignments aimed at detecting distant relationships.

18.2.4 ALGORITHMS FOR SEQUENCE ALIGNMENT

In principle, the only way to find homologous sequences to a given sequence is to align the query sequence against all sequences in the database and to estimate statistical significance for pairwise alignments that indicates whether the two sequences could be homologous or not.

18.2.4.1 Needleman–Wunsch and Smith–Waterman

Two commonly used algorithms that provide optimal alignment are Needleman–Wunsch and Smith–Waterman algorithms. Although both of them use computational method known as dynamic programming that guarantees optimal pairwise alignment between two sequences achieving the best alignment scores, the Needleman–Wunsch algorithm looks for global similarity between two sequences, while the Smith–Waterman algorithm looks for local alignment that includes only parts of the analyzed sequences. Both methods have their advantages and disadvantages, but it is generally believed that local alignments can be more effective because many homologous proteins share only parts of amino-acid sequences. The Smith–Waterman algorithm is used by several database search programs, such as SSEARCH, which is a part of the FASTA package (see the following text). The main limitation of dynamic programming algorithms is that they are slow when used to search large-sequence database and have a high computational cost.

18.2.4.2 FASTA and BLAST

Although dynamic programming algorithms should theoretically guarantee to find the best alignment between two sequences, alternative methods have been developed that are much faster. In 1988, William Pearson and David Lipman introduced FASTA, which achieved a comparable efficiency to the Smith–Waterman algorithm but was much faster. A few years later, the BLAST (Basic Local Alignment Search Tool) program was designed, which is currently the principal tool for comparing primary sequence information against sequences in public databases.

The principle of both FASTA and BLAST is based on the fact that two compared sequences should always contain short stretches of identical letters, called "words." Both FASTA and BLAST first search sequences in the database for words, which is very fast. Since the majority of sequences in the database are unlikely to contain matches to words, these sequences are skipped. Both programs then extend their matching segments in either direction in an attempt to produce longer alignments. In the final stage of the alignment, the FASTA uses dynamic programming for creating alignments of the regions with high scores, making it slower than BLAST. In the case of FASTA, the word length W is equivalent two amino acids, and there has to be exact match. In BLAST, the match has to score above a given threshold level T using the given substitution matrix.

The BLAST program provides the user with statistical parameters that indicate the probability that sequences in alignments were found by chance. The alignment score S is calculated by summing the scores for each aligned position and the scores for gaps. Gap scores are always negative, whereas the presence of the gap is more penalized than the length of the gap. The significance of the alignment is represented by P values and E values. The P value of an alignment score S is the probability that a score of at least S would have been obtained from comparisons of any unrelated sequences of the same length and composition as the query sequence. This means that the most significant matches are identified at P values close to 0. The E value, expectation value, is the number of different alignments with scores equivalent to or better than S that one should expect to find in a database search merely by chance. Thus, the E value can be any positive number, and more significant scores have lower E values.

Currently, BLAST2 is a principal tool for sequence alignments (http://blast.ncbi.nlm.nih.gov/Blast.cgi). As mentioned, the BLOSUM62 is the default substitution matrix in BLAST2. The BLAST has three programs for searching nucleotide

TABLE 18.1

BLAST Programs

BLAST Program	Compares
BLASTN	A nucleotide query sequence against a nucleotide sequence database
BLASTP	An amino acid query sequence against a protein sequence database
BLASTX	A nucleotide query sequence translated in all six reading frames against a protein sequence database
TBLASTN	A protein query sequence against a nucleotide sequence database dynamically translated in all six reading frames
TBLASTX	The six-frame translations of a nucleotide query sequence against the six-frame translations of a nucleotide sequence database

sequences similar to the query sequence and two programs that work for searching sequences in protein databases (Chapter 18, Table 18.1). As discussed, protein, rather than DNA, comparisons should be employed when searching for evolutionarily related sequences. An example of the BLAST search result is shown in Figure 18.1.

18.2.4.3 PSI-BLAST

Although BLAST is usually the first choice program when sequence similarity search is performed, pairwise sequence comparisons often fail to identify distantly related homologs. PSI-BLAST (Position-Specific-Iterative BLAST) is an iterative search that uses the BLAST algorithm to identify distant relatives of a protein. The

```
sp|P20856|MYG_CTEGU  Myoglobin
Length=154

Score = 41.2 bits (95),  Expect = 1e-06, Method: Compositional matrix adjust.
 Identities = 40/149 (26%), Positives = 58/149 (38%), Gaps = 22/149 (14%)

Query   1    MVLSPADKTNVKAAWGKVGAHAGEYGAEALERHF------------------DLSHGSA   41
             M LS  +   V   AWGKV     G +G E L R F                     D   S
Sbjct   1    MGLSDGEWQLVLNAWGKVETDIGGHGQEVLIRLFKGHPETLEKFDKFKHLKSEDEMKASE   60

Query   42   QVKGHGKKVADALTNAVAHVDDMPNALSALSDLHA--HKLRVDPVNFKLSHCLLVTLAAH   99
             +K HG   V  AL N +          L+ L+  HA  HK+ V  + F +S  ++   L +
Sbjct   61   DLKKHGTTVLTALGNILKKKGQHEAELAPLAQSHATKHHKIPVKYLEF-ISEAIIQVLESK   119

Query   100  LPAEFTPAVHASLDKFLASVSTVLTSKYR   128
              P +F      ++ K L      + +KY+
Sbjct   120  HPGDFGADAQGAMSKALELFRNDIAAKYK   148
```

FIGURE 18.1 An alignment of human α-globin (query sequence) and myoglobin (subject sequence) created by the BLAST search. The search was performed using the BLASTP program with human α-globin as a query sequence. Identical residues between the two sequences are listed in the middle, and conserved residues are represented by plus signs. Gaps are represented as dashes within the sequences. Note that, the sequences are different; despite that, both proteins are oxygen carriers (see Chapter 19 for further discussion).

program first performs a standard BLAST search of a protein query against a protein database. It then constructs a multiple sequence alignment, and from any significant local alignments, a profile called position-specific scoring matrix (PSSM), is created. The PSSM is generated by calculating position-specific scores for each position in the alignment. More conserved positions have high scores, and scores for less conserved positions are near 0. The generated profile is used for the second round of searching in the protein database, looking again for local alignments. A new multiple sequence alignment is constructed to create a profile that is used for the next search iteration. The process can be repeated for a defined number of cycles or until no more statistically significant sequences are detected. The PSI-BLAST is fast and identifies up to three times more related sequences than the standard BLAST. Its major methodological advance is the construction of PSSMs, which are more powerful than regular substitution matrices such as BLOSUMs and PAMs.

18.3   FUNCTION PREDICTION

The genomics era has provided us with a large number of potential protein-coding sequences, but the function of the majority of proteins is unknown. The quickest and simplest way to assign a probable function to a newly synthesized DNA or cDNA is to search for sequence homologs that have functions assigned already. The assumption is that the conserved sequence often indicates the conserved function. Although such analysis usually does not reveal the precise physiological function of the novel gene, it can often predict its biochemical function. For example, a match to a protein kinase gene reveals that the biochemical function of the newly identified gene is phosphorylation of target proteins, but its role in the cellular or whole-organism level remains unclear. Therefore, further experiments are needed for understanding the physiological roles of such genes with predicted biochemical function, but the information that the gene codes for a protein kinase is helpful for designing experimental strategies for studying its cellular role.

Although sequence alignment tools such as BLAST and FASTA are very powerful for pairwise comparisons, they are less effective for the identification of distantly but still functionally related proteins. Profile-based methods such as PSI-BLAST and RPS-BLAST (Reverse PSI-BLAST, opposite of the PSI-BLAST searching a query sequence against a database of profiles) can identify the relationships between protein families that are more distantly related. Therefore, these methods can be effective for the functional annotation of proteins and should be used when studying distant evolutionary relationships.

In addition to sequence comparisons, structures can be used for functional annotation because the primary structure of a protein is less conserved than the tree-dimensional structure. Furthermore, the three-dimensional structure of a protein is much more closely related to its function than the amino-acid sequence. Thus, structural comparisons can be used for protein function predictions. We return to this topic in Chapter 19.

FURTHER READING

1. Korf, I. et al. 2003, BLAST: An Essential Guide to the Basic Local Alignment Search Tool, O'Really & Associates, Inc., Sebastopol, CA.
2. Sharma, K. L. 2008, Bioinformatics: Sequence Alignment and Markov Models, McGraw-Hill, New York.
3. Dear, P. H. 2007, Bioinformatics (Methods Express), Cold Spring Harbor Laboratory Press, Cold Spring Harbor, New York.

19 Protein Structure Prediction

Tiit Land

CONTENTS

Determination of the three-dimensional structure of a protein can be considered as one of the most important goals in biochemistry. A variety of experimental methods exist that can be used for analysis of protein structures (described in part II of this book). Most of these techniques are slow, laborious, and expensive. At the same time, massive amounts of protein sequence data are produced by large-scale genome-sequencing projects, and knowing their three-dimensional structures could be very helpful for functional annotation. As we discussed in the Chapter 18, putative functions can sometimes be assigned to newly identified genes by sequence comparison methods. However, the three-dimensional structure of a protein rather than its amino-acid sequence carries out the biochemical function of a protein. Therefore, a tentative function can be assigned to a hypothetical protein by comparing its structure to previously solved structures of proteins with assigned functions. As an example, the importance of structural similarities between two functionally related proteins can be seen by comparing hemoglobin and myoglobin, which are both oxygen carriers. The solved three-dimensional structures of the human α-globin and myoglobin are very similar, but their amino-acid sequences are different (26% identity, 38% similarity; see Chapter 18, Figure 18.1).

One way to assign a tentative function to a newly identified gene is to express the gene in bacteria, purify the encoded protein, and examine its structure by using methods for protein structure analysis such as x-ray crystallography and nuclear magnetic resonance (NMR) spectroscopy. Both methods are notoriously slow and laborious and, further, require milligram amounts of very pure proteins. Therefore, computational approaches have been developed that predict protein structures and may provide some clues to functions of proteins even in the absence of detectable homologs. Protein structure prediction is aimed at the prediction of the three-dimensional structure of a protein from its amino-acid sequence. In recent years, it has become one of the ultimate goals in bioinformatics and theoretical chemistry, and has high importance for biotechnology and medicine.

In this chapter, we briefly describe methods for prediction of subcellular localization and structural features of proteins.

19.1 PREDICTION OF SUBCELLULAR LOCALIZATION OF PROTEINS

Most eukaryotic proteins are encoded in the nuclear genome and synthesized in the cytosol. In order to perform their functions, proteins must be localized at their appropriate subcellular compartments. In addition, determining protein subcellular localization is crucial for genome annotation and drug discovery. Important protein-sorting pathways are described in detail in chapter 10 of this book. Experimental methods for protein localization determination such as immunolocalization, and tagging proteins with green fluorescent protein, are accurate but slow and laborious. Several computational tools have been developed that provide fast and often accurate subcellular localization predictions for many organisms.

19.1.1 SIGNAL PEPTIDES

Signal peptides target proteins to the extracellular environment either directly through the plasma membrane in prokaryotes or through the endoplasmic reticulum in eukaryotes. The signal peptide is cleaved off from the resulting mature protein during the translocation process across the membrane. A popular program for signal peptide prediction is the SignalP that is developed from studies of biochemical properties of signal peptides. This program is available at the Web site of the Technical University of Denmark (http://www.cbs.dtu.dk/services/ SignalP). The SignalP predicts the presence and the location of cleavage sites of signal peptides from eukaryotes, Gram-negative bacteria, and Gram-positive bacteria. The original SignalP method was based on artificial neural networks for prediction of signal peptides and their cleavage sites (SignalP-NN). The current SignalP 3.0 program includes another prediction method based on hidden Markov models (SignalP-HMM). The user can choose to run either Signal-NN or Signal-HMM, or both.

19.1.2 TRANSMEMBRANE SEGMENTS

There are many methods for predicting transmembrane α-helices in proteins. Early methods used simple hydrophobicity scales searching for ~20 amino-acid stretches

TABLE 19.1

Programs for Prediction of Transmembrane Segments of Proteins

Program	Description, Web site
TMHMM	Prediction of transmembrane helices in proteins using Hidden Markov Model http://www.cbs.dtu.dk/services/TMHMM-2.0/
TopPred	Topology prediction of membrane proteins based on hydrophobicity analysis http://mobyle.pasteur.fr/cgi-bin/MobylePortal/portal.py?form=toppred
Tmpred	Prediction of transmembrane regions and protein orientation http://www .ch.embnet.org/software/TMPRED_form.html
HMMTOP	Prediction of transmembrane helices and topology of proteins using Hidden Markov Model http://www.enzim.hu/hmmtop/
PredictProtein	Advanced service for sequence analysis, structure, and function prediction http://www.predictprotein.org/
DAS	Prediction of transmembrane regions in prokaryotes based on the Dense Alignment Surface method http://www.sbc.su.se/~miklos/DAS/

with a predominance of hydrophobic amino-acid residues. Later methods include the prediction of membrane topology of the protein (which parts of the protein are located on the inside or on the outside of the membrane) using the "positive-inside" rule (described in Chapter 24 of this book), and more recent methods use different pattern-matching methods to search transmembrane helices based on experimentally determined transmembrane segments. Some of the widely used programs for predicting the location of transmembrane helices within a membrane protein are listed in Table 19.1.

19.1.3 PROTEIN TARGETING

Several computational tools for predicting protein localization from sequence data have been developed. The first widely used program for analysis of protein targeting in prokaryotes and eukaryotes was PSORT, developed by Kenta Nakai at the University of Tokyo. The PSORT (http://psort.ims.u-tokyo.ac.jp/) analyzes the input amino-acid-sequence based on known sorting signal motifs and predicts protein targeting to specific localizations by reporting the possibility of the protein sequence to be localized at particular sites. Recently, WOLF PSORT program (http://wolfpsort.org/) has been developed, which is an extension of PSORT II used for the analysis of animal, plant, and fungi sequences. In addition to known protein-sorting signals, WOLF PSORT makes predictions based on other sequence features such as amino-acid composition of proteins. A comprehensive list of subcellular localization prediction programs for both prokaryotes and eukaryotes is available at the PSORT.org Web site (http://www.psort.org/).

19.2 PREDICTION OF SECONDARY STRUCTURE

While the prediction of the secondary structure of a protein gives little insight into its function, it can be helpful for aligning distantly related proteins since protein evolution tends to proceed through insertions and deletions in less structured parts

TABLE 19.2

Selected Software Tools for Protein Secondary Structure Prediction

Program	Web site
PSIPRED	http://bioinf.cs.ucl.ac.uk/psipred/
JPRED	http://www.compbio.dundee.ac.uk/jpred_v2/
PredictProtein	http://www.predictprotein.org/
Predator	http://bioweb.pasteur.fr/seqanal/structure/
DSC	http://bioweb2.pasteur.fr/structure/
ZPRED	http://kestrel.ludwig.ucl.ac.uk/zpred.html

of polypeptide chains, whereas secondary structure elements are more conserved. Further, a reliable prediction of secondary structure elements within a protein can be used to generate a folded structure by packing together these elements. First, sequence-based secondary structure prediction methods were developed in the 1970s when relatively few three-dimensional structures were available. One of the early algorithms for secondary structure prediction was created by Chou and Fasman, which was derived from databases of known three-dimensional structures by calculating propensities of each amino acid residues to be found in helices, strands, and turns. Despite suffering from a lack of structural data, the Chou–Fasman algorithm has an accuracy rate of ~50%–60%. The main limitations of this algorithm is that it considers only local interactions, neglecting long-range orders, and makes no distinction between types of helices, types of turns, or orientation of β-strands. In addition, early predictions were performed on single sequences rather than on families of homologous sequences.

Recent methods for secondary structure predictions combine available structural data with sophisticated computational techniques such as neural networks, and achieve a 70%–75% accuracy. There are currently many software tools for secondary structure prediction (see Table 19.2). Several programs take a multiple alignment of protein sequences as an input, which increases prediction accuracy. Some programs accept a single sequence as an input, but first run PSI-BLAST and use the multiple alignment constructed by PSI-BLAST for secondary structure prediction.

19.3   PREDICTION OF THREE-DIMENSIONAL STRUCTURE

Predicting three-dimensional protein structures from only sequence information is a great challenge in theoretical chemistry and computational structural biology. The first prerequisite for predicting protein structures is understanding the physical basis of protein structural stability, which makes structure prediction a very difficult task. Second, we have to consider that the number of possible protein structures can be large. Nevertheless, any method that allows passing the determination of a protein structure experimentally by using x-ray crystallography or NMR is of high importance for biotechnology and is particularly attractive to the pharmaceutical industry.

The reliability of structural modeling methods is assessed every 2 years in the critical assessment of structural prediction (CASP) experiment in which a large number of structural predictions is rigorously compared with protein structures solved experimentally.

19.3.1 COMPARATIVE MODELING

Comparative modeling (also called homology modeling) can be used to predict protein structures when the sequence with unknown structure shows >25% identity to another sequence with known structure. This method is based on the assumption that two homologous proteins have similar structures. Further, since the three-dimensional structure of a protein is much more strongly conserved in evolutionary terms than the primary structure, a target sequence can be modeled on a relatively distantly related template. The first and critical step in homology modeling is the identification of the template. First, sequence alignment is carried out with the target protein and with one or more templates using either pairwise or multiple sequence alignment algorithms. Then the target is structurally aligned on the template. Regions of the target sequence that do not align on the template can be attempted to be modeled by loop modeling. However, the accuracy of regions that are predicted by loop modeling is much less than those of regions that are predicted using template-based modeling. The accuracy of modeling decreases with the number of amino acids in the loop, and the most accurate predictions are considered for loops of less than 8 amino acids.

MODELLER is a popular software tool for homology modeling. SWISS-MODEL (http://swissmodel.expasy.org/) is a fully automated homology modeling Web server.

19.3.2 THREADING

When the sequence identity between the target and template proteins is too low for comparative modeling, a method called protein threading or fold recognition can be used. The method is based on the observation that the tertiary structure of a protein is evolutionarily more conserved than the primary structure, and insertions and deletions occur during the evolution mostly in loop regions without affecting structural regions.

Briefly, this approach scans the target sequence against a library of known structural templates, and the template structure that is best compatible with the target sequence is selected. In other words, the sequence is "threaded" through a number of structures, and the method determines which structure fits best. To find the best fit, a series of scores is calculated, and the highest score is assumed to correspond to the structure that is adopted by the target sequence. The scoring system is based on finding the structure with the lowest energy. In that respect, the threading method shares some characteristics with the ab initio prediction method (see the following text). On the other hand, sequence alignment moment is a characteristic that threading shares with the comparative modeling method.

Some of the current threading tools are listed in Table 19.3.

TABLE 19.3
Selected List of Protein-Threading Tools

Program	Web site
Hhpred	http://toolkit.tuebingen.mpg.de/hhpred
3D-PSSM	http://www.sbg.bio.ic.ac.uk/~3dpssm/
Fugue	http://tardis.nibio.go.jp/fugue/
SAM-T02	http://www.soe.ucsc.edu/research/compbio/ HMM-apps/T02-query.html
Threader	http://bioinf.cs.ucl.ac.uk/threader/threader.html

19.3.3 AB INITIO MODELING

Ab initio protein modeling techniques ignore sequence homology and attempt to build three-dimensional protein models based on energetical or physicochemical properties of amino-acid residues of a protein. Thus, in contrast to homology modeling and threading techniques, ab initio modeling method does not compare the target sequence with previously solved structures. Ab initio modeling is based on the assumption that the protein conformation with the lowest energy corresponds to the native conformation of a protein.

Ab initio modeling methods require high computational power and long computing time, and the prediction results are much less reliable than those obtained by comparative modeling and threading techniques. Some recent hierarchic approaches that first build local structures and then more global structures is assembled, and seem to provide more accurate structure predictions. An alternative approach is a mini-threading method in which the matches between short fragments from resolved protein structures and the target sequence are used to build local structures, which are then employed in assembling the tertiary structure of a protein.

FURTHER READING

1. Branden, C. and Tooze, J. 1999, Introduction to Protein Structure, 2nd edition, Garland Science Publishing, New York.
2. Dear, P. H. 2007, Bioinformatics (Methods Express), Cold Spring Harbor Laboratory Press, Cold Spring Harbor, NY.
3. Lesk, A. 2008, Introduction to Bioinformatics, 3rd edition, Oxford University Press, New York.

20 Proteomics

Sherry Niessen and Benjamin F. Cravatt

CONTENTS

Proteomics is a large-scale initiative directed toward the discovery of the function of all proteins that are present in an organism. This postgenomic science aims to reveal protein function through the investigation of protein expression patterns, activities, interaction partners, and subcellular distributions. This is no small task, given that the proteome of a cell is not only under dynamic, spatial, and temporal control, but proteins themselves are posttranslationally regulated through molecular interactions and/or covalent modifications such as phosphorylation and glycosylation. Such cellular events are often not apparent from genomic information or mRNA abundance, highlighting the importance of proteomic information for a more complete understanding of biological processes.

One of the most challenging technical tasks addressed by proteomics is revealing significant differences in protein content that exist between two or more biological states. Identifying proteins that meet these criteria can then be followed up by functional experiments to examine their contribution to the biological process under analysis. For example, the field of cancer proteomics aims to identify proteins that are markers of malignancy or specific stage of disease progression as they represent potential diagnostic and therapeutic targets. This requires the use of technologies that are accurate, sensitive, reproducible, and provide a comprehensive analysis of protein content in biological samples; overcoming issues such as the high dynamic range of cellular protein expression, and the large differences in the physicochemical properties of proteins remains challenging. It has been estimated that the range in protein copy number is 7-8 orders of magnitude in human cells and up to 12 orders of magnitude in human serum samples. Toward the goal of functionally characterizing the proteome, several advanced technologies have been developed. These include methods such as differential protein gel staining (e.g., two-dimensional gel electrophoresis [2DE]), protein microarrays, mass-spectrometry (MS)-based profiling, and activity-based protein profiling (ABPP). Each of these proteomic methods will be discussed in turn.

20.1   TWO-DIMENSIONAL GEL ELECTROPHORESIS (2DE)

20.1.1   Overview

Two-dimensional gel electrophoresis (2DE) was developed in the late-1960s as a method for enhanced protein separation. Since that time the method has rapidly evolved with ever-increasing protein separation capabilities and protocols for protein detection. In 2DE, proteins are separated in two dimensions based on the intrinsic properties of each protein, which are then stained for visualization and quantification (Figure 20.1). By comparing the intensities of the stained protein spots, their relative amounts are estimated. Protein spots of interest that display differential levels or modified migration patterns in a biological comparison are excised from the gel, digested into peptides, and analyzed by MS to obtain protein identification.

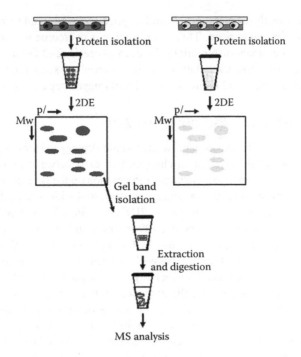

FIGURE 20.1 Two-dimensional gel electrophoresis (2DE). Proteins are separated on the basis of pI and molecular weight (Mw) and then stained for visualization and quantification. By comparing the intensities of the stained spots, the relative level of each protein is obtained. Protein spots that display differential levels or migration patterns are excised from the gel, digested into peptides, and analyzed by MS to obtain protein identification.

20.1.2   GENERAL METHOD OF 2DE

2DE achieves enhanced protein separation by combining two electrophoretic procedures, isoelectric focusing (IEF) and SDS–polyacrylamide gel electrophoresis (SDS-PAGE). In the first-dimension gel, proteins are separated according to their isoelectric point (pI) using immobilized pH gradient (IPG) strips. Proteins applied to IPG strips will move along the gel and accumulate at their isoelectric point; that is, the point at which their overall charge is neutral. This gel is then applied to the top of an SDS-slab gel and electrophoresed. The proteins in the first-dimension gel migrate into the second-dimension gel where they are separated on the basis of their molecular weight.

Silver and Coomassie Brilliant Blue stains are two historical methods for in-gel protein detection. However, these stains are not ideal as they display poor detection sensitivity (Coomassie Brilliant Blue), result in diminished peptide recovery from the gel (Silver), and have an overall limited detection range (Silver and Coomassie Brilliant Blue). As such, these methods have been largely replaced by fluorescent stains, including several from the SYPRO family. To detect abundance differences or changes in migration patterns between two samples, their stained images are compared. Because of significant variations and irreproducibilities between gels, no two gel images are directly superimposable and require computational software packages to overlay them for comparison. This complication is overcome with the use of fluorescence two-dimensional differential gel electrophoresis (2-D DIGE). In 2-D DIGE, each protein sample is covalently labeled with a fluorophore such as Cy2, Cy3, or Cy5, mixed together, and resolved on the same gel and imaged separately for comparison.

20.1.3 CHARACTERIZING PROTEINS USING 2DE

2DE is most widely used as a comparative proteomic strategy to identify protein expression or posttranslational modification changes that exist between two systems under analysis. As this method provides no direct information on the type of posttranslational modification, the position of this modification, or even the identity of the protein, it is often combined with MS for these purposes. 2DE in combination with MS has been successfully used to identify proteins potentially involved in biological processes, including cancer. In a recent breast cancer study by Ou and colleagues, 2DE identified more than 80 overexpressed and 30 underexpressed proteins in aggressive breast cancer cell lines compared to their nonaggressive counterparts, a small number of which were validated by additional approaches. A 2DE study performed by Shen, Tian, and colleagues comparing primary pancreatic cancers and corresponding normal tissues resulted in the identification of several proteins with altered levels including protein disulphide isomerase and galectin-1. Further functional studies are required to determine whether such proteins may represent useful biomarkers and therapeutic targets.

20.1.4 LIMITATIONS OF 2DE

There are many intrinsic limitations of 2DE that have resulted in its being superseded by other proteomic technologies. First, on average, only 3000 spots can be separated and stained in a typical 2DE experiment. As a result the comigration of proteins in the same spot is not uncommon, making accurate quantification by fluorescence challenging. For example, Gygi and colleagues identified six different proteins in the same faint spot. Second, 2DE has a small detection range and is therefore unsuitable to analyze low abundance proteins. Third, many proteins, including membrane bound and nuclear, display very poor solubility in the buffers used during IEF leading to their absence in the analysis. The addition of detergents such as Trition-X114 and ASB14 can partially help in the detection of some, but not all, membrane proteins. Finally, the range of protein pIs and molecular masses found in nature exceed what is normally analyzed by 2DE, therefore excluding small, large, acidic, and basic proteins from analysis.

20.2 PROTEIN MICROARRAYS

20.2.1 OVERVIEW

Protein microarrays provide a second basis for characterizing protein function. There are two main types of protein microarrays that are currently in use: analytical microarrays and functional microarrays (Figure 20.2). Analytical microarrays are high-density arrays of affinity reagents such as antibodies or aptamers that detect proteins in a complex mixture. These types of arrays are used to profile protein

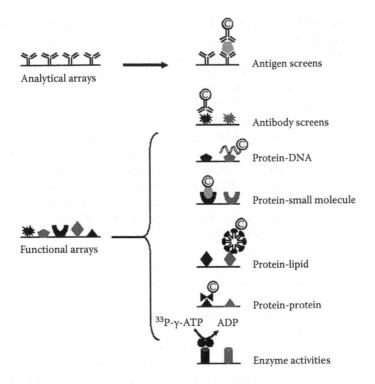

FIGURE 20.2 Use of microarrays in defining protein function. There are two main types of protein microarrays: analytical arrays and functional arrays. Analytical arrays screen for protein interactions with affinity reagents such as antibodies. Functional arrays composed of proteins or protein domains can be used to screen for a wide variety of biochemical functions, including antibodies screens, enzymatic activities, and biomolecular interactions. In this figure, (c) denotes a biomolecule labeled with a fluorescent dye for visualization and quantification. (Adapted from Hall, D. A., Ptacek, J. and Snyder, M. (2007). Protein microarray technology. Mech. Ageing Dev. 128, 161–167. With permission.)

expression patterns across biological samples. Functional microarrays are composed of a large number of full-length functional proteins or protein domains. They are used to study the biochemical activity of an entire proteome by examining antibody interactions, enzymatic activities, and biomolecular interactions. Analytical and functional protein arrays are generated by printing antibodies/proteins onto a solid surface in a known ordered pattern providing specific binding sites for target molecules (e.g., antigens, proteins, lipids). The interaction between the protein on the slide and its target is then visualized and quantified by various detection methods.

20.2.2   GENERAL METHOD OF PROTEIN MICROARRAYS

To perform reproducible and reliable assays using microarrays, it is important to immobilize surface antibodies/proteins in such a way that they are still folded and functional. Several types of slides are commonly used to print proteins including glass slides with or without a polymer coating, as well as nitrocellulose membranes.

Proteins are immobilized to these surfaces through adsorption (poly-L-Lysine), covalent linking (aminosilane, epoxysilane, or aldehyde), affinity attachment (nickel, avidin, protein A/G), or physical entrapment (agrose, polyacrylamide, or supermolecular hydrogel). One of the best ways to maintain proteins in their native conformation is through affinity attachment, where proteins are fused with an epitope tag and then linked to the surface of the slide through this tag. Both label and label-free methods are available for the purposes of detecting and quantifying protein binding to the array. One example of a label-free approach involves surface plasmon resonance (SPR). For SPR, proteins are spotted on to a gold surface, which is then incubated with the target proteome mixture. The interaction between a target protein and the slide is revealed by changes in the refractive index giving an SPR signal. Advantages of the SPR method include its ability to measure the amount of bound target protein and determine affinity constants and kinetic data. A second label-free method involves performing a sandwich ELISA. This approach, commonly used with analytical microarrays, employs a second tagged antibody (fluorescence or chemiluminescence) to recognize the antigen of interest. Although this method is very sensitive, it requires that two antibodies that recognize different epitopes are available for each antigen. Moreover, it limits the number of proteins that can be assessed at the same time due to antibody cross reactivities. The most commonly used reagents for covalently labeling of proteins are fluorescent dyes like Cy3 or Cy5. Cell extracts or individual biomolecules are labeled with the covalent tag, applied to the protein array, and spots are visualized by fluorescence intensity. One of the main advantages of direct protein modification is the ability to analyze hundreds of target proteins on the same slide.

20.2.3 CHARACTERIZING PROTEINS USING MICROARRAYS

Protein microarrays have been used to screen proteome-wide expression patterns, protein posttranslational modifications, detection of circulating antibodies, enzyme activities, and protein-biomolecular interaction patterns. Zhu and colleagues developed a functional protein microarray that represents 5800 of the known yeast proteins and applied this array to the analysis of protein–protein and protein–lipid interactions. Its potential to identify protein–protein interactions was demonstrated using a calmodulin probe through the detection of six known and 33 novel calmodulin-binding proteins. When applied to the identification of phosphoinositide-binding proteins, 150 novel lipid-binding proteins were identified, of which, 45 were membrane associated. In addition, Hudson and colleagues recently generated a human functional protein microarray containing over 5000 purified proteins from insect cells to screen autoantibodies from patients with ovarian cancer and identified 94 proteins that displayed enhanced reactivity to cancer patient sera. Microarrays are also used to examine enzyme activities, including kinases, serine hydrolases, and metalloproteinases. To assess the activity of clinically relevant serine hydrolases and metalloproteinases, an analytical microarray platform for activity-based protein profiling (ABPP) (Section 20.4) was introduced by Sieber and colleagues. In this approach, proteomes are treated with tagged chemical probes that selectively react with active target enzymes in solution, and then the labeled enzymes are captured and visualized on glass slides displaying an array of antibodies. This method permitted the analysis of prostrate specific antigen (PSA), a clinical prostrate cancer biomarker, in the range of the endogenous levels of this enzyme. Finally, microarrays are applied to evaluate protein expression patterns especially in human cancer cells and tissues. Studies have identified proteins displaying differential expression in malignant samples such as eIF-4E, MAPK7, annexin XI, cytokines such as IL-8, and several proteins involved in resistance to chemotherapeutic agents.

20.2.4 LIMITATIONS OF PROTEIN MICROARRAYS

One of the major limitations of protein microarrays is preparation of functional folded proteins and antibodies to apply to the array for screening. Misfolded proteins or inaccessible protein binding domains may lead to false-negative or false-positive results. Moreover, given the complexity of the human proteome (posttranslational modifications, proteolytic events, etc.), a complete functional analysis will require larger numbers of antibodies and functional purified proteins than are currently available. One interesting option to achieve enhanced proteome wide coverage includes the use of cell-free *in situ* transcription and translation systems of DNA to form the protein array. A second major limitation is that proteins/enzymes normally operate as parts of larger networks that may be lost in this artificial noncellular environment.

20.3 MASS SPECTROMETRY (MS)-BASED APPROACHES

20.3.1 OVERVIEW

MS-based approaches characterize protein function on a global scale through in-depth identification and quantification. In combination with other biochemical techniques, including affinity enrichment, cellular fraction, or biological perturbation, MS can offer a valuable approach to discovery a protein's role in biological systems. Most MS-based approaches involve the analysis of peptides generated through the enzymatic digestion of protein samples to obtain the identity and quantity of the parental proteins in the proteome. Given that there are many types of MS instruments, for simplicity we will use the example where peptides are evaluated with an ion trap mass analyzer through an electrospray device. Figure 20.3 illustrates two approaches that are commonly used in MS-based proteomics for protein identification. In Scheme A, a proteome is resolved by one-dimensional gel electrophoresis (1DE), stained, and sliced into sections. Proteins from each section are enzymatically digested into peptides and extracted. Prior to MS analysis, peptides are fractionated by high-resolution liquid chromatography (LC) using a single-phase column composed of a C18 reverse phase (RP) resin for separating peptides based on their hydrophobic character. In Scheme B, the entire proteome is digested and analyzed by multidimensional protein identification technology (MudPIT). As this mixture of peptides is very complex, it is best resolved using LC in combination with a biphasic column. The biphasic column integrates a strong cation exchange (SCX) and a C18 resin. Peptides interact with the SCX resin, as they are positively charged and are eluted as batches into the C18 resin by a mobile phase containing ammonium acetate. Similar to the single-phase column, peptides are eluted from the C18 resin into the MS with an organic gradient. If the proteome under analysis is complex, then a more gradual increase in ammonium acetate concentration can be used to elute smaller amounts of

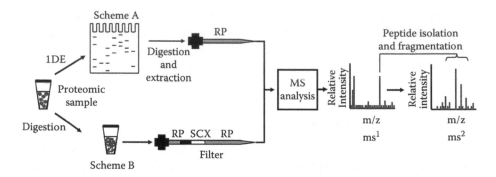

FIGURE 20.3 Overview of MS-based approaches for protein identification. In Scheme A, proteins are separated by 1DE, stained, and sliced into sections. Proteins from each section are enzymatically digested into peptides and extracted. In Scheme B, the entire proteome is digested and analyzed by multidimensional protein identification technology (MudPIT). Peptides are fractionated by liquid chromatography using either a single phase or a biphasic column. Eluted peptides are ionized and transported into the mass analyzer. In a typical data-dependent MS experiment, a repeating cycle is performed that consists of a full survey scan (ms^1), followed by a number of product ion scans where peptides are isolated for tandem mass spectrum (ms^2) generation for peptide sequencing and identification.

peptides from the SCX resin into the C18 resin for better separation. Eluted peptides are ionized by electrospray ionization (ESI) and transported into the mass analyzer. In a typical data-dependent MS experiment, a repeating cycle is performed that consists of a full survey scan (ms^1) followed by a number of product ion scans where peptides are isolated for tandem mass spectrum (ms^2) generation that permits peptide sequencing and identification. In data-dependent acquisition, computational control directs specific peptide ion isolation based on their mass to charge (m/z) ratio and relative peptide abundance in the total ionized mixture. The selected ions are typically fragmented by collision-induced dissociated (CID) to generate an ms^2 spectrum, although a number of alternative fragmentation techniques are available. For example, electron transfer dissociation (ETD) is a useful fragmentation method for posttranslational modification studies. Peptide identifications are yielded from the ms^2 spectrum by searching each against a protein database using a search algorithm such as SEQUEST or Mascot. One key development in using MS to define protein function is through its ability to be performed in a quantitative fashion. To achieve protein quantification by MS, several approaches are available. They are broadly divided into two classes—those that employ some sort of differential stable isotope for the generation of a specific mass tag and those that are label free.

20.3.2 USE OF STABLE ISOTOPES FOR PROTEIN QUANTIFICATION

20.3.2.1 Metabolic Labeling Strategies: Stable Isotope Labeling by Amino Acids in Cell Culture (SILAC)

Mass tags can be introduced *in vivo* metabolically or *in vitro* by chemical or enzymatic steps and are generally employed to compare relative levels of proteins between biological samples. One of the most widely used metabolic labeling strategies is stable isotope labeling by amino acids in cell culture (SILAC). In this method, isotopically labeled amino acids are used as supplements in dialyzed growth medium and are incorporated into proteins during normal cell growth and protein turnover. Protein labeling can be achieved in excess of 95% in five to eight cell passages. Several isotopically labeled amino acids are available, allowing the comparison of up to five biological states that contain ^2H instead of ^1H, ^{13}C instead of ^{12}C, or ^{15}N instead of ^{14}N. $^{13}C_6$-arginine and $^{13}C_6$-lysine (bearing six ^{13}C atoms) are the most common employed isotopically label amino acids (either alone or together), as their use ensures that all tryptic peptides from a protein carry at least one labeled amino acid (except the C-terminal peptide). Although ^2H-labeled amino acids are available, they should be used with caution due to changes in retention time characteristics during reverse phase chromatography of the labeled versus nonlabel peptides.

The basic SILAC procedure involves growing one population of cells in a media containing the natural form of the amino acid ("light") and the second population in a media containing a stable isotope-labeled analogue ("heavy") (Figure 20.4). The light and heavy cells are mixed in equal ratios and prepared for MS analysis, which ensures that errors introduced during biochemical preparation and MS analysis affect both populations of cells equally. The digested peptides from the labeled and unlabeled cells coelute during LC-MS analysis and are distinguished from each other by the mass difference introduced by the labeled amino acids. Relative protein quantification is determined from the ms^1 survey scans by comparing the relative signal intensity peak areas of the heavy and light peptides. Most available computational software programs select isotopic peptide pairs for extraction based on successful ms^2 peptide identification. The ratios of all peptides from each individual protein are then averaged to give a protein ratio. In addition, as ms^2 spectra are normally gathered for both the light and heavy peptides, the introduction of the heavy amino acid also reveals structural information that increases confidence in protein identification.

20.3.2.2 *In Vitro* Labeling Strategies: Isotope-Coded Affinity Tagging (ICAT) and Isobaric Tags for Relative and Absolute Quantification (iTRAQ)

A prototypical strategy for *in vitro* chemical labeling is isotope-coded affinity tagging (ICAT) developed by Gygi and colleagues. The first-generation ICAT reagent consisted of an iodoacetamide group that reacts with the thiol functionality of cysteine residues, a linker that contains heavy or light isotopes, and a biotin affinity tag. The biotin affinity tag facilitates the purification of labeled peptides using avidin. Several second-generation ICAT tags have been described to overcome issues concerning tag fragmentation during ms^2 generation. For example, an acid-cleavable moiety was introduced between the isotope linker and the biotin group so that the affinity handle could be cleaved off before MS analysis. In ICAT, two biological states are compared by labeling one protein sample with the isotopically light reagent and the second protein sample with the isotopically heavy reagent (Figure 20.5). Labeled proteins are combined, subjected to enzymatic digestion, and labeled peptides are enriched by

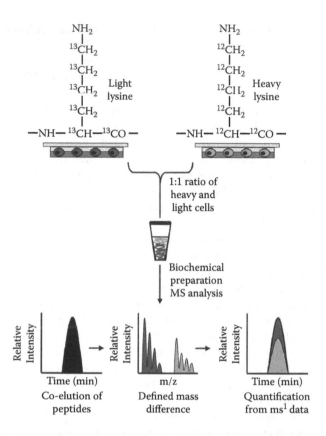

FIGURE 20.4 Stable isotope labeling by amino acids in cell culture (SILAC). One population of cells is grown in a media containing the natural form of the amino acid ("light") and the second population in a media containing a stable isotope-labeled analogue ("heavy"). The light and heavy cells are mixed in equal ratios and prepared for MS analysis. The digested peptides from the labeled and unlabeled cells coelute during LC-MS analysis and are distinguished from each other by the mass difference introduced by the labeled amino acids. Relative protein quantification is determined from the ms^1 survey scans by comparing the relative signal intensities peak areas of the heavy and light peptides.

avidin chromatography and analyzed by LC-MS. The procedure of enriching tagged cysteine peptides leads to a significant reduction in the complexity of the sample for MS analysis, which can increase the depth of proteome coverage. However, since this method involves enrichment of cysteine-containing peptides it is not entirely suited for the identification of posttranslational modifications. Similar to quantification by SILAC, isotopically labeled peptides will coelute during the LC-MS analysis and can be distinguished by the mass difference of the introduced ICAT tag. Quantification and peptide identification are determined from the ms^1 and ms^2 spectra as described for SILAC. One exception, relative protein quantification, is not based on every peptide from a protein but only those that contain cysteine residues.

One issue with the ICAT tagging and enrichment strategy is that some proteins do not contain cysteine residues (it has been estimated that one out of seven proteins does

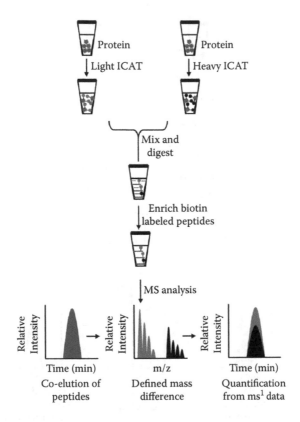

FIGURE 20.5 Isotope-coded affinity tagging (ICAT). Protein extracts are labeled with either a light or heavy ICAT tag. The light and heavy labeled proteins are mixed, digested, and the labeled peptides are enriched by avidin chromatography and analyzed by LC-MS. The digested peptides from the labeled and unlabeled cells will elute together during LC-MS analysis and are distinguished from each other by the mass difference introduced by the ICAT tag. Relative protein quantification is determined from the ms[1] survey scans by comparing the relative signal intensities peak areas of the heavy and light peptides.

not contain a cysteine residue) or have a small number of MS-compatible cysteine-containing peptides and are therefore not monitored accurately by this method. An alternative approach to ICAT is isobaric tags for relative and absolute quantification (iTRAQ) (Applied Biosystem™). iTRAQ applies chemical isobaric mass tags to label all peptides through N-terminal and lysine amine groups (Figure 20.6). Currently, eight different isobaric mass tags are available, supporting the comparative analysis of up to eight different biological samples. These isobaric mass tags consist of a reporter group, a balancer group, and a peptide reactive group. Although they introduce the same mass to each peptide they can be distinguished from each other during peptide fragmentation, as each mass tag breaks down to generate a unique reporter ion. In this approach, labeled samples are combined, analyzed by LC-MS, and, based on database searching of the fragmentation data, peptides are identified and quantified by comparing the extracted ion currents of each unique peptide tag reporter ion.

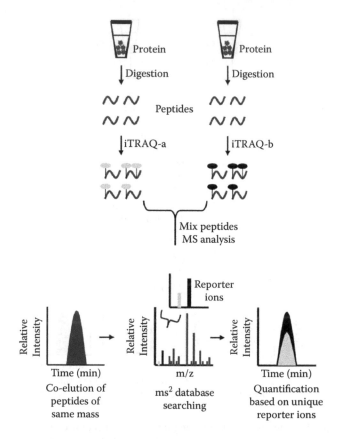

FIGURE 20.6 Isobaric tags for relative and absolute quantification (iTRAQ). Protein extracts are digested and peptides labeled with one of the different iTRAQ reagents. Labeled peptides are combined, analyzed by LC-MS where they will co-elute, and based on database searching of the fragmentation data, peptides are identified and quantified by comparing the extracted ion currents of each unique peptide tag reporter ion.

20.3.3 Characterizing Proteins Using Isotopic Labeling Strategies

Isotopic labeling strategies in combination with MS have been used for the identification of protein interacting partners, phosphotyrosine-binding proteins, protein posttranslational modifications, rates of protein synthesis/turnover, and in general quantitative protein profiling experiments. SILAC has shown particular promise as a method to identify and quantify protein posttranslational events. Kruger and colleagues used SILAC labeling in combination with phosphotyrosine immunoprecipitation to characterize the tyrosine phosphoproteome in response to insulin signaling. The authors found that the activation of insulin signaling leads to the tyrosine phosphorylation of over 40 proteins and provided temporal activation profiles for insulin-induced tyrosine phosphorylation events. In a similar study, Zhang and colleagues used iTRAQ reagents to identify and quantify the tyrosine phosphoproteome in response to epidermal growth factor receptor (EGFR) activation, where they identified 58 phosphorylated proteins and provided temporal profiles for EGF-induced tyrosine phosphorylation events. In a clever application of SILAC technology, "light" and "heavy" methionine was used to label and quantify cellular methylated proteins. The heavy methionine is converted intracellularly to S-adenosyl methionine, the cell's sole methyl donor. The labeled S-adenosyl methionine transfers the heavy methyl group to methyl acceptors proteins, which can then be identified by MS.

A second arena where isotopic labeling studies are particular effective is the large-scale analysis of protein expression patterns in biological systems. In a SILAC study of cellular membranes isolated from aggressive and nonaggressive human breast cancer cell lines, Liang and colleagues identified over 800 membrane or membrane-associated proteins. Several of these proteins displayed altered expression when comparing the aggressive to nonaggressive breast cancer cell lines, of which half were reported in the literature as having an association with cancer. In a series of ICAT papers aiming to identify protein biomarkers for pancreatic cancer, several biological sources were profiled, including pancreatic cancer juice, tumor specimens, chronic pancreatitis specimens, and normal pancreas. Authors were able to define two potential biomarkers, annexin A2 and insulin-like growth-factor-binding protein 2 (IGFBP-2), that were up-regulated in pancreatic tumor specimens compared to chronic pancreatitis specimens and normal tissue.

20.3.4   Use of Label-Free Strategies for Protein Quantification

To complement isotopic mass tagging strategies, a number of label-free quantification methods are available. These approaches offer several advantages, including increased throughput, low cost, and methodological simplicity (i.e., no additional chemistries or sample manipulations are required). Moreover, when examining biological samples, such as human tumors, the numbers of samples that need to be analyzed to detect statistically relevant protein changes far out number the available mass tags. Label-free strategies are broadly divided into two groups: those that rely on the survey scans (ms^1) for quantification and those that are based on product ion scans (ms^2) scans for quantification.

Methods have been described for determining protein abundance by directly comparing peptide peak areas of the same m/z and retention time extracted from the ms^1 data without any isotopic labels to that of to a known amount of spiked-in internal peptide standard (e.g., AQUA strategy). These methods either use peptide identification from ms^2 data to direct peptide quantification in the ms^1 data or directly use the ms^1 data alone. Direct quantification from ms^1 data requires that additional experiments are performed using a targeted ms^2 approach to identify peptides that are differentially expressed. To quantify from ms^2 data, the MS is set either in data-dependent acquisition mode or data-independent acquisition mode. In data-independent acquisition the MS sequentially directs the isolation and fragmentation of a defined m/z window throughout a mass range. In data-dependent and independent methods, peptides are identified through database searching and quantified by comparing the total ion current for all ions or for selected ions in the ms^2 spectra. Quantification from ms^2 spectra offers several benefits, including a decrease in chemical noise and increased specificity and sensitivity.

More recently, the label-free method of spectral counting has gained considerable popularity for comparative quantification of protein levels in biological samples. Spectral counting relies on the observation that there is a linear relationship between a protein's abundance in a complex proteome sample and its level of ms^2 sampling during an MS run. Using this approach, the number of ms^2 spectra that are generated during the course of an LC-MS and correspond to a particular protein are summed and compared between samples. To take into consideration the size of the protein, the number of spectral counts can be normalized by the theoretical number of peptides that are generated from the protein upon digestion (protein abundance index (PAI)).

20.3.5 CHARACTERIZING PROTEINS USING LABEL-FREE APPROACHES

Asara and colleagues recently described a data-dependent acquisition method where they compared averaged total ion currents for all ms^2 spectra that were obtained from a protein. In a screen for phosphotyrosine-binding proteins, the authors identified several known and novel interactors, including pyruvate kinase M2, a glycolytic enzyme critical for cancer cell proliferation. Further studies demonstrated the importance of this embryonic form of pyruvate kinase in cancer cell metabolism and tumor growth. Spectral counting has been successfully used to compare the relative abundance of proteins in many biological samples. For example, this method was used by Jessani and colleagues to quantify active serine hydrolase enzyme activities dysregulated in human breast tumor samples compared to normal breast tissues. Briefly, using MudPIT analysis in combination with activity-based protein profiling (see Section 20.4.), serine hydrolase enzyme profiles were obtained for breast tumor and tissue samples, demonstrating the presence of more than 50 enzyme activities. Using spectral counting, several of these enzyme activities, including KIAA1363, fibroblast-activating protein (FAP), and platelet-activating factor acetylhydrolase 2 (PAF-AH2) were identified as being highly up-regulated in a subset of aggressive breast tumors.

20.3.6 LIMITATIONS OF USING MS

In addition to the unique limitations of individual MS-based proteomic methods highlighted in their respective sections, there are also several issues that are shared by most of these methods that should be considered. First, when quantifying from ms^1 or ms^2 data, if only one or two peptides contribute to these values, then errors in quantification will have a large effect on protein abundance ratios. This problem is especially important for methods such as ICAT, which only quantify a small subset of total possible peptides (i.e., cysteine containing peptides). Second, in complex proteomic samples, coelution of the labeled or unlabeled peptides with any unrelated peptides of similar mass may lead to isotope cluster overlap, which can hinder the accuracy of quantitative measurements. Such issues will be most prevalent when whole proteomes are under analysis. The use of functional enrichment strategies can overcome these limitations by reducing the complexity of the proteome.

20.4 ACTIVITY-BASED PROTEIN PROFILING (ABPP)

20.4.1 OVERVIEW

The proteomic methods that have been described thus far are an excellent starting point to define protein function in biological systems. However, these methods typically record changes in protein abundance, which provide only an indirect assessment of protein activity. Protein activity is regulated by a plethora of posttranslational events in vivo, including protein–protein interactions, protein processing, and post–translational modification. Enzyme classes, such as proteases, are known to be tightly regulated by multiple posttranslational processes. For example, a typical protease is first translated as an inactive zymogen, (proenzyme), which requires proteolytic cleavage to become active. Second, the activity of the protease is spatially and temporarily controlled by endogenous protein inhibitors. As such, assessing protease levels in a cell using mRNA or protein abundance may hold very little relevance to the actual amount of catalytically active enzyme in the system. To address the need for higher-order functional proteomic methods, a chemical technology referred to as activity-based protein profiling (ABPP) has been introduced.

20.4.2 GENERAL METHOD OF ABPP

ABPP utilizes small molecule chemical probes to monitor the activity of many enzymes in parallel within the confines of the proteomes in which they are naturally expressed. Activity-based probes selectively label active enzymes in a proteome but not their inactive forms (e.g., zymogen or inhibitor bound) (Figure 20.7B). ABPP studies have succeeded in revealing changes in enzyme activity that do occur in the absence of similar changes in protein or mRNA levels, demonstrating the importance of this functional proteomics method. Most activity-based probes often target a large, but manageable, fraction of the enzyme proteome by exploiting shared catalytic or structural features in enzyme active sites. To date, activity-based probes have been developed for more than 20 enzyme classes, including serine-, cysteine-, aspartyl-, and metallo-hydrolases, histone deacetylases, kinases, cytochrome p450s, glycosidases, and oxidoreductases.

A typical activity-based probe consists of three chemical elements: (1) a reactive group (RG) for binding and covalently labeling the active site of a given enzyme class (or classes), (2) a linker region, and (3) a direct or latent reporter tag for the rapid and sensitive detection, enrichment, and identification of labeled enzymes from proteomes (Figure 20.7A). The probe RGs can be founded on mechanism-based inhibitors, protein-reactive natural products, or general electrophiles. A direct reporter tag can be a rhodamine group for quantification by fluorescent scanning or a biotin group for the rapid isolation and identification of labeled proteins using avidin chromatography in combination with LC-MS. Latent reporter tags, such as alkynes or azides, provide sterically benign substitutes for bulky rhodamine or biotin groups, which can be conjugated to rhodamine/biotin following the probe-labeling step using bio-orthogonal reactions such as click chemistry (CC) or the Staudinger ligation. Probes bearing latent tags often show improved affinity for their enzyme targets, as well as allowing ABPP experiments to be performed in living systems. Labeled enzyme activities are revealed through two approaches (Figure 20.7C). In the first approach,

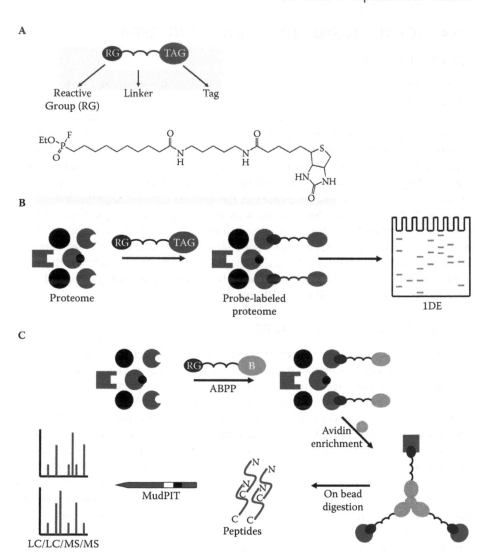

FIGURE 20.7 Activity-based protein profiling (ABPP). (A) (top) A general activity-based probe consists of a reactive group (RG), a linker region, and a reporter tag (Tag) (bottom). The fluorophosphonate activity-based probe with biotin as the reporter tag is shown as an example. (B) Activity-based probes selectively label a class/family of active enzymes (open circles versus open squares) within a given proteome, but not their inactive (zymogen [closed circles] or inhibitor-bound [open circles filled]) forms. Labeled activities can be resolved by 1DE-PAGE and revealed by fluorescence. (C) ABPP-MudPIT involves the isolation of biotin-labeled enzyme activities by avidin affinity. Enriched activities are next digested off the beads and analyzed by MudPIT technology.

1DE is used to resolve labeled enzyme activities, which, when conjugated to a fluorescent probe, can be visualized and quantified by in-gel fluorescence scanning. This approach is widely used, as it is very robust, requires minimal amounts of proteome, and has high throughput. The second approach is a gel-free MS-based method called ABPP-MudPIT, where biotin-labeled enzyme activities are enriched using avidin chromatography and then directly analyzed by MudPIT through an on-bead trypsin digestion. The ABPP-MudPIT analysis, not only reveals the identities of the probe-labeled enzymes, but also their relative quantity through spectral counting. This platform, although having enhanced resolution over the 1DE approach and providing much higher information content, requires a significant increase in the amount of proteomic material required for analysis. In an extension of the ABPP-MudPIT technology, a tobacco etch virus (TEV) protease cleavage site was introduced between the RG and the affinity handle for the selective release of probe-modified peptides allowing enzyme active site profiling.

One important application of ABPP technology is for the identification of potent and selective enzyme inhibitors. Such inhibitors are invaluable tools in defining protein function in living systems. In this assay, known as competitive ABPP, libraries of candidate inhibitors are tested for their ability to block probe labeling of an enzyme. This approach has several advantages compared to more traditional techniques relying on substrate assays, including the ability to (1) screen inhibitors in native proteomes, avoiding the need for recombinant expression and purification of enzymes, (2) identify inhibitors for uncharacterized enzymes that lack known substrates, and (3) determine both the potency and selectivity of inhibitors. Competitive ABPP has been applied to identify both reversible and covalent enzyme inhibitors.

20.4.3 Characterizing Proteins Using ABPP

ABPP has explored protein function on many frontiers by revealing conserved catalytic residues in enzyme active sites, providing a screening platform for the generation of selective inhibitors, assigning enzymes to a specific mechanistic class, and profiling enzyme activities in human tissues and cell lines. An ABPP study by Jessani and colleagues of the serine hydrolase family in a panel of human cancer cell lines permitted the classification of these lines into functional subtypes based on their tissue of origin and states of invasiveness. ABPP in combination with MS revealed that the novel serine hydrolase enzyme KIAA1363 was particularly up-regulated in the invasive cancer cell lines compared to their noninvasive counterparts. Further ABPP studies demonstrated that KIAA1363 was also up-regulated in breast and ovarian human tumors compared to normal tissue counterparts together, suggesting an important role for this enzyme in human cancer. To define the endogenous biochemical function for KIAA1363, competitive ABPP technology was employed to identify a potent and selective inhibitor. Using this inhibitor, in combination with a global metabolite discovery platform (DMP), KIAA1363 was shown to regulate the endogenous levels of a family of ether lipid metabolites known as monoalkylglycerol ethers (MAGEs) in aggressive ovarian cancer cell lines.

20.4.4 LIMITATIONS OF USING ABPP

One of the biggest limitations of ABPP is that activity-based probes have not been designed for all enzyme classes in the proteome. Moreover, for those enzyme classes for which probes have been developed, an open question remains in certain cases as to how extensively the probes cover the entire enzyme family and the accuracy with which they report on all facets of enzyme activity. Nonetheless, given the well-recognized technical challenges confronting the field of proteomics, the increased sensitivity, resolution, and information content provided by function-based approaches such as ABPP promise to keep them at the forefront of proteomic technology development and application.

20.5 CONCLUSIONS

As described in this chapter, several quantitative proteomic methods have been developed to enhance our understanding of protein function. Using these technologies, researchers are able to identify pertinent changes in protein expression, activity, interactions, and the posttranslational modification state in complex biological systems, which can enhance biological understanding through follow-up functional experiments. Methods for quantitative proteomics, although ever advancing, are still far from having the potential to characterize whole proteomes comprehensively. As such, methodologies that able enrichment or prefractionate the proteome are gaining favor as a means to profile less abundant proteins. One example of such a functional enrichment strategy is ABPP, which in combination with DNA microarrays and MS-based proteomic techniques can provide a rich set of data for functional exploration of uncharacterized portions of the proteome, leading to the discovery of enzymatic pathways involved in different biological processes.

FURTHER READING

1. Rabilloud, T. (2002). Two-dimensional gel electrophoresis in proteomics: old, old fashioned, but it still climbs up the mountains. Proteomics, 2, 3–10.
2. Hall, D. A., Ptacek, J., and Snyder, M. (2007). Protein microarray technology. Mech. Ageing Dev. 128, 161–167.
3. Bantsheff, M., Schirle, M., Sweetman, G., Rick, J., and Kuster, B. (2007). Quantitative mass spectrometry in proteomics: a critical review. Anal. Bioanal. Chem. 389, 1017–1031.
4. Ong, S. E. and Mann, M. (2005). Mass spectrometry-based proteomics turns quantitative. Nat. Chem. Biol. 1, 252–262.
5. Cravatt, B. F., Wright, A. T., and Kozarich, J. W. (2008). Activity-based protein profiling: From enzyme chemistry to proteomic chemistry. Annu. Rev. Biochem. 77, 383–414.

Introduction to Part IV

Ülo Langel

Genes encode multiple, possibly hundreds of thousands, of proteins and peptides. It has been suggested that tens of thousands of protein families exist, although it is difficult to exactly define a protein family. It seems that most of endogenous peptides and proteins carry biological functions, making the peptide/protein field extremely important for understanding the functioning of organisms, as well as for development of therapeutic strategies.

In Part IV, Chapters 21–27, we have selected six classes of proteins for an introduction, together with several bioactive peptides. We admit that the choice between multiple possibilities was not an easy task, however, we believe that the introduction to enzymes, nucleic acid interacting proteins, receptors, membrane proteins, antibodies, and fibrous proteins is a good start in learning about proteins. Also, some bioactive peptide classes are introduced in order to lead the reader into this quickly developing field of research.

In these chapters, the authors concentrate on biochemical mechanisms of the protein's actions and, if possible, indicate the therapeutical or biotechnological possibilities for these peptides and proteins. It is our understanding that today there exist novel ways to apply the overall wide knowledge, which, in the future, can lead to novel drugs, materials and research tools, and deeper understanding of Nature. Hence, Part IV tries to explain the rules and applications of several representative classes of peptides and proteins.

21 Enzymes

Matjaž Zorko

CONTENTS

Enzymes are proteins that are able to catalyze chemical reactions, not only in cells but also outside the cells in extracellular liquids (such as blood), and even in vitro, if given the appropriate conditions. Historically, in the 1890s E. Büchner (Nobel Prize 1907) was the first to demonstrate the catalytic activity of enzymes in a cell extract, thus refuting Pasteur's thesis that life processes such as fermentation can operate only within living cells, and implying that enzymes are molecules that can be extracted and characterized. Only in the 1920s was it proven by Sumner, Northrop, and Stanley (Nobel Prize 1946) that physiological catalysts were proteins, although catalytic activity had been proposed to be associated with proteins much earlier. Today, enzymes are regarded as the central protein class that enables chemical reactions that take place in organisms under very mild conditions to run with physiologically required rates. Additionally, enzymes not only constitute the key points of metabolic regulation but are also very important in biotechnology for producing a variety of biomolecules and drug precursors, and in medicine for diagnostics and therapy. The idea of producing an enzyme with designed specificity and activity led to their construction based on antibodies. These "artificial" enzymes are called *abzymes* or catalytic antibodies, and are discussed in Chapter 30. Besides proteins, some RNA molecules also have catalytic properties. Catalytic RNAs are called ribozymes and have an important role in the maturation of certain RNA molecules and in protein synthesis, as discussed in Chapter 16.

21.1   ACTIVE SITE CONCEPT

The active site, the region where an enzyme binds its substrate, is only a small part of the enzyme molecule. It recognizes the substrate molecule, in whole or in part, by establishing several usually weak bonds with appropriately oriented functional groups of the amino acid residues within the active site. Binding energy is the only energy available to the enzyme for converting substrate into the product along an alternative reaction pathway that is faster than that in the absence of enzyme. How this is done is the subject of the catalytic mechanism and will be discussed later in this chapter. Recognition of substrate by enzyme was for a long time regarded as static. In 1890, Fischer postulated the lock-and-key principle in which enzyme active site was compared to the lock and substrate to the key that precisely fitted the lock without any changes in the enzyme structure. Koshland, in 1958, put forward the opposing theory of "induced fit," in which the enzyme adapts to the substrate by conformational changes in the active site (see Chapter 17, Figure 17.2). Knowing today the detailed structure of many enzymes and their catalytic mechanisms, it appears that most enzymes change the conformation of at least some active site amino acid residues during substrate binding and conversion. Aldose reductase is an example of an enzyme with a very flexible active site that can accommodate substrates of very different shape and size, ranging from small aliphatic aldehydes to steroids. Acetylcholinesterase is an enzyme with a partly rigid active site that sits at the bottom of an approximately 20-Å-deep gorge. The structure of the gorge is considered to be very rigid, but it has been shown that some amino acid residues change their conformation on binding of substrate to the peripheral active site at the entrance of the gorge, initiating a conformational change at the catalytic site near the bottom of the gorge, where substrate is hydrolytically cleaved. DNA polymerase is an enzyme with a very rigid active site that displays extremely high selectivity for a specific base pair according to size and geometry. It is proposed that such active site rigidity is the chief determinant of the high fidelity of DNA replication.

Catalytic events take place in the active site that usually comprises less than 10% of total number of amino acids in the enzyme. The question arises as to why the remaining part of the enzyme molecule is needed. The main reasons for this are the following:

- An active site can be formed only by mutual interactions of many amino acid residues that bring together amino acid residues with appropriate properties that may be far away from each other in the protein sequence and keep them in the proper three-dimensional structure inside the active site cleft or pocket.
- Adaptation of the active site conformation to the structure of the substrate, via conformational changes, sometimes demands the involvement of many amino acid residues outside the enzyme active site.
- Substrate distortion, leading to stabilization of the transition state, must be compensated by distortion in the enzyme molecule that has to be distributed among many amino acid residues in order to retain enzyme stability.
- Conformational changes in allosteric enzymes are transferred from subunit to subunit over a long distance, affecting a number of amino acids outside the active site.
- Regulation of enzyme activity, particularly by allosteric modulators, involves additional regulatory binding site(s) in the enzyme molecule and a number of complex interactions of many amino acid residues outside the active center are needed for the formation of regulatory binding sites and for providing communication between sites.

21.2 SPECIFICITY, ACTIVITY, AND TURNOVER

Binding between substrate and enzyme is usually specific. An enzyme can exhibit absolute or group specificity. An absolutely specific enzyme can bind only one particular substrate. Several such enzymes are known, such as carbonic anhydrase, succinate dehydrogenase, and lactase. Group-specific enzymes can accept several, usually structurally related, molecules as substrates. Proteases provide well-known examples, such as trypsin, chymotrypsin, and elastase, that are able to catalyze the hydrolysis of peptide bonds in many different proteins. This does not mean that they lack specificity. This depends on the structure of the enzyme active site and mainly on the amino acid residue in the position P_1 in the substrate protein, which means the first residue at the N-terminal site of the peptide bond to be cleaved. Trypsin accepts substrates proteins with Lys or Arg in this position, chymotrypsin with Phe, Tyr, or Trp and elastase with small residues like Ala. A special type of group specific enzymes comprises bond specific enzymes, such as amylase that catalyses the a(1 → 4) O-glycosidic bond in several oligo- and poly-saccharides. As generally accepted, the specificity of enzymes requires at least a three-point attachment of the substrate to the enzyme active site. This is particularly true for the enzymes that recognize only one enantiomer of the substrate molecule. It is, for instance, not possible to understand how D-amino acid oxidase can differentiate between L- and D-amino acids if the substrate amino acid is not bound to an active site with at least three functional groups.

Enzyme activity is the rate by which enzyme converts substrate into product. It can be monitored either by following the degradation of substrate or formation of product in time. The proper unit of enzyme activity in the SI system is the katal (kat), which is defined as the amount of enzyme that converts 1 mol of substrate per second. Katal is seldom used in practice because the activity of 1 kat is unrealistically large. The alternative unit is the enzyme unit (U), later denoted as the international unit (IU). This is the amount of enzyme that converts 1 µmol of substrate per minute. In use, both the international unit and the katal must be defined by specifying the assay conditions. Another unit in use in enzymology is the specific activity. This is the enzyme activity per milligram of total protein. It can be expressed for instance in kat per mg protein. Specific activity is practical in monitoring the purity of the enzyme during the purification procedure. In all the above cases, the activity of the enzyme directly depends on the amount of enzyme present in the assay solution, and the obtained activity does not reflect enzyme efficiency. To avoid this, one can use the turnover number, which is defined as the number of substrate molecules that are converted to product by a single enzyme molecule or, more precisely, per enzyme active site, per unit of time. Turnover number is equal to the catalytic rate constant k_{cat}, as will be explained in Section 21.4.

21.3 THERMODYNAMICS AND THE MECHANISM OF ENZYME REACTION

In an enzyme reaction, substrate binds to the enzyme, forming the enzyme–substrate complex; substrate is converted into product, which is released from the unchanged enzyme. If the simple reaction

$$S \rightleftharpoons P \tag{21.1}$$

is catalyzed by an enzyme, the most fundamental presentation of this reaction would be

$$E + S \underset{k_{-1}}{\overset{k_1}{\rightleftharpoons}} ES \underset{k_{-2}}{\overset{k_2}{\rightleftharpoons}} E + P \tag{21.2}$$

where S is substrate, P product, E enzyme, and ES enzyme–substrate complex; rate constants are denoted by k. ES, which is frequently referred as the Michaelis complex, is a true complex, though unstable, and should not be confused with the transition states, which are not shown in this equation, although they must appear during the course of the reaction, one before and another after ES formation (see Figure 21.1). As illustrated in Figure 21.1, enzyme leads to the conversion of substrate to product over a modified reaction path in which ΔG, and thus the equilibrium of the reaction, is not changed, but the activation energy $\Delta G*$ is lowered and the reaction rate is increased.

21.3.1 BINDING OF SUBSTRATE(S)

Considering the thermodynamics of the enzyme catalyzed reaction, it is convenient to divide the reaction into two consecutive steps. The first step is the binding of substrate(s) to the enzyme and, second, the conversion of substrate(s) into product(s). Binding of substrates to the enzyme is entropically unfavorable because of loss of translational and rotational freedom. The translational entropy for a small molecule in 1 M solution

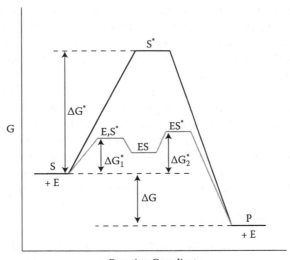

FIGURE 21.1 Energy profile of the catalyzed (lower profile) and noncatalyzed reaction (upper profile).

(standard state) is around 120 J/mol K and increases with the decreasing concentration. Rotational entropy for a large organic molecule is also around 120 J/mol K, but it is not dependent on concentration. When two molecules, each with three degrees of translation entropy and three degrees of rotation entropy, are associated, the formed complex has three degrees of translation entropy and three degrees of rotation entropy; three degrees of each are lost. The calculated entropy loss for the small molecule is up to 190 J/mol K. The loss might be partly compensated by increased internal vibration and rotation entropy but the contribution of these two is usually small. The loss of entropy must be compensated by binding energy, which is provided by the decrease of enthalpy arising from bonds formed between substrates and enzyme. When the binding includes a hydrophobic interaction between enzyme and substrate, partial desolvation of the participating molecules results in release of water molecules, which can provide additional compensation for the entropy loss.

21.3.2 CONVERSION OF SUBSTRATE(S) TO PRODUCT(S)

After binding of the substrate(s) to the enzyme, the second step of enzyme reaction starts. This comprises all phases of conversion of substrate(s) to product(s). The essential role of the enzyme is to stabilize the transition state of the conversion reaction and thus increase the reaction rate. Enzymes use different mechanisms to force the reaction of bound substrate(s) to proceed faster than in the absence of enzyme. The most important mechanisms will be discussed briefly here.

21.3.2.1 Proximity and Orientation

Because substrate molecules are already bound to the enzyme they do not need to find each other at random in the solution. Additionally, the encounter of more than two reacting substrates that are free in solution is highly improbable. This is not so in an enzyme catalyzed reaction where a number of substrates can be bound to the enzyme active site, and ternary and quaternary complexes are not uncommon. In these complexes the functional groups of the substrates are also appropriately oriented for the reaction to take place. The orientation factor plays an important obstacle for the reaction in random encounters in solution, particularly with more complex reactants. With enzyme-bound substrates, the reaction between substrate functional groups is no more intermolecular but becomes intramolecular, or in the most simplified case, the bimolecular reaction essentially proceeds as an unimolecular reaction. The concept of the effective concentration (see Chapter 2.5) can be applied here, and the increase of the rate of reaction can be regarded as the consequence of the increased effective concentration of the reacting groups.

21.3.2.2   Acid-Base Catalysis

Acids act as catalysts by donating protons and bases by accepting protons. In both cases the transition state is stabilized by transfer of proton—either from acid to substrate or from substrate to base. When the proton donor or proton acceptor is a water molecule, this is called general acid-base catalysis in order to distinguish it from specific acid-base catalysis, in which the proton donor or proton acceptor is another acidic or basic group. Such basic and acidic amino acid residues frequently reside in enzyme active sites.

21.3.2.3   Covalent Catalysis

In covalent catalysis, a transient covalent bond between substrate and the corresponding amino acid residue in the enzyme active site is formed in order to stabilize the transition state. Functional groups of amino acids are normally not sufficiently reactive to covalently bind the substrate. Therefore, they are helped by the functional groups of adjacent amino acid residues to gain the needed reactivity, usually by becoming stronger nucleophiles. A well-known example is the charge relay system in esterases and proteases that consists of one acidic and one basic amino acid residue acting together to make a Ser–OH group or a Cys–SH group a strong nucleophile that can attack the substrate. The following charge relay system, also called a catalytic triad, has been identified in chymotrypsin:

$$\tag{21.3}$$

Suitable groups that can be activated and used as nucleophile reagents are –OH of Ser, Thr, and Tyr; -SH of Cys; -NH$_2$ of Lys; -COO$^-$ of Asp; and imidazole of His. In covalent catalysis, some enzymes use metal ions like Zn^{2+} and coenzymes like pyridoxal phosphate and thiamine pyrophosphate to covalently bind substrates.

Since stabilization of the transition state is the central event in all enzyme catalytic mechanisms, one would expect the enzyme active site to be adapted to bind substrate in the form of a transition state. In very effective enzymes, this is not so. Because the equilibrium between the ground state and transition state is shifted almost entirely to the ground state, it is reasonable for the enzyme to recognize substrate in the ground state and to bind it weakly by noncovalent interactions. In the next step, the enzyme changes its conformation, which provides additional interactions between substrate and the active site. These interactions stabilize the substrate transition state and thus shift the equilibrium between the ground state and transition state toward the transition state. In practice, enzymes usually combine different catalytic mechanisms in accelerating the rate of the reaction (see Figure 21.6). Enzyme reactions are often complex and proceed via several intermediate states. This can also contribute to the acceleration of the reaction rate. More intermediates are included the more they are structurally similar to each other, and the activation energy needed to reach the transition states between them is lower.

21.4   BASIC ENZYME KINETICS

21.4.1   SINGLE SUBSTRATE KINETICS

At the beginning of the twentieth century, Brown was the first to explain the frequently observed hyperbolic curve showing the initial rate of enzyme reaction as a function of substrate concentration (Figure 21.2). The initial rate, v_0, is the rate at $t = 0$. At this point, the concentration of substrate is precisely defined and the back reaction cannot proceed because no product is present; v_0 increases with increasing concentration of substrate, reaching a plateau called the maximal rate, V_{max}, which cannot be exceeded by further increase of substrate concentration. Brown proposed

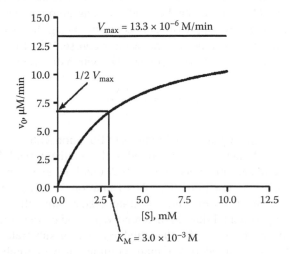

FIGURE 21.2   Dependence of initial rate (v_0) on the concentration of substrate [S] for an enzyme obeying Michaelis–Menten kinetics with indicated parameters K_M and V_{max}.

that in the course of the reaction an enzyme-substrate complex is formed that is further converted into product and regenerated enzyme (Equation 21.2). At very high substrate concentrations, all enzymes are occupied by substrate and a further increase of substrate concentration has no effect. V_{max} is thus not dependent on the concentration of substrate but on the total concentration of enzyme $[E_0]$:

$$V_{max} = k_2[E_0] \tag{21.4}$$

where k_2 is the rate constant for the decomposition of ES into E and P. This concept was elaborated by Leonor Michaelis and his coworker Maud Menten. Starting from the scheme shown in Equation 21.2 and taking into the account the initial rate concept ($[P] = 0$ and k_{-2} not needed) and full equilibrium between E, S, and ES, they developed the equation that is known today as the Michaelis–Menten equation:

$$v_0 = \frac{V_{max}[S]}{[S] + K_M} \tag{21.5}$$

K_M is called Michaelis constant and represents the dissociation constant of ES into E and S

$$K_M = \frac{[E][S]}{[ES]} = \frac{k_{-1}}{k_1} \tag{21.6}$$

K_M thus shows the affinity of substrate to the enzyme. This holds only if k_2 is so small in comparison to k_{-1} that it can be neglected (in this case, K_M is frequently denoted by K_s); this has been shown to be true for some enzymes but not for all. The modified approach of Briggs and Haldane made the Michaelis–Menten equation more generally valid by introducing the steady-state assumption, which says that the concentration of ES remains approximately constant during the reaction, except for the very beginning and end of the enzyme reaction course. With this assumption, the same equation as Equation 21.5 was obtained, but with a different meaning of K_M:

$$K_M = \frac{k_{-1} + k_2}{k_1} \tag{21.7}$$

Clearly, when $k_2 \ll k_{-1}$, Equation 21.7 reduces to Equation 21.6.

The Michaelis–Menten equation provides the mathematical background to understanding the hyperbolic plots (Figure 21.2) obtained experimentally for most single substrate enzyme reactions that obey Equation 21.2. V_{max} and K_M provide valuable information on an enzyme-substrate system, and their values can be extracted from experimental data in different ways. Today, the simplest approach is to feed data on v_0 measured for the suitable range of substrate concentrations into the computer and to apply one of many commercially or freely available programs that will fit the curve, according to the Michaelis–Menten equation, to the experimental data using the least square method. The program will also calculate values of K_M and V_{max}. An alternative is to use different graphical procedures. The most known is the Lineweaver–Burk plot that uses the reciprocal form of the Michaelis–Menten equation:

$$\frac{1}{v_0} = \frac{K_M}{V_{max}} \cdot \frac{1}{[S]} + \frac{1}{V_{max}} \tag{21.8}$$

$1/v_0$ is a linear function of $1/[S]$. K_M and V_{max} can easily be obtained from the slope of the line, and its intercepts with the coordination axes from the graph constructed by plotting $1/v_0$ against $1/[S]$. In spite of many disadvantages, this plot is still much in use, particularly as a diagnostic plot in inhibition studies (see Section 21.6). Many other plots exist, also named after their inventors—for instance, the Eade–Hofstee plot, Hanes plot, Eisenthal–Cornish–Bowden plot, to list only the most known. They are all based on the Michaelis–Menten equation linearized in different manners. Whichever procedure one chooses for the extraction of K_M and V_{max}, it should be used cautiously and the reliability of the obtained result should be critically evaluated.

How informative are values of K_M and V_{max}? K_M is related to the affinity of enzyme for the substrate; the lower the value of K_M, the higher the affinity. Its precise meaning is dependent on the kinetic scheme used and relative values of rate constants included (see Equations 21.6 and 21.7). With more complex kinetic schemes that comprise more intermediates and rate constants, K_M also becomes more complex. As shown in Figure 21.2, K_M equals the concentration of S where $v_0 = \frac{1}{2} V_{max}$. This is the most general definition of K_M and, although rudimentary, it tells us at which substrate concentrations an enzyme is effective. It is not surprising that values of K_M measured for many enzyme-substrate systems correspond well to the cellular concentrations of the substrates. V_{max} reflects the catalytic efficacy of the enzyme–substrate system. For the simple kinetics shown in Equation 21.2 and initial rate conditions, V_{max} depends only on k_2 and $[E_0]$. k_2 is equal to k_{cat}, and can be calculated from Equation 21.4 when total concentration of the enzyme is known. With more complex kinetics k_{cat} depends on all rate constants that lead the decomposition of complex ES into product(s). For the simplified scheme, assuming rapid equilibrium between E, S, and ES:

$$E + S \underset{k_{-1}}{\overset{k_1}{\rightleftharpoons}} ES \xrightarrow{k_2} EP \xrightarrow{k_3} E + P \tag{21.9}$$

k_{cat} depends on the rate constants k_2 and k_3:

$$\frac{1}{k_{cat}} = \frac{1}{k_2} + \frac{1}{k_3} \tag{21.10}$$

For efficient catalysis the enzyme–substrate system should operate with large k_{cat} and small K_M. These two parameters are joined in the ratio k_{cat}/K_M, called the *specificity constant*, which is often used to illustrate and compare the efficiency of different enzyme–substrate systems. In Table 21.1 values for K_M, k_{cat} and k_{cat}/K_M are given for some selected systems. Note the differences in values obtained for different substrates with the same enzyme.

Introduction to Peptides and Proteins

TABLE 21.1

Values of k_{cat}, K_M, and k_{cat}/K_M for Some Selected Enzyme-Substrate Systems

Enzyme	Substrate	k_{cat} (s^{-1})	K_M (M)	k_{cat}/K_M (M^{-1}s^{-1})
Superoxide dismutase	O_2^-	1.0×10^6	3.6×10^{-4}	2.8×10^9
Acetylcholinesterase	Acetylcholine	1.4×10^4	9.0×10^{-5}	1.6×10^8
Fumarase	Fumarate	8.0×10^2	5.0×10^{-6}	1.6×10^8
	Malate	9.0×10^2	2.5×10^{-5}	3.6×10^7
Chymotrypsin	N-acetyltyrosine ethyl ester	1.9×10^2	6.6×10^{-4}	2.9×10^5
	N-acetylvaline ethyl ester	1.7×10^{-1}	8.8×10^{-2}	1.9
	N-acetylglycine ethyl ester	5.1×10^{-2}	4.4×10^{-1}	1.2×10^{-1}
DNA polymerase I	dNTPs[a]	1	2.5×10^{-5}	4.0×10^4
Lysozyme	(NAG-NAM)$_3$[b]	5.0×10^{-1}	6.0×10^{-3}	83

[a] dNTPs = deoxynucleotide triphosphates.
[b] NAG-NAM = N-acetyl-(2-deoxy-2-aminoglucopyranosyl)-β(1,4)-N-acetylmuramic acid.

21.4.2 MULTISUBSTRATE KINETICS

Most enzymes deal with the reactions of more than one substrate. Consider the two-substrate reaction scheme where A and B are substrates, and P and Q are products:

$$E + A + B \rightleftharpoons EAB \rightleftharpoons E + P + Q \tag{21.11}$$

In two-substrate reactions, a functional group or electron(s) is usually transferred from A to B. For these reactions, the Michaelis–Menten equation can be applied only under some special conditions. If, for instance, one of the substrates is water—which very frequent occurs in biological systems—the water concentration is not changed during the reaction course, and hence does not affect the reaction rate. Reaction runs in accordance with the pseudo first-order mechanism, and the Michaelis–Menten equation is adequate. Similarly, if none of the substrates is water but the concentration of one of them is kept constant, the Michaelis–Menten equation can be used to provide valuable information. However, the K_M and V_{max} obtained are valid only for the given concentration of the fixed substrate. Graphical procedures based on the Lineweaver–Burk modification of the Michaelis–Menten equation with one substrate concentration fixed are often used to differentiate between different kinetic schemes obeyed by two-substrate mechanisms.

A general equation for a two-substrate enzyme catalyzed reaction based on the steady-state assumption was first derived by Alberty:

$$v_0 = \frac{V_{max}[A][B]}{K_M^B[A] + K_M^A[B] + [A][B] + K_S^A K_M^B} \tag{21.12}$$

where

V_{max} is the maximal rate when both substrates A and B are saturating,

K_M^B is the concentration of B giving $\frac{1}{2}V_{max}$ when A is saturating,

K_M^A is the concentration of A giving $\frac{1}{2}V_{max}$ when B is saturating,

K_S^A is the dissociation constant of $E + A \leftrightarrow EA$.

The two-substrate reaction shown in Equation 21.11 can proceed by different mechanisms and for each mechanism Equation 21.12 can be modified correspondingly. These mechanisms are usually divided into two groups: one in which the ternary complex EAB is formed and the other where this complex does not exist. The basic mechanisms with a ternary complex are random and ordered sequential mechanisms, and the basic mechanism without ternary complex is the ping-pong mechanism. Schemes for these mechanisms are shown below in Cleland's notation.

In the random sequential mechanism

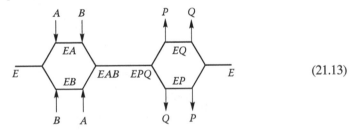

$$(21.13)$$

substrates A and B bind to the enzyme E one after another but E does not care which one binds the first. Similarly, products P and Q are released ad libitum. In contrast to this, in the ordered sequential mechanism

$$\begin{array}{cccc} A & B & P & Q \\ \downarrow & \downarrow & \uparrow & \uparrow \\ \hline E & EAB & EPQ & E \end{array}$$

$$(21.14)$$

substrate B can bind only if A is already bound and the sequence of the release of products, first P and then Q, is also fixed. In both sequential mechanisms, the group that is exchanged between substrates is transferred directly from A to B. In the ping-pong mechanism

$$\begin{array}{cccc} A & P & B & Q \\ \downarrow & \uparrow & \downarrow & \uparrow \\ \hline E & EA \quad FA \quad F & FB \quad EB & E \end{array}$$

$$(21.15)$$

the functional group is first transferred from A to the enzyme, which is temporarily modified (denoted by F), and the first product P is released. Only then does substrate B bind to the enzyme and accept the functional group, thus becoming product Q, which is released and the enzyme recovered. Diagnostic plots that enable the

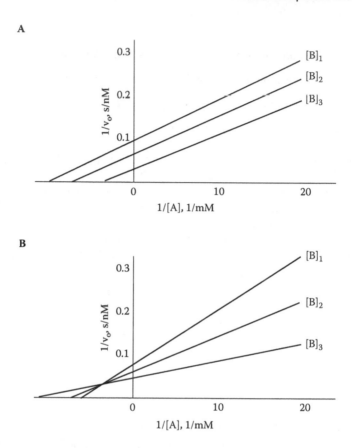

FIGURE 21.3 Diagnostic plots for two-substrate mechanisms: A: ping-pong mechanism; and B: random sequential mechanism. [B] is constant for each line, $[B]_1 < [B]_2 < [B]_3$. In these two mechanisms, $1/v_o$ versus $1/[B]$ plots appear analogous to the shown $1/v_o$ versus $1/[A]$ plots. In ordered sequential mechanism, plot $1/v_o$ versus $1/[A]$ is similar to that shown in B, whereas in $1/v_o$ versus $1/[B]$ plot lines intersect on the $1/v_o$ axis.

differentiation between the mechanisms presented in Equation 21.13–21.15 are shown in Figure 21.3.

Enzymes with more than two substrates are known, as, for instance, the three-substrate enzymes glutamine-dependent NAD^+ synthetase, and pyruvate carboxylase. Mechanisms of these reactions are complex and will not be discussed here.

21.4.3   ALLOSTERIC KINETICS

In all the enzymes discussed so far, only one active site or independent multiple active sites were assumed. Where there is more than one active site, binding of a substrate to one active site does not affect binding and conversion of substrate at the other active site. Allosteric enzymes always comprise more than one active site and cooperativity exists between the sites (see Chapter 2.5). The functional background for cooperativity is a conformational change of the enzyme that occurs after binding of substrate. This affects binding and conversion of the substrate at the other active site. Generally, allosteric enzymes are multisubunit proteins and each active site resides in a different enzyme subunit. Subunits may or not be identical. Cooperativity can be homotropic or heterotropic. Homotropic cooperativity will be briefly presented here and heterotropic in Section 21.6.3. Enzymes exhibiting positive homotropic cooperativity between active sites show typical sigmoidal v_0 versus $[S]$ plots (Figure 21.4A). This plot was explained by Hill on the level of

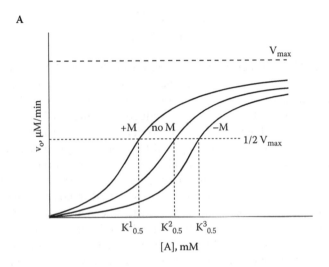

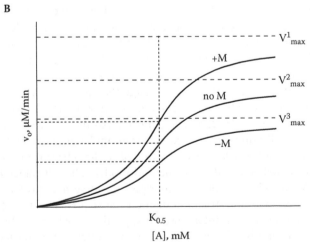

FIGURE 21.4 Dependence of initial rate (v_0) on the concentration of substrate $[S]$ for an allosteric enzyme with a positive homotropic allosteric effect of substrate in the presence of the heterotropic allosteric modulators—effect on $K_{0.5}$ (A) and on V_{max} (B). +M and −M denote an allosteric activator and an allosteric inhibitor, respectively.

cooperative binding of ligand, L, to protein with n monomers, each comprising one binding site:

$$Y = \frac{K_b[L]^n}{1 + K_b[L]^n} \tag{21.16}$$

where Y is fractional saturation and K_b the overall binding constant ($K_b = \frac{[M_nL_n]}{[M_n][L]^n}$, M is protein monomer). In terms of initial rate, the Hill equation can be expressed as

$$v_0 = \frac{V_{max}[S]^n}{[S]^n + K_{0.5}^n} \tag{21.17}$$

where n is the Hill coefficient and $K_{0.5}$ the concentration of S at $\frac{1}{2}V_{max}$. Positive cooperativity, resulting in sigmoidal curve of v_0 versus $[S]$, is obtained with a Hill coefficient greater than 1 and negative cooperativity with n below 1. When $n = 1$, no cooperativity is observed, Equation 21.17 reduces to Equation 21.5, and $K_{0.5}$ is K_M. The physical basis for homotropic cooperativity can be described by two mechanisms. The concerted mechanism proposed by Monod, Wyman, and Changeux is called the MWC model. It assumes an oligomeric structure of the enzyme with the existence of two conformational states in equilibrium. The predominant state is state T (tense) of the free enzyme. The second state is state R (relaxed), which has higher affinity for the substrate. This model is also called the symmetric model because of the symmetry assumption that predicts identical affinity for the substrate in all protomers of the same state. By binding to the enzyme, substrate shifts the equilibrium toward state R for which it has higher affinity. This results in gradual increase of the overall affinity of the enzyme for the substrate and gives the sigmoidal curve in the plot v_0 versus $[S]$. Negative homotropic cooperativity cannot be described by this model. Another mechanism based on Koshland's idea of induced fit was proposed by Koshland, Nemethy, and Filmer and is thus called the KNF model. Initially, all protomers of the free enzyme are in the same conformation. Binding of substrate to one promoter induces a conformational change that alters this and the neighboring protomers. Conformationally modified neighboring protomers show altered affinity for the binding of the substrate; higher affinity leads to positive and lower to negative cooperativity. Binding of the next substrate molecule induces further conformational change that is sequentially transmitted to the other protomers. Because of this, the KNF model is called the sequential model. The advantage of this model over the MWC model is its ability to interpret negative homotropic cooperativity, in good agreement with the structural data on the sequence of the conformational changes revealed for some allosteric proteins, particularly hemoglobin. Eagan later unified the MWC and KNF models in a general model that is, for most cases, too complicated for the use in practice.

Besides active sites, allosteric enzymes frequently contain one or more regulatory sites where allosteric modulators can bind and modulate enzyme activity. Modulators control the activity of allosteric enzymes via the heterotropic allosteric effects, and this makes many allosteric enzymes crucial regulatory points in metabolism. They are briefly discussed in Section 21.6.3.

21.5 COFACTORS, COENZYMES, AND PROSTHETIC GROUPS

Many enzymes are assisted by cofactors. These are ions or small no-protein molecules that are required for the catalytic process. The term cofactor is not precisely defined. In the case of ions, only those that are not ubiquitously present in the enzyme surroundings are considered as cofactors. These are most frequently metal ions. They are usually tightly and specifically bound to the enzyme. If removed, the enzyme becomes inactive. Enzyme without cofactor is called an apoenzyme; a cofactor plus an apoenzyme gives a holoenzyme. Some enzymes require more than one cofactor. Cytochrome c oxidase, a large multisubunit enzyme complex with 13 subunits comprises two Fe ions in two hemes and two Cu ions in two copper centers. Examples of metal ion cofactor and some enzymes for which they are required are listed in Table 21.2. Metal ions shown in this table appear to be needed by all organisms, including human, but some organisms, particularly microorganisms, also require other ions, including ions of Cd, V, and W.

Coenzymes and prosthetic groups are organic molecules that can temporarily bind specific functional group or electron(s) during the catalytic process in which this group is transferred between substrates. Prosthetic groups are tightly and permanently bound to the protein, and are an integral part of the enzyme where they accept a functional group or electron(s) in one catalytic process and release it in another. In contrast to this, coenzymes are promiscuous. They bind transiently and rather weakly to the enzyme, where they accept a functional group from the substrate. Then they are released and subsequently bound to another enzyme where they pass the functional group to another substrate and are recovered. In this, coenzymes resemble substrates and are sometimes treated as such. Indeed, in the two- and three-substrate enzyme reactions discussed above (see Section 21.4.2), one or sometimes even two substrates are actually coenzymes. An example is shown in Equations 21.18–21.20

TABLE 21.2
Some Examples of Metal Ions as Enzyme Cofactors

Metal Ion	Enzymes
Fe^{2+}/Fe^{3+}	Cytochrome oxidase, catalase, nitrogenase, hydrogenase
Zn^{2+}	Carboxypeptidases, alcohol dehydrogenase, carbonic anhydrase
Cu^{2+}	Cytochrome oxidase, hexose oxidase, nitrite reductase
Mg^{2+}	Hexokinase, pyruvate kinase, glucose-6-phosphatase
Mn^{2+}	Ribonucleotide reductase, arginase, water-splitting enzyme (photosynthesis)
Ni^{2+}	Urease, carbon monoxide dehydrogenase
Se	Glutathione peroxidase, formate dehydrogenase
Mo	Dinitrogenase, sulfite oxidase, nitrate reductase

Note: Hundreds of enzymes need metal ions as cofactors.

TABLE 21.3

Selection of Common Coenzymes and Prosthetic Groups

Coenzyme/Prosthetic Group	Transferred Group	Precursor
Thiamine pyrophosphate	Aldehydes	Thiamine
Flavin adenine dinucleotide (FAD)	Electrons and protons	Riboflavin
Flavin mononucleotide (FMN)		
Nicotinamide adenine dinucleotide (NAD)	Hydride ion (2 electrons and 1 proton)	Nicotinic acid
Coenzyme A	Acyl groups	Pantothenic acid
Pyridoxal phosphate	Amino groups	Pyridoxine
5-Deoxyadenosylcobalamin (coenzyme B12)	H atoms and alkyl groups	(Cyano)cobalamin
Biocytin	CO_2 (–COOH)	Biotin
Tetrahydrofolic acid	One-carbon groups	Folic acid
Lipoic acid	Electrons and acyl groups	Lipoic acid

below (Cyt stands for cytochrome) of the role of one coenzyme, NADPH, and two prosthetic groups, heme and FAD, in the transfer of electrons in the process of introducing a double bond into saturated fatty acid:

$$NADPH + H^+ + Cytb_5\text{-reductase}(FAD) \rightarrow NADP^+ + Cytb_5\text{-reductase}(FADH_2) \tag{21.18}$$

$$Cytb_5\text{-reductase}(FADH_2) + 2\ Cyt\ b_5(Fe^{3+}) \rightarrow Cyt\ b_5\text{-reductase}(FAD) + 2\ Cytb_5(Fe^{2+}) + 2\ H^+ \tag{21.19}$$

$$2\ Cyt\ b_5(Fe^{2+}) + 2\ H^+ + O_2 + \text{-}CH_2\text{-}CH_2\text{-} \rightarrow 2\ Cyt\ b_5(Fe^{3+}) + 2\ H_2O + \text{-}CH=CH\text{-} \tag{21.20}$$

NADPH is the reduced form of coenzyme nicotinamide adenine dinucleotide phosphate, which passes two electrons and one proton to FAD; flavin adenine dinucleotide, using an additional proton from the solution, is reduced to $FADH_2$. FAD is the prosthetic group of the enzyme $Cytb_5$-reductase. Electrons are transferred from the reductase to Fe^{3+} in the prosthetic group heme of two $Cytb_5$s. Eventually, electrons are delivered to an oxygen molecule, which takes two protons from the solution and two protons and two electrons from the group $-CH_2\text{-}CH_2\text{-}$, transforming the latter into $-CH = CH\text{-}$; $-CH_2\text{-}CH_2\text{-}$ is the part of the fatty acid alkane chain where a double bond is formed.

Vitamins are precursors of many prosthetic groups and coenzymes. Table 21.3 shows the main coenzymes and prosthetic groups, together with the transferred functional groups.

21.6   ENZYME INHIBITION

Molecules that bind to an enzyme and totally or partially reduce its activity are called inhibitors. Extrinsic inhibitors are usually toxic and some of most effective toxins, including components of snake and insect venoms, are of peptide or protein origin. Intrinsic inhibitors, on the other hand, are usually synthesized inside the cell where they act and are important regulators of enzyme activity, particularly of proteases.

21.6.1 IRREVERSIBLE INHIBITION

Inhibitors can be divided into reversible and irreversible, according to the mode of binding to the enzyme. The latter bind irreversibly to the active center of the enzyme by forming a covalent bond with a specific amino acid residue that is important for the binding of substrate and/or catalyzing the conversion of substrate to product. The enzyme is thus permanently inactivated. Most irreversible inhibitors have an electrophilic group that will attack a nucleophilic functional group such as sulfhydryl in Cys or hydroxyl in Ser, Thr, or Tyr in the enzyme active center. Organophosphates, carbamates, and sulfonate are well-known irreversible inhibitors that bind to the hydroxyl group of Ser and powerfully inhibit many enzymes with Ser in the active center, including acetylcholinesterase and serine proteases. Some of these inhibitors are used for chemical warfare and others as insecticides. The latter are made selective to insect acetylcholinesterase and are much less toxic to mammals. Aspirin and penicillin are examples of useful irreversible inhibitors. Aspirin functions by acylating Ser in the active center of cyclooxygenase, which is required for the synthesis of prostaglandins and thromboxanes. The extensively used antibiotic penicillin is also an acylating agent that irreversibly modifies Ser in the active center of bacterial transpeptidases, the enzymes that participate in the synthesis of the bacterial cell wall, particularly in Gram positive species. Penicillin is a member of the special class of irreversible inhibitors called suicide inhibitors. They originate from substrate analogs that are converted by the catalytic reaction into an irreversible inhibitor that subsequently blocks the enzyme. Penicillin mimics the dipeptide D-Ala-D-Ala, which is part of the transpeptidase substrate. Penicillin is therefore recognized by the transpeptidase active site where it binds. There the penicillin β-lactam ring is cleaved by the enzyme, and the transformed penicillin is covalently attached to the active site Ser. Some irreversible inhibitors are shown in Table 21.4.

21.6.2 REVERSIBLE INHIBITION

Reversible inhibitors are classified, according to the mode of binding to the enzyme, as competitive, uncompetitive, and mixed. Competitive inhibitors are structurally similar to substrates, and they compete with substrates for binding to the active center of the enzyme. A competitive inhibitor, I, can bind to E but not to ES, as shown in a scheme valid for the initial rate conditions:

$$E + S \; \underset{}{\overset{K_s}{\rightleftharpoons}} \; ES \; \xrightarrow{k_2} \; E + P$$
$$+$$
$$I$$
$$\Big\updownarrow K_i$$
$$EI$$

$$(21.21)$$

TABLE 21.4
Some Irreversible Inhibitors

Inhibitor	Origin and Role
DIFP, sarin, soman, VX, malathion	Synthetic organophosphates—some used in chemical warfare and as insecticides, others in protein isolation and purification to prevent degradation by serine proteases; bind to serine –OH in serine proteases, cholinesterases, and other esterases
Aspirin	Natural methyl salicylate (many synthetic variants); as an antiinflammatory, analgesic and anticoagulatory drug; acts by acylating active site serine of cyclooxygenases, preventing synthesis of prostaglandins and thromboxanes
Penicillin	Natural antibiotic (many synthetic variants); binds to active site serine of transpeptidase preventing cross-linking of peptidoglycans in bacterial cell wall
Sulbactam	Synthetic analog of penicillin; binds to serine in active center of β-lactamase, enzyme expressed by some bacteria that inactivates penicillins
Vinyl sulfone inhibitors	Family of synthetic sulfonylated peptidomimetics; bind to active site cysteine of cysteine proteases in parasitic protozoa—a potential drug for malaria, sleeping sickness and similar parasitic diseases

K_i, the inhibition constant, is the dissociation constant of complex EI. The Michaelis–Menten equation in the presence of competitive inhibitor is:

$$v_0 = \frac{V_{max}[S]}{[S]+\alpha K_M}, \text{ where } \alpha = \left(1+\frac{[I]}{K_i}\right) \tag{21.22}$$

and the corresponding Lineweaver–Burk transformation is:

$$\frac{1}{v_0} = \frac{\alpha K_M}{V_{max}} \cdot \frac{1}{[S]} + \frac{1}{V_{max}} \tag{21.23}$$

From these equations values of α and K_i can be obtained numerically or graphically (see Figure 21.5). V_{max} is not affected by competitive inhibitors because, at constant $[I]$ and increasing $[S]$, substrate is displacing inhibitor from the active site and v_0 is approaching V_{max}. This is employed in intoxication with competitive inhibitors, where increased concentrations of substrate can be administered in order to reverse inhibition. Transition state analogs of substrates are very potent competitive inhibitors. They mimic the transition state of the substrate that is bound to the active site of the enzyme with very high affinity. Transition state analogs are discussed in Chapter 25.

Uncompetitive inhibitors bind only to the complex ES and not to the free enzyme. The reason for this lies either in a conformational change induced by the binding of the substrate that shapes the binding site for the inhibitor, or in the additional functional group or hydrophobic surface provided by the substrate molecule to effectively

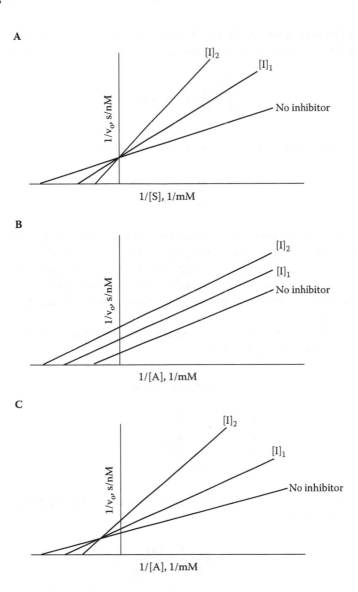

FIGURE 21.5 Diagnostic Lineweaver–Burk plots for different types of reversible inhibition: (A) competitive inhibition; (B) uncompetitive inhibition; and (C) mixed inhibition. [I] denotes concentration of the inhibitor; $[I]_1 < [I]_2$.

bind the inhibitor. The kinetic scheme is as follows:

$$E + S \xrightleftharpoons{K_s} ES \xrightarrow{k_2} E + P$$

$$+$$

$$I$$

$$\big\Updownarrow K_i'$$

$$ESI$$

(21.24)

K_i', the inhibition constant, is the dissociation constant of complex ESI. The Michaelis–Menten equation for v_0 in the presence of uncompetitive inhibitor is:

$$v_0 = \frac{V_{max}[S]}{\alpha'[S] + K_M}, \text{ where } \alpha' = \left(1 + \frac{[I]}{K_i'}\right) \qquad (21.25)$$

and the corresponding Lineweaver–Burk transformation:

$$\frac{1}{v_0} = \frac{K_M}{V_{max}} \cdot \frac{1}{[S]} + \frac{\alpha'}{V_{max}} \qquad (21.26)$$

Although not directly apparent from the Equations 21.26 and 21.27, it can easily be demonstrated that both parameters, K_M and V_{max}, are equally affected by the inhibitor and are decreased by being divided by α'. Procedures to determine K_i' are similar to those with the competitive inhibitor. Uncompetitive inhibition is very rare in single-substrate reactions, but it appears more frequently in multiple substrate processes.

Mixed inhibition appears as a combination of competitive and uncompetitive inhibition. The corresponding scheme is

$$
\begin{array}{ccccc}
E + S & \xrightleftharpoons{\quad K_s \quad} & ES & \xrightarrow{\ k_2\ } & E + P \\
+ & & + & & \\
I & & I & & \\
\Big\updownarrow K_i & & \Big\updownarrow K_i' & & \\
EI + S & \xrightleftharpoons{\quad K_s' \quad} & ESI & &
\end{array}
\qquad (21.27)
$$

K_s' being the dissociation constant for the decomposition of ESI into EI and S. The Michaelis–Menten equation is

$$v_0 = \frac{V_{max}[S]}{\alpha'[S] + \alpha K_M} \qquad (21.28)$$

and the Lineweaver–Burk equation

$$\frac{1}{v_0} = \frac{\alpha K_M}{V_{max}} \cdot \frac{1}{[S]} + \frac{\alpha'}{V_{max}} \qquad (21.29)$$

A mixed inhibitor will decrease V_{max} and also affect K_M. K_M will be decreased if $K_i > K_i'$ and increased in the opposite situation. A special case of mixed inhibition is noncompetitive inhibition, in which $K_i = K_i'$ and $\alpha = \alpha'$. Noncompetitive inhibitor will decrease V_{max} and will not affect K_M. Apparently, inhibitor does not interfere with the binding of substrate, but it decreases the ability of enzyme for catalysis. Different inhibition patterns as revealed by the Lineweaver–Burk plots for different types of reversible inhibitors are shown in Figure 21.5. These and many other plots are used to classify reversible inhibitors and to determine the values of the kinetic parameters K_i, K_i', α, and α'. All types of reversible inhibition shown in this figure and discussed above are fundamental and are members of the so-called linear inhibitors, named so because the corresponding Lineweaver–Burk plots are linear. Many more complex inhibition patterns are known to give nonlinear Lineweaver–Burk plots. Such plots are characteristic of several types of partial inhibition, substrate inhibition, product inhibition, and allosteric inhibition. The latter is discussed below.

21.6.3 ALLOSTERIC INHIBITION

Allosteric modulators are molecules that bind reversibly to the regulatory site of allosteric enzymes and induce conformational changes in the enzyme which results in modified activity. When the enzyme activity is increased, the allosteric modulators are called allosteric activators and when it is decreased, allosteric inhibitors. Only enzymes that show positive homotropic allostery toward substrate, resulting in a sigmoidal v_0 versus $[S]$ curve, will be discussed here. As discussed in Section 21.4.3, it is assumed that the enzyme can be in two different states, R and T. Two general types of allosteric inhibition are distinguished, denoted by K and V. In K type, substrate has higher affinity for the R than for the T state, but the catalytic ability (k_{cat}) of the two states is the same. Contrary to substrate, inhibitor binds with higher activity to the T than to the R state. Binding of inhibitor to the enzyme decreases the overall affinity of the enzyme for the substrate, either by shifting the equilibrium between the R and T states toward the T state, or by inducing a conformational change in the R state that eventually results in alteration of R into T. Either way, the decreased affinity of the enzyme for substrate reduces the enzyme activity. This is seen in the plot of v_0 versus $[S]$ as a shift of the curve to the right and an increase of the value of $K_{0.5}$ (see Figure 21.4A). In V type allosteric inhibition, affinity for the substrate is the same in each state of the enzyme, but k_{cat} of R state is much higher. An inhibitor that binds to T state with higher affinity than to R state shifts the equilibrium toward state T, thus decreasing the apparent V_{max} without affecting the binding of the substrate (Figure 21.4B). Pure K and V inhibitions are two extremes in a broad spectrum of realistic situations in which both types are mixed in different proportions.

21.7 REGULATION OF ENZYME ACTION

All processes in organisms are strictly regulated by many different regulatory mechanisms that are linked and integrated into a unified regulatory system. Regulation of enzyme action is an important part of this system. Main strategies for the regulation of the enzyme activity are

- Modulation of enzyme concentration
- Activation by proteolytic cleavage
- Covalent modification
- Allosteric modulation of enzyme activity
- Synthesis of specific inhibitors

According to Equation 21.4, activity of the enzyme is linearly dependent on enzyme concentration, which is regulated at the level of enzyme biosynthesis. Some enzymes are constitutive and their concentrations in the cell are steady. Other enzymes are inducible. Their substrates are not always present in the cell, and during this time, the concentration of the inducible enzymes is very low. When substrate appears, nutrient or xenobiotic for instance, it is detected by the signal transduction system, which activates the expression machinery and the concentration of the enzyme is increased. Inducible enzymes are numerous, particularly in bacteria that are exposed to constantly changing environments.

Activation by proteolytic cleavage is one of the mechanisms of preventing harmful action of some enzymes at the site of their synthesis and in tissue where they are not supposed to act. It is not regarded as proper regulation because it is not reversible. Proteolytic activation is very common in proteases. Pancreatic proteolytic enzymes trypsin, chymotrypsin, elastase, and carboxypeptidase are all synthesized in pancreatic cells in inactive forms called *zymogens*. Only after being secreted into the intestinal tract do they become active by part of their sequence being cleaved off, resulting in the functional three-dimensional structure. The principal cleavage occurs in the N-terminal part of the zymogen. Trypsin plays a key role in this process.

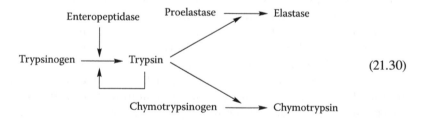

$$(21.30)$$

Trypsin is synthesized in exocrine pancreas in acinar cells as inactive trypsinogen. After being secreted in the duodenum, its activation is started by enteropeptidase, which is a serine protease synthesized in the duodenum wall cells. It converts a small amount of trypsinogen into active trypsin by removing the N-terminal hexapeptide Val-(Asp)$_4$-Lys. The specificity of trypsin is similar to that of enteropeptidase and is able to activate the remaining trypsinogen. Activation of chymotrypsinogen is more complex:

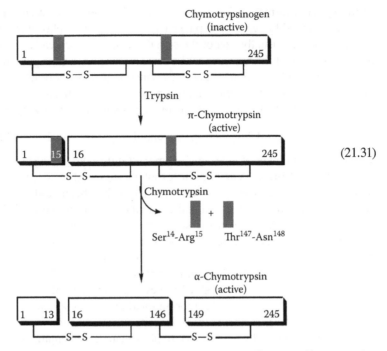

The first step is cleavage of the peptide bond between Arg^{15} and Ile^{16} by trypsin, resulting in active π-chymotrypsin that, in the next step, autolytically clips out two dipeptides, Ser^{14}-Arg^{15} and Thr^{147}-Asn^{148}. The active α-chymotrypsin obtained consists of three polypeptide chains linked together by two disulfide bonds. Proteolytic activation is also important in the blood-clotting system, where a series of serine proteases are activated sequentially to permit the production of an insoluble fibrin network. This type of regulation is not limited only to enzymes. It participates in maturation of other proteins and peptides. A well-studied example is the multistep proteolytic maturation of the peptide hormone insulin (see Chapter 5).

Reversible covalent modification is a convenient type of long-term regulation of enzyme activity, in which two enzymes cooperate. One covalently attaches the modifying group to the target regulatory enzyme, and the other is used to remove it. Modification that occurs away from the active site alters the activity of the target enzyme, typically by shifting the enzyme into a new conformational state. A huge number of covalent protein modifications are known (see Chapter 5), but only a few of them, including phosphorylation, methylation, ADP-ribosylation, adenylation, and uridylation, are more frequently used for enzyme regulation. By far the most common regulatory covalent modification is reversible phosphorylation. A phosphoryl group is transferred from ATP to the hydroxyl group of Ser, Thr, or Tyr by protein kinases. This brings into the protein structure a new bulky, negatively charged group that prevents existing, and introduces new interactions between amino acid residues, resulting in conformational change of the regulated enzyme. When appropriate, this effect is reversed by removal of phosphate by the specific phosphatase. Kinases and phosphatases are usually part of the signal transduction cascades controlled by hormones. Binding of the hormone to the corresponding receptor governs the phosphorylation/dephosphorylation of many enzymes and induces other cellular events in order to control complex physiological processes (see Chapter 23). Glycogen phosphorylase catalyses the release of glucose-1-phosphate from glycogen:

$$\text{glycogen(Glu)}_n + P_i \overset{1}{\rightleftharpoons} \text{glycogen(Glu)}_{n-1} + \text{glucose-1-phosphate} \qquad (21.32)$$

where n is the number of glucose monomers (Glu) in the glycogen molecule.

Being one of the key enzymes in carbohydrate metabolism in muscles and liver, its activity must be under very strict control; it has been studied extensively. The enzyme is a homodimer that can undergo extensive conformational change, shifting the enzyme from the less active T state (phosphorylase b) to the more active R state (phosphorylase a) and vice versa. Besides allosteric modulation via ATP, AMP, and glucose-6-phosphate, the main regulatory mechanism is phosphorylation of Ser^{14}. Upon phosphorylation, the N-terminal part (residues 5 to 22) adopts the structure of a distorted 3_{10} helix and establishes new contacts with the neighboring subunit. This is followed by a series of conformational changes that lead to activation of the enzyme by improved access to the active site. The activation is reversed by dephosphorylation catalyzed by phosphorylase phosphatase. Schematically:

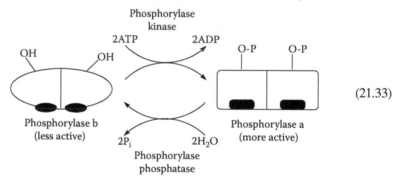

$$(21.33)$$

Phosphorylase kinase is activated indirectly by glucagon and epinephrine, and phosphorylase phosphatase by insulin.

Allosteric regulation is mediated by allosteric inhibition or allosteric activation of key regulatory enzymes that usually catalyze the first reaction in the metabolic pathway. Fundamental mechanism of this regulation is feedback, by which a regulatory enzyme can be temporarily shut down by the final product of the regulated pathway. In this way, the product acts as a heterotropic allosteric inhibitor:

$$A \overset{\times}{\rightarrow} B \longrightarrow C \longrightarrow D \longrightarrow E \longrightarrow F \qquad (21.34)$$

Product F inhibits the first enzyme of the pathway A–F that catalyzes conversion of A to B. This stops further production of F when the required concentration of F is achieved. The stationary concentration of F, $[F]_{set}$, in the system is set by the affinity constant for the binding of F to the regulatory enzyme. If $[F]_{set}$ is much greater than the concentration of the regulatory enzyme, half the available regulatory enzyme molecules will bind F and will be inhibited when the concentration of F reaches $[F]_{set}$. If F is used by other processes in the cell and its concentration drops below $[F]_{set}$, F will dissociate from the enzyme, whose activity will again increase, resulting in increased production of F. An example of such system is the feedback regulation of aspartate transcarbamoylase. This complex enzyme consists of two sets of catalytic trimers and three sets of regulatory dimers. It catalyzes the first irreversible step (called a *committed step*) of pyrimidine synthesis and is inhibited by the end product CTP. At the appropriate concentration, CTP molecules can bind to the enzyme, one to each regulatory subunit. This stabilizes the much less active T state of the enzyme and decreases its activity in order to stop further increase of CTP concentration. Kinetic schemes of the feedback inhibition can be very different. One of the more complex examples is shown below:

$$A \xrightarrow{X} B \longrightarrow C \longrightarrow D \longrightarrow E \begin{matrix} \nearrow F \\ \searrow G \end{matrix} \tag{21.35}$$

This mechanism permits separate regulation of the concentrations of two products, F and G. The ATP/ADP system in energy metabolism operates in a similar fashion. The pathways that are included in providing the cell with the energy, that is, the synthesis of ATP, are down-regulated via feedback inhibition by ATP as soon as the required concentration of ATP is reached. When the concentration of ATP is low, and ADP concentration is thus high, energy metabolism is no more inhibited, and, moreover, it is accelerated by ADP that acts as allosteric activator.

Some enzymes, particularly proteases, can be down-regulated by endogenous inhibitors that are small proteins. More than 20 structurally diverse families of inhibitory proteins are known for serine proteases. On the basis of the mechanism of action they are grouped into three classes: canonical inhibitors, noncanonical inhibitors, and serpins. Canonical inhibitors bind to the enzyme through an exposed binding loop, which is complementary to the active site of the protease. Noncanonical inhibitors use their N-terminal segment for interaction with the active site and establish additional interactions outside it. Serpins (serine protease inhibitors) bind to proteases similarly to canonic inhibitors and substrates, but are subsequently cleaved without being released due to their tight folding and numerous interactions with protease. Binding affinities of inhibitors are very high; for instance, bovine pancreatic trypsin inhibitor (BPTI) binds to trypsin with a constant of 10^{13} M^{-1}. Inhibitors similar to those targeting serine proteases are known also for other types of proteases. Protease inhibitors are very common in nature, and can be found in bacteria, archaea, and eukaryotes. In human blood more than 10% of all proteins present are protease inhibitors, including more than 35 members of serpins.

21.8   ENZYMES IN MEDICINE AND TECHNOLOGY

Enzyme applications in medicine are extensive. Pancreatic enzymes have been used in digestive disorders since the nineteenth century. Many enzymes are used extracellularly. Collagenase, for instance, is applied topically as an ointment on the skin to clean dead tissue from skin ulcers, burns, and other wounds, and can be also injected to reach deeper in tissue. Rhodanase can be used as an anticyanide drug that converts cyanide to thiocyanate, which is excreted normally in urine and is 100 times less toxic than cyanide. In disorders within the blood circulation system, streptokinase is used as thrombolytic agent for acute myocardial infarction. Urokinase is an activator of plasminogen and is also used to dissolve blood clots in the lungs (pulmonary embolism). Asparaginase is used in cancer to reduce the level of free Asp in the bloodstream. This retards tumor growth because tumor cells are unable to synthesize Asp, whereas normal cells are not affected. In diagnostics, the release of enzymes from damaged or decaying tissue is an important tool. Common enzymes used for clinical diagnosis include acid phosphatase, alanine aminotransferase, alkaline phosphatase, amylase, angiotensin converting enzyme, aspartate transaminase, creatin phosphokinase, γ-glutamyltransferase, lactate dehydrogenase, renin, and many others. Early diagnosis of myocardial infarction is performed routinely by measuring the activity of creatine phosphokinase, aspartate transaminase, and lactate dehydrogenase in blood. In cancer, many enzymes released in the circulation can be detected in blood, and the results are used to support diagnosis and for following carcinogenesis. Well-known examples are aspartate aminotransferase indicating liver cancer, acid phosphatase in prostatic cancer, and alkaline phosphatase in most cancers. Mutations in genes for enzymes lead to genetic diseases. The importance of enzymes is shown by the fact that a lethal illness can be caused by the malfunction of just one type of enzyme out of the thousands of types present in our bodies. Detection of mutated genes is today pertinent for the prenatal diagnostics of some genetic diseases, for instance, phenylketonuria, in which adequate treatment can help to provide survival and better quality of life.

Enzymes are used in the chemical industry and other industrial and home applications when highly specific catalysts are required. However, enzymes in general are limited in the number of reactions they catalyze and by their lack of stability in organic solvents and at high temperatures. Some enzymes can be used directly in solution—for instance, proteases, amylase, and cellulase in detergents; alkaline proteases and pancreatic enzymes in the leather industry; and amylase in the production of glucose syrup from starch, to mention only a few applications. An improvement was achieved by immobilizing enzymes on different technologically acceptable matrices, enabling their repeated use and other advantages, such as use in nonaqueous solutions and in continuous production systems, allowing simpler separation and purification of products. Examples of immobilized enzymes include the following: lactase immobilized on cellulose triacetate fibers, used to convert lactose to glucose and galactose in milk, which is vital for patients with lactose intolerance; penicillin amidase immobilized in different ways, used in production of penicillin and cephalosporins; and nitrile hydratase excreted by bacteria immobilized in polyacrylamide gel, which catalyzes the addition of water to acrylonitrile in the production of acrylamide. Disadvantages of immobilization are the additional cost and the fact that immobilization can reduce the stability and efficacy of an enzyme, although in most cases the stability of enzymes increases following immobilization.

Enzymes find applications as diagnostic enzymes. The term refers to enzymes used directly or as a component of an assay system for the determination of many substances ranging from simple chemicals (such as phosphates, ammonia, ethanol, and acetic acid, and metabolites such as glucose, urea, creatinine, and ATP) to toxic substances such as pesticides, herbicides, and heavy metals; proteins such as viral antigens, antibodies, and serum proteins, and DNA drugs; vitamins; and hormones. The usefulness of enzymes in technological processes has led to attempts to create new enzymes with novel properties by the modification of existing enzymes by methods of protein engineering or by designing totally new biocatalysts, as for instance, by developing catalytic antibodies (see Chapter 25).

21.9   ENZYME CLASSIFICATION

Enzyme nomenclature and classification was developed by the Nomenclature Committee of the International Union of Biochemistry and Molecular Biology. Later, with the participation of the International Union of Pure and Applied Chemistry, the Joint Commission on Biochemical Nomenclature (JCBN) was established. Each newly discovered enzyme is submitted to JCBN for classification, which is based on the reactions that it catalyzes. Enzymes are divided into six classes (Table 21.5). Each enzyme is given an enzyme classification number (EC number) consisting of four digits corresponding to four levels of classification. Peroxidases, for example, are given the EC number 1.11.1.X. The first digit, 1, represents oxidoreductases, one of the general classes of enzymes, as shown in Table 21.5. The second digit, number 11, denotes the oxidoreductases, which act on peroxides as donors. The third digit, 1, identifies the enzyme as a peroxidase. The fourth digit (X) would identify the exact type of peroxidase. Horseradish peroxidase, for instance, is identified by EC number 1.11.1.7.

Historically, enzymes were identified by trivial names that are still widely in use today. To avoid confusion, systematic names have been introduced in parallel to the EC number. These names end with suffix –ase and are based on the substrate that is converted and reaction that is catalyzed. Although precise, systematic names are

TABLE 21.5
Enzyme Classification

EC Number	Class	Catalyzed Reaction
1	Oxidoreductases	Oxidation/reduction (transfer of electrons)
2	Transferases	Transfer of functional groups
3	Hydrolases	Hydrolysis of various bonds
4	Lyases	Addition to double bond or formation of double bond by removal of functional groups
5	Isomerases	Isomerizations within a single molecule
6	Ligases	Formation of bonds by condensation reaction coupled to ATP cleavage

sometimes very complex and inconvenient. The trivial name of the enzyme that catalyzes the first reaction of glycolysis is hexokinase. The systematic name of this enzyme is ATP:glucose phosphotransferase. This and other similar cases fuel further use of simpler trivial names in practice.

21.10   PROTEASES

Proteases are important members of the hydrolases that catalyze hydrolytic cleavage of peptide bonds in peptides and proteins. They are also termed peptidases, proteinases, and proteolytic enzymes; in general, all these names are synonyms. Proteases are numerous enzymes found in all species and also in viruses. Just one class of proteases, serine proteases, has about 1000 entries of classified and characterized enzymes in the MEROPS proteinase database (http://merops.sanger.ac.uk/). More than 2% of all genes in the human genome have been identified as coding for proteases. They are grouped, on the basis of mechanism of catalysis, into serine, cysteine, threonine, aspartic, glutamic, and metallo catalytic types (and a few unknown catalytic types). The assignment refers to the amino acid residue or metal ion that acts as or promotes the nucleophile in the active site and is crucial for the catalysis. Parallel to this, proteases are also classified on the basis of sequence homology and tertiary structure by grouping sets of homologous peptidases into families and sets of related families into clans. Finally, proteases have been grouped also by the position of the cleaved peptide bond. For instance, endopeptidases cleave internal peptide bonds away from the N- and C-termini; exopeptidases cleave at or near free N- or C-termini, or both; aminopeptidases liberate single amino acids from the free N-terminus; carboxypeptidases do the same at the free C-terminus.

Proteases catalyze the cleavage of a peptide bond

$$(21.36)$$

By a mechanism in which a reactive nucleophile group of the enzyme e attacks the peptide carbonyl group. One way to make such a nucleophile is by a catalytic triad, where usually a histidine residue is used to activate Ser, Cys, or Thr as a nucleophile, as shown for chymotrypsin in Equation 21.3. A detailed study of the mechanism of the serine protease chymotrypsin reveals that it combines several catalytic strategies, as shown in Figure 21.6. In the first step, His acts as a base which, assisted by Asp, accepts a proton from a serine −OH group, making this group a strong nucleophile that attacks the carbonyl carbon in the target peptide bond of the substrate molecule and establishes a temporary covalent bond. The scissile peptide bond is cleaved and the first product, a peptide ranging from this bond to the C-terminus, is released. This results in a covalent catalysis combined with an acid-base catalysis. The second step is an acid-base catalytic process in which water is introduced and assisted by protonated His, which now acts as a proton donor, reestablishing the original state

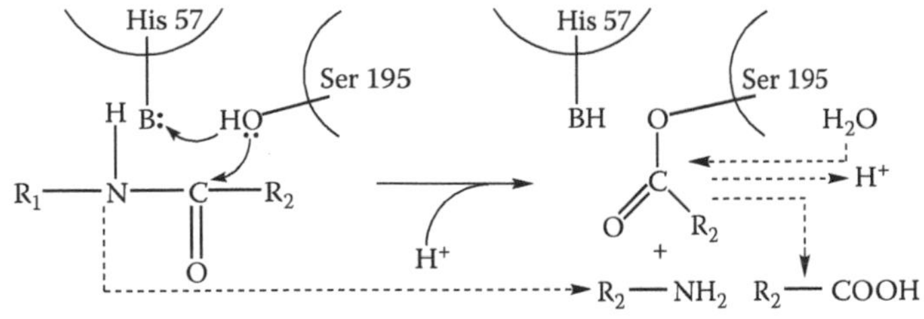

FIGURE 21.6 Schematic presentation of chymotrypsin mechanism: the first step, acylation of Ser195, is a covalent catalysis; the second step, hydrolytic deacylation is an acid–base catalysis.

of the active site. This is accompanied by release of the second product, a peptide ranging from the N-terminus to the scissile bond. The Cys- and Thr-proteases, with activated –SH and –OH nucleophile groups, function in essentially the same way. In aspartic, glutamic, and metallo proteases, the nucleophile reagent is water, activated by the concerted action of carboxyl groups of Asp and Glu or by a metal ion, usually Zn^{2+}. The activated water molecule is not part of the enzyme structure, and transient covalent attachment of the reaction intermediate to the active site is thus not possible. The covalent catalytic step is here replaced by an acid-base catalysis.

Proteases exhibit diverse functions. They are not only important food digestive enzymes and cleaners of redundant and damaged peptides and proteins in the organism (see Chapter 8), but are currently regarded as important signaling molecules, involved in a number of key physiological processes, including blood clotting and lysis of the clot, cell-cycle progression, cell proliferation, immune response, and cell death. By directed and strictly controlled cleavage of proteins, proteinases serve as highly effective switches that rapidly switch the activity of other proteins on and off. In many processes, proteases are organized in cascade systems, which result in rapid and effective amplification of an organism's response to a physiological signal.

FURTHER READING

1. Fersht, A., 2002. Structure and Mechanism in Protein Sciences. New York: W. H. Freeman and Company.
2. Smith, H. J., 2005. Enzymes and Their Inhibition (Drug Development). Boca Raton: CRC Press.
3. Segel, I. H., 1993. Enzyme Kinetics. New York: John Wiley & Sons.
4. Cornish-Bowden, A., 1995. Fundamentals of Enzyme Kinetics. London: Portland Press Ltd.
5. Jitrapakdee, S., St Mazrice, M., Rayment, I., Cleland, W. W., Walace, J. C., and Attwood, P. V., 2008. Structure, mechanism and regulation of pyruvate carboxylase. Biochem. J. 413, 369–387.
6. Cleland, W. W., Frey, P. A., and Gerlt, J. A., 1998. The low barrier hydrogen bond in enzymatic catalysis (Minireview). J. Biol. Chem. 273, 25529–25532.
7. Cannon, W. R. and Benkovic, S. J., 1998. Solvation, reorganization energy, and biological catalysis (Minireview), J. Biol. Chem. 273, 26257–26260.
8. Warshel, A., 1998. Electrostatic origin of the catalytic power of enzymes and the role of preorganized active sites (Minireview), J. Biol. Chem. 273, 27035–27038.

22 Nucleic Acid–Binding Proteins

Ülo Langel

CONTENTS

Cells react to environmental stimuli with short-term responses (occuring rapidly as a result of modifications of already-existing proteins) or longer-term responses (usually through changes in gene transcription). Transcription is regulated by the cell's transcription factors and other proteins determining the potentially transcribable genes at any given time. Such regulation of gene expression is the key process in cells, and, hence, the understanding of this is essential in examining the multiple complicated molecular interactions that control protein production. In this chapter, we briefly describe and categorize the various proteins known to regulate information transfer from transcription to translation.

Broadly, one can differentiate between (1) oligonucleotide, DNA, and RNA interacting proteins, recognizing them by chemical interactions (e.g., Zn-fingers and Leu-zippers—cf. below), and (2) transcription- and translation-regulating proteins, defined by their biological function. The situation is often complicated since classification of these protein classes is sometimes overlapping.

One way to illustrate the possible classification of transcription factors is presented by the simplified scheme in Figure 22.1, where different signaling schemes

Cell Surface Receptors

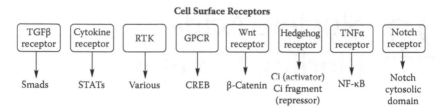

Signaling Transcription Factors

FIGURE 22.1 Classification scheme for transcription factors based on their cell surface receptor coupling and signaling. Different receptors signal after specific, extracellular ligand binding followed by activation of the transcription factor (often phosphorylation in cytosol) followed by the translocation and DNA interactions in the nucleus. RTK—receptor tyrosine kinases, GPCR—G-protein-coupled receptors, (Modified from Lodish, H. et al., *Molecular Cell Biology,* 6th ed., 2008. W. H. Freeman. New York. With permission.)

by cell surface receptors yield the activation of different transcription factors, leading to multiple cellular events as suggested by H. Lodish and colleagues (cf. below). The multiple intracellular signaling pathways in this process involve the actions of protein kinases, the disassembly of multiprotein complexes, and other ways for creation of the active transcription factor. These pathways are highly regulated in order to control the level and duration of signaling in gene expression. Such classification is relatively novel and could serve as additional way to understand the complicated world of transcriptional regulators.

22.1 DNA POLYMERASE-BASED CLASSIFICATION AND MECHANISMS OF TRANSCRIPTION FACTORS

In eukaryotic cells, three RNA polymerases—RNA polymerase I, II, and III—are responsible for transcription of genes, requiring several additional transcription factors (although sharing one of them, the TATA-binding protein, TBP) to associate with specific promoter sequences (Figure 22.2). One way to classify transcription factors is by their interaction with their specific RNA polymerases.

RNA polymerase I is responsible for the transcription of tandem repeat ribosomal RNA genes. Other factors, including UBF (upstream binding factor) and SL1 (selectivity factor 1), cooperatively recognize the promoter and, subsequently, recruiting polymerase I and forming the initiation complex. The SL1 transcription factor is composed of four protein subunits, one of which is TBP.

RNA polymerase II is responsible for the synthesis of mRNA from protein-coding genes. It has been demonstrated in 1979 by Robert Roeder's group that RNA polymerase II is able to initiate transcription only if additional proteins (distinct initiation factors) are added to the system. Basic transcription machinery constitutes polymerase II promoters and additional transcription factors bind to DNA sequences that control the expression of individual genes and are thus responsible for regulating gene expression (Figure 22.2). Initiation of transcription by RNA polymerase II involves several general transcription factors. One of them, TFIID, binds to the

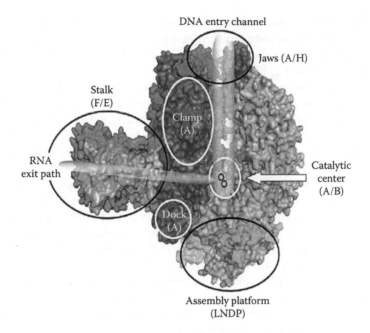

FIGURE 22.2 Structural organization of archaeal RNA polymerase. The RNA polymerase (RNAP, RNAPII, 1WCM) subunits are shade-coded according to function; the two magnesium ions in the active site are shown as metallic spheres. The major DNA entry channel and the exit path of the RNA transcript are highlighted as grey-transparent cylinders. (From Werner F., *Mol. Microbiol.* 2007, 65(6), 1395–1404. With permission.)

TATA-box and is involved in the transcriptional initiation. TFIID itself is a multiprotein complex composed of TBP and 10–12 other proteins, called TBP-associated factors (TAFs). TBP recognizes another general transcription factor (TFIIB) forming a TBP-TFIIB complex at the promoter. In addition, TFIIB forms complex with a third factor, TFIIF. Two additional factors, TFIIE and TFIIH, are required for initiation of transcription.

The genes for tRNAs, 5S rRNA, and some of the small RNAs involved in splicing and protein transport are transcribed by polymerase III, involving transcription factors TFIIIA, TFIIIC, and TFIIIB. The TFIIIB is composed of multiple subunits, one of which, again, is the TATA-binding protein, TBP.

22.2 GENERAL TRANSCRIPTION FACTORS

General transcription factors (GTFs) in eukaryotes are proteins controlling whether genes are transcribed (TFIIA, TFIIB, TFIID, TFIIE, TFIIF, and TFIIH where TF denotes transcription factor, and II denotes polymerase II). Some of these GTFs are part of the large multiprotein transcription complex that interact with RNA polymerase and do not bind DNA directly. However, many recognize regulatory DNA sequences using DNA-binding domains and, by directly controlling transcription,

are central to the regulation of gene expression. TFs function either alone or by interacting with multiprotein complexes by increasing (activators) or inhibiting (repressors) transcription in the presence of a RNA polymerase, one of the best-studied
transcriptional activators.

GTFs are present in all cells at all times and most are constitutively active.
However, one large class of TFs requires activation prior to initiation of events
and are, consequently, these are called conditionally active. These TFs involve
subclasses of developmental, signal-dependent (extra- and intracellular ligand-
dependent, e.g., p53), cell membrane receptor dependent (CREB) and latent cytoplasmic (NFκB) TFs.

TFs generally consist of two domains, one specifically recognizing DNA and the
other independently activating transcription by interacting with other components of
the transcriptional machinery. The activation domains of transcription factors may
have differential structural features such as acidic activation domains that are rich in
negatively charged residues Asp and Glu, or in Pro and Gln residues. The activation
domains often stimulate transcription by interacting with general transcription factors such as TFIIB or TFIID, thereby facilitating the assembly of the transcriptome
on the promoter. In these interactions, additionally, various activators can bind to
different general transcription factors, yielding the combined action of multiple factors synergistically stimulating transcription.

22.3 TRANSCRIPTIONAL REGULATORY PROTEINS

Approximately 5%–10% of the mammalian genome encodes regulatory proteins for
gene transcription. Genes are regulated by multiple gene regulatory proteins; each
of those proteins is the product of a gene that is in turn regulated by multiple other
proteins, whereas these regulatory proteins are additionally influenced by extracellular signals, adding even more complexity.

The events that control transcription, recombination, and replication (cf. Chapter
4) involve multiprotein-DNA complexes (Figure 22.3). Besides the "direct" regulation of gene transcription (e.g., transcriptional activators or repressors), the role of
the multiple proteins is often to bring the requisite proteins into close proximity in

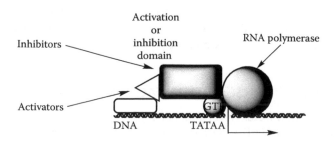

FIGURE 22.3 Schematic illustration of multiprotein–DNA complexes, with activation
domain controlling events of transcription, recombination, and replication.

order to induce conformational changes in the DNA (involving architectural proteins, such as TATA-binding protein TBP, interacting exclusively with the minor groove of DNA). Analysis of protein binding to specific DNA has defined the correlation of protein-binding sites with the regulatory elements of enhancers and promoters, indicating that these sequences generally constitute the recognition sites of specific DNA-binding proteins. The association of the complexes is often initiated by the presence of a specific DNA sequence, which can cause sufficient affinity for cooperative DNA recognition for two gene regulatory proteins with a weak affinity. This, in turn, leads to protein dimerization, enabling recognition by a next set of transcriptional regulators.

Eukaryotic repressors bind to specific DNA sequences and inhibit transcription or, in some cases, they interfere with the binding of other transcription factors to DNA (Figure 22.3). Often, the repressors compete with activators for binding to specific DNA regulatory sequences, yielding inhibition of the activator binding, and thereby inhibiting transcription. Many active repressors contain specific functional domains that inhibit transcription via protein–protein interactions and serve as critical regulators of cell growth and differentiation.

22.4 DNA SEQUENCE SPECIFIC AND NONSPECIFIC PROTEIN BINDING

Interactions of proteins with nucleic acids are, in general, between linear sequences of amino acid and nucleotide residues in aqueous solvent. Nonspecific protein–oligonucleotide interactions are largely electrostatic due to the polyelectrolyte character of DNA. Specific base-pair sequence interactions involve several increments such as structural, coding and thermodynamic components, which can be characterized in separate experiments.

Proteins recognize a specific sequence by interacting with functional groups at nucleobases within the major and minor grooves in the DNA double helix, influencing the conformation of each region of the helix and, indirectly even protein binding. Noncovalent interactions between DNA and its binding proteins determine the specificity. Hydrogen bonds within DNA major grooves, and hydrophobic interactions within minor grooves are responsible for these interactions, as exemplified in Figure 22.4 by interactions between side chains of Arg and Asn with base pairs G-C and A-T, respectively.

Specific recognition of DNA sequence by its binding protein is possible despite the availability of nonspecific sites of interaction, suggesting involvement of the greatest possible number of DNA-protein contacts, which, in turn, explains the favorable fit of DNA-binding motifs to DNA major and minor grooves as well as induced fit of both interaction parts. One way to maximize contacts yielding higher specificity is to insert multiple DNA-binding domains within a single protein, or to assemble dimers of DNA-binding proteins as seen in the case of, for example, many helix-turn-helix and zinc-finger proteins where both DNA-binding motifs of the two proteins are able to access the helix.

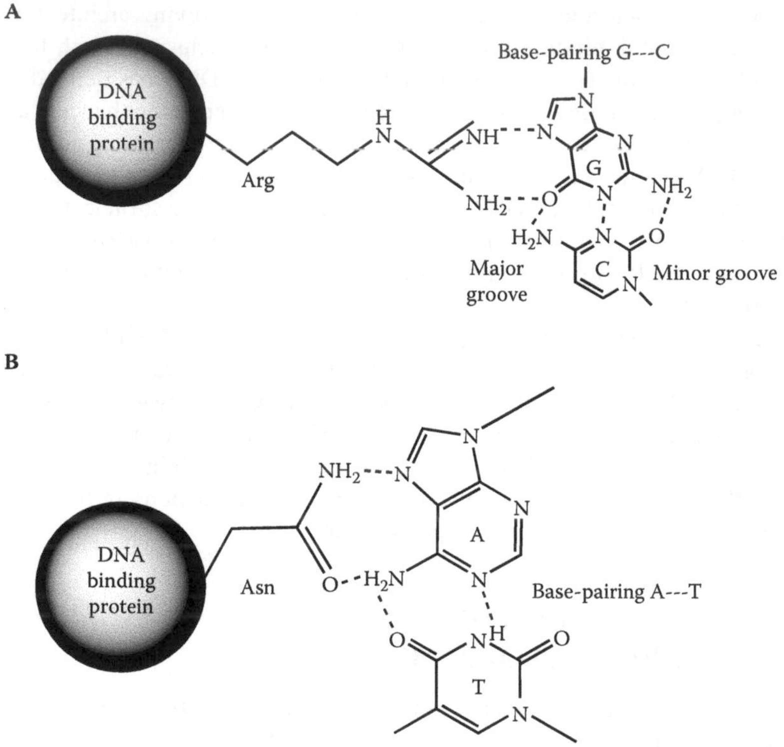

A

B

FIGURE 22.4 Specific interactions between proteins and oligonucleotides: (A) Arg and G-C; (B) Asn and A-T.

22.5 RNA-BINDING PROTEINS

Posttranscriptional regulation of gene expression—that is, events occurring at the level of RNA—is relatively poorly understood as compared to transcriptional regulation. RNA-binding proteins (RBP) are key regulators of RNA splicing, transport, and translation. RBPs also determine the route of action for bound pre-mRNAs and RNAs. By these actions, they may be key players in human disease such as cancer. However, even though many RBPs have critical roles in the translation in various tissues, it is yet largely unclear how RBPs control these events, primarily due to the difficulty in identifying their RNA targets. The list of proteins with identified RNA-binding properties is around 800 (not counting the alternative splice variants) in human genome, and this number is constantly growing.

Multiple RBPs can be classified according to their type of targets or their expression pattern. One can distinguish between ribonucleases, adenosine deaminases, aminoacyl-tRNA synthetases, ribosomal structural proteins, and translation factors among well-known RBPs. In their RNA targets, RBPs may recognize specific sequences, structures, or both. However, the RNA-binding specificity of most RBPs is unknown, mainly due to the lack of identified RNA targets. Nonetheless, knowledge of the RNA binding specificity of some RBPs is beginning to emerge. In several cases, RNA-binding domains predict the molecular function of the RBP, such as DEAD/DEAH box for RNA helicase activity and the PAZ (Piwi/Argonaut/Zwille) domain for short, single-stranded RNA-binding in RNAi or miRNA processes.

Several RBP domains include elements being associated with DNA-binding properties of transcription factors such as zinc fingers (mainly C-x8-C-x5-C-x3-H type) and homeodomains such as fingers 4-6 in TFIIIA. Multiple specific RBP motifs are known, additionally, such as the dominating ribonucleoprotein (RNP, also known as RRM and RBD) domain comprising four β-strands and two α-helices in the order β-α-β-β-α-β; the double-stranded RNA-binding domain (dsRBD) with the structure α-β-β-β-α such as dsRBD3; the κ-homology (KH) domain with the structure β-α-α-β-β-α such as the Nova 1 KH3, DEAD/DEAH box; the Pumilio/PBF (PUF) domain; PAZ domain, Sam domain such as Vts1, and others. Many RBPs have one or more copies of the same RNA-binding domain, whereas others have more distinct domains.

RBPs often contain multiple repeats of a few general modular units yielding higher specificity, affinity, and versatility of interactions with RNA molecules by combining multiple functionalities and weaker interactions of the individual units. In Figure 22.5A, this is exemplified by a schematic modular structure of RBP Dicer, comprising, among other motifs, of dsRBD and endonuclease catalytic domain which, respectively, enable recognition of double-stranded RNA and Dicer-specific interaction and cleavage of specific RNA in RNA-interference pathway (cf. Chapter 16).

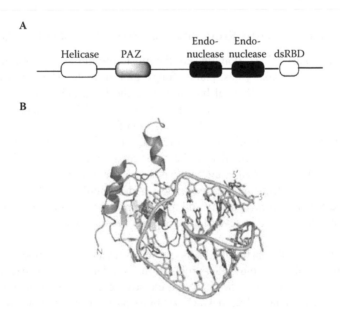

FIGURE 22.5 Interactions of RBPs with RNA. A: Modular structure of RNA-binding protein Dicer. PAZ—Piwi/Argonaut/Zwille domain, dsRBD—double strand RNA-binding domain. B: Structure of the N-terminal RNA-recognition motif (RRM) of human U1A bound to RNA. (From Macmillan Publishers Ltd.: Lunde, B.M., 2007. Nat. Rev. Mol. Cell *Biol.* 8, 479–490. With permission.)

In Figure 22.5B, an example of RBD, U1A, bound to RNA, is presented. In this structure, and in many other RRM–RNA complexes, single-stranded bases are specifically recognized through the protein β-sheet and through two loops that connect the secondary structure elements. Binding is mediated, in most cases, by conserved residues of one Arg or Lys, forming a salt bridge to the phosphodiester backbone, and two aromatic residues making stacking (π-π) interactions with the nucleobases. These three amino acids reside in highly conserved RNP-motifs located in the two central β-strands.

The modular organization of RBPs allows for their simultaneous interactions with other proteins besides RNA, where the simplest example is the dimerization of RBPs. Such simultaneous interactions cause additional synergy and complexity in RBP networks.

22.6   STRUCTURE OF DNA-BINDING PROTEINS

Multiple DNA-binding proteins have been identified; among them, one can name some well-known classes: transcription factors, TATA-binding protein, RNA polymerases, transcriptional activators and repressors, histones, RecA (in DNA recombination), DNA glycosylases, nucleases, origin recognition proteins, DNA polymerases and ligases, single-strand binding proteins, DNA topoisomerases, and restriction endonucleases. The structural features for some of these are summarized below and illustrated in Figure 22.6.

22.6.1   Helix-Turn-Helix Regulatory Proteins, Homeodomain Proteins

The helix-turn-helix (HTH) domains are the most common motifs in multiple transcription factors in prokaryotes, plants, fungi, and animals. The core of the HTH domain comprises of a trihelical bundle where the third helix of the HTH motif interacts mostly with the DNA major groove (Figure 22.6A), whereas the second helix stabilizes the interaction. HTH transcription factors demonstrate considerable structural variety (e.g., simple trihelical or tetrahelical structures) and participate in various functions (e.g., mediating the substrate interactions of different DNA-operating enzymes, and regulating both basal and specific transcription regulators, depending on their DNA-binding properties). Additionally, the HTH domains can demonstrate RNA-binding capacity and mediate protein–protein interactions. HTH modules have been found in several transcription factors such as TFIIB and TFIIE. (cf. to follow.)

In eukaryotic cells, helix-turn-helix proteins include the homeodomain proteins (Figure 22.6B), playing critical roles in the regulation of gene expression during embryonic development. Some of the earliest recognized Drosophila mutants resulted in development of flies in which one body part was transformed into another; for example, in the mutant called Antennapedia, legs rather than antennae outgrowth in the head of the fly were observed. Analysis of these mutant genes indicated that they contain "homeoboxes," conserved sequences of 180 base pairs encoding the DNA-binding domains (homeodomains) of transcription factors. A wide variety of homeodomain proteins have been identified in mammals, fungi, and plants.

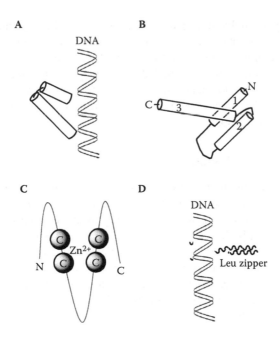

FIGURE 22.6 Some structural features of DNA-binding proteins, α-helical domains of proteins are indicated by cylinders. (A) Schematic interaction of second and third helices from helix-turn-helix motif with DNA; (B) schematic drawing of a homeodomain protein with indication of helices 1-3; (C) topology and structures of ZnF domain of classical GATA (1GAT) domain; (D) leucine zipper motif and its interaction with DNA.

22.6.2 Zinc Finger Regulatory Proteins

Zinc finger domains are defined as Cys and His containing repeats (either CCHH or a CCHC), coordinating Zn^{2+} ions and folding into DNA-binding looped structures ("fingers," ZnFs; cf. Figure 22.6C). These domains are common in the polymerase III transcription factor TFIIIA (cf. upcoming text). The TFIIIA-type ZnFs are predicted to exist in around 1000 different proteins in humans, and altogether in more than 15,000 domains. The C-terminal ZnF domain of GATA-1 (Figure 22.6C) has additionally been demonstrated to mediate interactions with several transcriptional regulators, including CBP, Fli-1, Sp1, EKLF, and PU. It has been additionally demonstrated that a major function of many classes of ZnFs is to function as protein recognition modules, thereby regulating transcription by mechanisms distinct from ZnF-DNA interactions. Zinc finger domain containing transcription factors are also represented by the gene transcription regulating steroid hormone receptors, recognizing, for example, steroid hormones.

22.6.3 Leucine Zipper Motifs

Leucine zipper proteins contain DNA-binding domains formed by dimerization of two polypeptide chains (cf. Figure 27.6D). The leucine zipper proteins contain sequences of 4-5 Leu residues separated by seven amino acids, and, consequently, exposing their hydrophobic side chains to one side of a helical cylinder structure and yielding the dimerization of the two protein subunits. A DNA-binding region rich in positively charged amino acids, such as Lys and Arg, follows the leucine zipper. The leucine zipper regulatory proteins include TFs such as c-fos and c-jun, important regulators of normal development which, when mutated or overexpressed, can generate cancer. These DNA interacting homo- or heterodimeric proteins are also called basic zipper proteins (bZips).

22.6.4   OTHER DNA-BINDING MOTIFS

Additional DNA-binding motifs have been discovered in different proteins, and one can find several ways to classify them in the literature. One of the better studied is TATA-binding protein (TBP), which is defined by its specific binding to TATA-box by recognition of β-sheet structure, and the main contacts are within the minor groove of the DNA molecule. The basic domain containing protein recognizes DNA by the interaction-induced α-helical structure containing a high number of basic amino acids such as Arg. An additional structural motif, the ribbon-helix-helix motif, found in bacterial proteins, involves a ribbon (i.e., two strands of a β-sheet) and makes contact with the major groove of DNA.

FURTHER READING

1. Lodish, Harvey, Berk, Arnold, Kaiser, Chris A. et al. 2008. Molecular Cell Biology, 6th ed. New York: W. H. Freeman and Company.
2. Fletcher, S. and Hamilton, A. D. 2005. Protein surface recognition and proteomimetics: mimics of protein surface structure and function. Curr. Op. Chem. Biol. 9(6): 632–638.
3. Lunde, B. M., Moore, C., and Varani G. 2007. RNA-binding proteins: modular design for efficient function. Nat. Rev. Mol. Cell. Biol. 8(6): 479–490.
4. von Hippel, P.H. 2007. From "simple" DNA-protein interactions to the macromolecular machines of gene expression. Annu. Rev. Biophys. Biomol. Struct., 36, 79–105.

23 Cell Surface Receptors and Signaling

Ülo Langel

CONTENTS

23.1 CELL SURFACE RECEPTORS

The (cell surface) receptor concept answers the question of how information can be transmitted to the cell: receptors are "small, discrete (protein) area(s) on the cell membrane or within the cell with which molecules or molecular complexes (e.g., hormones, drugs, and other chemical messengers) interact." The origins of the receptor concept date back to the 1878 when Paul Ehrlich and John Newport Langley first mentioned the idea. The receptor concept was initially based on biological response data and, subsequently, on radio ligand-binding properties. The development of receptor characterization began with receptors that mediate response by coupling to G-proteins, also known as G protein-coupled receptors (GPCRs; cf. following).

In addition to cell surface receptors, there are nuclear and endoplasmic reticulum (ER) receptors, which are located within the interior of cells. Nuclear receptors are responsible for the expression of specific genes acting as DNA-binding proteins and regulating transcription in response to hormones and in concert with other proteins. ER receptors recognize intracellular second messengers such as IP3, identified by the inositol triphosphate receptor, which is an ER membrane glycoprotein complex acting as Ca^{2+} channel. These two receptor classes are not described in detail here.

Pharmacological term "receptors" usually refers to proteins that are "signal-transducing receptors," according to the IUPHAR (International Union of Pharmacology) classification. "A receptor is a cellular macromolecule, or an assembly of macromolecules, that is concerned directly and specifically in chemical signaling between and within cells." Transduction of extracellular signals across the plasma membrane to the intracellular environment is achieved by the interaction of regulatory molecules with specific membrane-spanning cell surface receptors. Interaction of an appropriate activating ligand with the receptor at the external face of the cell results in the generation of an intracellular signal. Cell surface receptors are able to distinguish their specific ligands from the multitude of other bioactive factors in the extracellular milieu. In response to specific activation, such receptors function in the transmission, amplification, and integration of extracellular signals through a variety of intracellular mechanisms to control cellular functioning. Ligand-activated receptors are internalized and in this way, the signal initiated by them can be transferred to different cellular compartments.

Based on structural and functional criteria, three broad categories of cell surface receptors that recognize specific molecules may be defined. These comprise (1) ion channels and ligand-gated ion channels, (2) G-protein coupled (7-transmembrane domain) receptors, and (3) enzyme-associated receptors with subunits having one transmembrane domain. However, this classification of receptors is not overwhelming since it does not include several important, additional cell-signaling proteins like cytokine receptors (interleukin receptor family, tumor necrosis factor receptor family, etc.), the immunoglobulin superfamily, Wnt-activated Wnt receptor complex, NGF receptors (p75 or low-affinity nerve growth factor receptor), and others.

Cells express receptors that operate through different signaling mechanisms applied by these different receptors, and cross talk often occurs between intracellular signaling pathways. Hence, the individual signaling mechanisms of particular receptors are highly regulated and integrated by cross talk, which explains the high complexity of how the cell surface receptors function.

23.1.1 Ion Channels

IUPHAR defines 9 subclasses of ion-channel and ligand-gated ion channel receptors, such as the Cys-loop superfamily (anion channels like $GABA_AR$ and cation channels like nAChR), glutamate-gated cation channels (NMDAR and non-NMDAR), epithelial Na^+ channels, voltage gated cation channels (Na^+, Ca^{2+}, K^+), chloride channels, ATPase-linked transporters, and transporters related to neurotransmitters.

23.1.1.1 Structural Features of Ion Channel Receptors

Receptors in this class all contain homologous polypeptides expressed by homologous genes or resulting from alternative splicing and, hence, an array of receptors

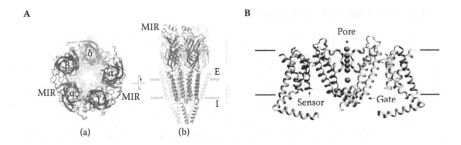

FIGURE 23.1 Structural organization of two representative ion channel receptors. (A) Ribbon diagrams of the whole nAChR from Torpedo at 4 Å resolution. receptor, as viewed from the synaptic cleft (left) and parallel with the membrane plane (right). (From Unwin, N. J. 2005. *Mol. Biol.*, (346): 967–989. Reprinted with permission from Elsevier.) (B) Structure of a voltage-gated potassium channel, side view showing two of the pore-forming domains with the associated ion binding sites. (From Corry, B. 2006. *Mol. BioSystems*, 2: 527–35. Reprinted with permission of the Royal Society of Chemistry, U.K.)

can be constructed using different combinations of these gene products. Often, these receptors contain subunit structures, each of these subunits comprising of 2–12 trans-membrane domains, depending on the receptor type. The assembly of the structure can yield an oligomer surrounding a membrane pore. The subclasses of transmitter-gated ion channels vary largely in the membrane topologies of their subunits.

Nicotinic acetylcholine (nACh) receptors (Figure 23.1A) have historically served as a source of structural information for this class of receptors because of their high content in skeletal muscles and electric organs. nAChR have important role in neuronal signal transmission, nociception, learning, memory, and addiction. The secondary structure of Torpedo nAChR is presented in the atomic-scale model at 4 Å resolution in Figure 23.1A. nAChR is a pentameric protein (stochiometrically subunits $\alpha_2\beta\gamma\delta$, each subunit spanning the plasma membrane four times; neuronal receptors contain only α- and β-subunits) that encloses a central ion channel of 7–9 nm, which is permeable to cations such as Na^+, K^+, or Ca^{2+}. Various combinations of subunits form subtypes of the nAChR. The ligand, ACh, is recognized by an interface of the α-subunit and its neighbor in the pentameric structure; the number of ACh binding sites in each receptor can be two.

Among multiple ion channel receptors, the potassium channel (K^+ channel) is another important example, being responsible for moving K^+ ions in and out of nerve cells and thus allowing rapid nerve signaling (Figure 23.1B). K^+ channels are able to rapidly transport ions, and this is achieved by a relatively wide pore. To be able to select between ion types, however, the pore must be appropriately narrow. The structure determination of K^+ channels demonstrates that such a compromise can be achieved by utilizing a short, narrow section of the pore (selectivity filter), while keeping the rest of the pore wider to aid rapid diffusion. The voltage gated K^+ channels provide an example of channel gating as a number of structures exist for the channels in different functional states. The crystal structure studies showed that the voltage sensing domain extends into the lipid in a paddle-like structure (Figure 23.1B).

23.1.1.2 Ion Channels and Signaling

Transduction of the signal through these receptors occurs via the opening of a cation or an anion channel. These receptors are often activated by recognition of neurotransmitter molecules initiating rapid signaling between neuronal cells excited by the passage of an electrical action potential. Binding of the ligand, extra- or intracellularly, to the receptor changes the ion permeability of the plasma membrane as the receptor undergoes a conformational change that opens or closes an ion channel followed by a rapid return of the receptor to initial state. Transmembrane voltage-gated channels, such as those for Ca^{2+} and K^+, are sensitive for voltage changes in the cellular membrane potential, and regulating action potentials by returning the depolarized cell to a resting state.

In case of nAChR, the receptor can exist in three different conformations in equilibrium: resting, active, and desensitized, or temporary inactive states. Agonists bind to the active state, but show higher affinity for the desensitized one owing to a conformational change produced by the agonist. In the desensitized state, the channel does not respond to further stimuli, and the desensitization-causing agonist behaves like an antagonist (Figure 23.6A).

23.1.2 G-Protein-Coupled Receptors

A wide range of neurotransmitters, neuropeptides, polypeptide hormones, inflammatory mediators, and other bioactive molecules transduce their signal to the intracellular environment by specific interaction with a class of receptor that relies upon interaction with intracellular guanine nucleotide (GTP or GDP)-binding proteins (G-proteins) for activation of intracellular effector systems. In recent years, molecular cloning approaches have allowed the identification of more than 800 discrete G protein-coupled receptor (GPCR) molecules. It has been established that approximately 80% of known hormones and neurotransmitters activate cellular signal transduction mechanisms by activating GPCRs, and that 40%–50% of prescription drugs today target GPCRs. It is predicted that the genome may code 1500–2000 receptors in total. This renders the GPCR superfamily the largest single class of eukaryotic receptors. Three subclasses of these receptors are defined as rhodopsin-like, secretin/calcitonin receptor-like, and as metabotropic glutamate-like receptors, cf. below.

23.1.2.1 Structural Features of GPCR

GPCRs are integral membrane proteins that occur in a wide range of organisms and interact with a diversity of ligands (Figure 23.2A). All members of the GPCR superfamily described to date comprise a single polypeptide chain that characteristically contains seven stretches of mostly hydrophobic residues of 20 to 30 amino acids, linked by hydrophilic domains of varying length; accordingly, they are often called 7TM receptors. This structure was originally proposed on the basis of the similarity of hydropathic features exhibited by GPCRs bovine rhodopsin (Figure 23.2B) and the α-adrenergic receptor. A proposed structure consistent with the projection map of rhodopsin was obtained by electron crystallography of two-dimensional crystals and suggests that the general three-dimensional arrangement of TM helices is similar for these membrane proteins. In 2007, the group of Brian K. Kobilka reported on crystal structure of the human β_2 adrenergic receptor at 3.4 and 2.4 Å resolution when bound to different ligands (Figure 23.2C) in what can be considered as breakthrough in the field. Although similar to rhodopsin structure on first sight, several differences are seen when compared to the structure of rhodopsin;

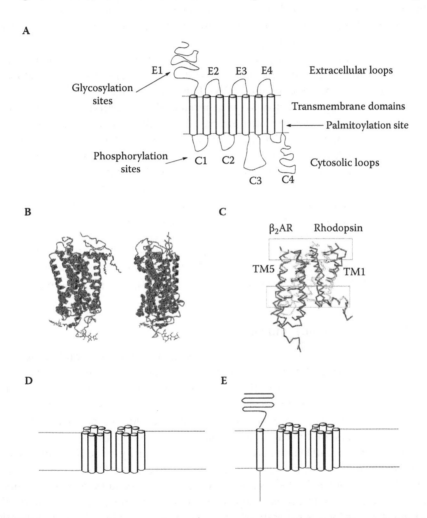

FIGURE 23.2 G-protein coupled receptors. (A) Simplified scheme of structural features of G protein coupled receptors, 7TM helices assemble to bundle as illustrated in schemes B and C. (B) Crystal structure model of bovine rhodopsin at 2.2 Å. (From Okada, T. 2004. *J. Mol. Biol.,* 342(2): 571–583. Reprinted with permission of Elsevier.) (C) Comparison of β_2AR and rhodopsin structures, β_2AR is superimposed with the homologous structure of rhodopsin. (From Rasmussen, S. G. F. 2007 *Nature,* 450(7168): 383–387. Reprinted with permission from Nature Publishing Group, Macmillan Publishers Ltd.) (D) GPCR multiprotein complexes are often formed as homo- or heterodimers as illustrated here by dimer of 7-helix bundles. (E) GPCR can form multiprotein complexes with receptor associated accessory proteins, RAMP.

for example, the weaker interactions occur between cytoplasmic ends of TM3 and TM6, involving the conserved E/DRY sequences.

Almost all GPCRs contain Cys residues in the first and second extracellular loops (Figure 23.2A). Mutation of these Cys residues results in altered function of rhodopsin, α-adrenergic, and muscarinic acetylcholine receptors. These data are consistent with the existence of intramolecular disulfide bond(s) between the adjacent first and second extracellular loops in native receptors that are critical for maintaining the tertiary structure of the receptor proteins.

Most receptors also exhibit conservation of a Cys residue in the C-terminal segment of the cytoplasmic C-terminal domain. This is the site of palmitoylation in rhodopsin and the β-adrenergic receptor (Figure 23.2A). This posttranslational modification would result in the anchoring of receptors to the membrane, thereby introducing an additional intracellular loop (or short helix) into the molecular architecture of the receptor.

Another form of posttranslational modification during biosynthesis of GPCRs is Asn-linked (N-linked) glycosylation, which results in the covalent attachment of carbohydrate residues to extracellular domains of the protein. A further form of posttranslational modification is seen in the phosphorylation of particular residues located in intracellular domains of receptors. Receptor phosphorylation by multiple kinases is of fundamental significance in desensitization processes.

The molecular understanding of GPCRs has been conceptualized as a monomeric protein until relatively recently. The current understanding is that several GPCRs may exist as homo- or heterodimers (Figure 23.2D), or associate with accessory proteins such as RAMPs (Figure 23.2E), yielding alterations in their pharmacological properties.

23.1.2.2   GPCR Families and Subtypes

Based on structural diversities (the size of cytoplasmatic tail/loops and intracellular tail) of mammalian GPCRs, they are grouped into three major families (Figure 23.3). Considerable variation is seen in the overall length of the receptors. For example, the mature metabotropic glutamate receptor mGluRla is composed of approximately 1180 amino acid residues. Adrenergic, dopamine, serotonin, and muscarinic acetylcholine receptors are typically on the order of 350 to 600 amino acids in length. Variability in the overall length of receptors occurs primarily in the N-terminal extracellular domain, which is between 550 and 575 residues in mGluR1-7 as compared to less than 10 residues in the adenosine A_{2A} and A_{2B} receptors. Size variability is also seen in the third intracellular loop and the C-terminal intracellular domain. The third intracellular loop is generally shorter in receptors that bind protein or peptide ligands or lipid mediators, relative to receptors for the bioactive amines.

Family 1 receptors respond to a wide range of stimuli: light, odorants, peptide hormones, glycoproteins, nucleotides, and chemokines, and can be subdivided according to the ligand and receptor homology; they include receptors such as the β2-adrenergic receptor (Figure 23.3). Family 2 receptors all contain a larger, 100–150-residues-long N-terminal tail, and include the Ca^{2+}-sensing receptor (Figure 23.3). Family 3 receptors involve a huge N-terminal tail (500–600 residues; Figure 23.3).

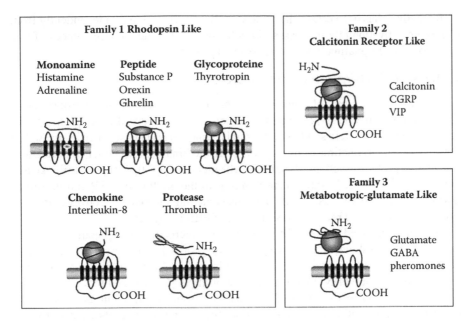

FIGURE 23.3 Schematic models of the three GPCR subfamilies with their typical ligands. (From Lang, M. 2006. *Curr. Protein Peptide Sci.*, 7: 335–353. With permission of Bentham Science Publishers Ltd.)

Orphan receptors comprise novel receptor sequences whose assignment to the G protein-coupled receptor superfamily is currently based on sequence features alone, but whose activating ligand remains unidentified. Such receptors have been identified mainly by application of low-stringency hybridization and PCR amplification, using degenerate oligonucleotide primers. Endogenous ligands have been identified only for relatively few receptors originally isolated as orphan sequences.

23.1.2.3 G-Protein and Effector Coupling

The G-protein-coupled receptor-mediated activation of enzymes, such as adenylyl cyclase, guanylate cyclase, phospholipases C and A2, and phosphodiesterases, results in the generation of intracellular second-messenger molecules such as cyclic AMP (cAMP), cGMP, diacylglycerol (DAG), inositol 1,4,5-trisphosphate (IP3), and arachidonic acid and associated metabolites. This sets in motion a cascade of events that continues through the capability of second messenger molecules to elicit a variety of subsequent effects (Figure 23.6B). These may include alteration of the activity of various protein kinases or the mobilization of calcium from intracellular stores, which may in turn affect the activity of ion channels or other enzymes.

Mutagenesis studies on adrenergic, muscarinic acetylcholine, and neuropeptide receptors have established that all of the intracellular domains are involved in efficient functional coupling of receptors to G-proteins. There is little primary sequence homology within these domains, and it is their secondary structure that is thought to be important in mediating interactions with G-proteins.

Recently, it has been established that GPCRs can even signal independently from G-protein activation in a few cases, though, the physiological consequences of this unconventional signaling have not yet been explored.

23.1.2.4 G-Proteins and G-Protein Cycle

G-proteins comprise part of a superfamily of proteins that bind and hydrolyze GTP. In addition to trimeric G-proteins, the larger group includes the elongation factor Ef-Tu, involved in protein synthesis; the small monomeric GTPases such as ras, which convey signals originating from receptor tyrosine kinases; and larger cytoskeleton-associated proteins such as dynamin. All members of this superfamily switch between "on" and "off" states corresponding to the binding of GTP and its subsequent hydrolysis. In the case of GPCRs, this switching mechanism conveys information from an occupied receptor to some sort of effector system.

G-proteins coupled to the 7TM receptors are heterotrimeric molecules, comprising a variable α-subunit and relatively invariant β- and γ-subunits (Figure 23.4B and Table 23.1). On activation by agonist A ([1]; Figure 23.4A), guanosine-5'-diphosphate (GDP), which is normally bound to the α-subunit, is released and replaced by GTP (2). The α-subunit, with bound GTP, separates from the βγ subunits (3), and this generates a signal within the cell, since both the α- and βγ-subunits are able to bind to effector systems in the membrane (4), to either stimulate or inhibit their activity. The specificity of signal transduction is determined largely by the specificity of G-protein coupling, since different G-protein subunits preferentially stimulate particular effectors. Subsequently, the G-protein α subunit, which contains an integral GTPase activity, hydrolyses GTP (5), reverts to the GDP-bound, inactive state, and dissociates from the ion channel or enzyme, and reassociates with the βγ-subunit. Signal amplification occurs as a consequence of the ability of a single receptor to activate many G-protein molecules and from the initiation of several subsequent catalytic cycles of effector enzymes.

It has been generally proposed that a given receptor always interacts with a particular G-protein or with multiple G-proteins within one family. However, for several GPCRs, it is generally accepted that simultaneous functional coupling with distinct unrelated G-proteins can be observed, leading to the activation of multiple intracellular effectors with distinct efficacies and/or potencies, often referred to as promiscuous coupling.

23.1.3 Enzyme-Associated Receptors with Subunits Having One Membrane-Inserted Domain

These receptors involve four subclasses of receptors. First, receptors with intrinsic tyrosine kinase (TK) activity (e.g., PDGF, EGF, insulin, and NGF receptors) are single-subunit proteins with or without extracellular Ig domains, multiple subunit TK receptors formed by post-translational cleavage, or Trk receptors for neurotrophins. Second, nonenzyme-containing receptors associating with extrinsic TK (e.g., IL1, GH, GDNF receptors) comprise a wide range of multisubunit receptors with a ligand-specific subunit and a subunit for signal transduction. Third, there are the receptors with Ser or Thr kinase activity (e.g., TGF receptors). Last, there are the intrinsic cyclase receptors, such as those with guanylate cyclase activity.

A

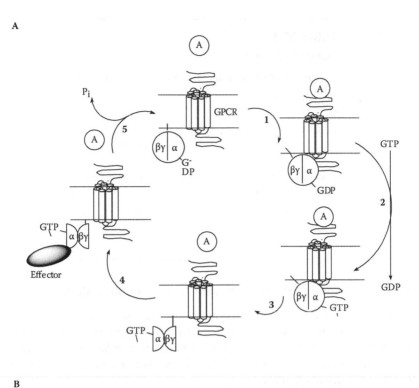

B

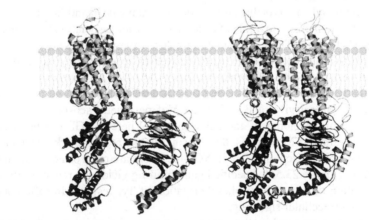

FIGURE 23.4 G-protein-coupled receptors and G-proteins. (A–G) protein cycle; (B) models of the receptor–G-protein complex. Two representations of receptor–G-protein complexes are shown. On the left (R1·Gα(O) βγ complex) is a representative model of classical GPCR–G-protein interactions in which one GPCR interacts with one G-protein. On the right is a representation of an alternative model that depicts the ability of a heterotrimeric G-protein to interact with a GPCR dimer complex (R2·Gα-βγ-complex). (From Holinstat, M., Oldham, W. M., and Hamm, H. E. 2006 *EMBO Reports* 7(9): 866–869. Reprinted with permission from Nature Publishing Group, Macmillan Publishers Ltd.)

TABLE 23.1

List of G-Proteins

G$_i$ Subfamily, Pertussis Toxin Sensitive	Gq Superfamily
Gα_{i1}, Gα_{i2}, Gα_{i3}	Gα_q
Gα_{oA}, Gα_{oB}	Gα_{11}
Gα_{t1}, Gα_{t2}	Gα_{14}
Gα_{gust}	G$\alpha_{15/16}$
Gα_z	
G$_s$ subfamily, cholera toxin sensitive	G$_{12}$ superfamily
Gα_{olf}	Gα_{12}
Gα_{sL}	Gα_{13}
Gα_{sS}	
XLGα_s	

23.1.3.1　Structural Features of Enzyme-Associated Receptors

This class of receptors is defined structurally by referring to subunits with a single transmembrane spanning helix (or membrane-inserted segment of nonpeptide, such as a glycolipid). Most of these receptors are heteromers where one subunit within a group of receptors is often shared. When single subunits are involved, the ligand causes their dimerization or clustering.

Extracellular domains of these receptors can contain the ligand-binding site, whereas intracellular domains have a role of transduction or for association with other subunit types in a heteromer. Three important examples of these are presented as follows.

Nerve growth factor (NGF) receptors recognize specifically trimeric glycoprotein NGF, the best-characterized member of family of 14 kD neurotrophins (other known representatives are BDNF, NT-3, and NT-4), with ability to stimulate growth, differentiation, and survival and maintenance of peripheral sensory and sympathetic neurons during development and nerve injury. Whereas NGF receptors consist of two distinct single TM proteins, the tyrosine kinase receptor (trkA^{NGF}) and the p75 pan-neurotrophin receptor (p75^{NTR}), NGF receptors are most often active in homodimeric form (Figure 23.5A). Specific ligand binding yields the receptor dimerization and activation of intracellular phosphorylation of Tyr residues in different proteins leading to intracellular signaling.

One mechanism used by the immune system to detect invasion by pathogens is through the Toll-like receptors, TLRs. Stimulation of different TLRs induces different patterns of gene expression leading to the activation of the acute immune response and the development of antigen-specific acquired immunity. TLRs are integral membrane glycoproteins, being the members of a larger superfamily that includes the interleukin-1 receptors, IL-1Rs based on considerable homology in the cytoplasmic region. The extracellular region of the TLRs contains leucine-rich repeats, whereas in IL-1Rs are three Ig-like domains (Figure 23.5B). TLRs and IL-1Rs have a conserved 200 aa region in their cytoplasmic tails known as Toll/Il-1R (TIR) domain, which are responsible for signaling.

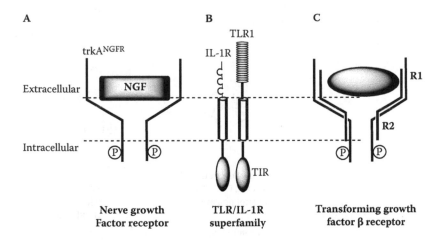

FIGURE 23.5 Simplified examples of enzyme-associated cell surface receptors with possible phosphorylation sites (P) indicated. (A) Heterodimeric nerve growth factor receptor trkA; (B) Toll/interleukin-1 receptor complex; (C) hetero-tetrameric transforming growth factor-β receptor.

Transforming growth factor-β, TGF-β, signals receptor Ser/Thr kinases whose activity is located in intracellular part, and Smad effectors, to inhibit proliferation and induce apoptosis in various cell types. This signaling pathway yields interactions with oncogenic pathways in humans: TGF-β acts at tumor suppressor in early stages of tumorigenesis, but can promote advanced tumor cell invasiveness and metastasis. TGF-β is a common name for more than thirty members of the TGF-β superfamily that share a common machinery of signaling. TGF-β is a disulphide-linked homodimeric polypeptide bound to, for example, latent TGF-β binding proteins (LTBPs). This mature ligand binds directly to the protein core of a hetero-tetrameric transmembrane proteoglycane receptor, TGF-βR (Figure 23.5C), consisting of two types of receptors, R1 and R2. Receptor trans-phosphorylation results in a conformational change of R1 and activation of its catalytic center, leading to signaling by Smad effectors. Recently, it has been demonstrated that the TGF-β superfamily ligands can even activate MAP kinase and Act pathways of signaling.

23.1.3.2 Enzyme-Associated Receptors and Effector Coupling

This class of receptors is involved in very wide set of receptor functions in immune and nervous systems, and elsewhere. The following three examples illustrate these diverse actions.

The first is one in which, upon binding and following internalization, NGF triggers a cascade of biochemical events. trkA^{NGFR} is a transmembrane glycoprotein belonging to the class of receptor tyrosine kinases. Its signaling cascade is complicated, including MAPK-Ras-Erk pathway with several protein phosphorylation events involved. Most of the NGF-mediated biological activities are due to trkA^{NGFR} auto-phosphorylation and subsequent activation of several signaling pathways (Figure 23.6C).

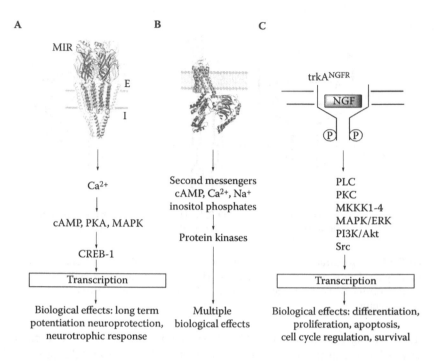

FIGURE 23.6 Simplified examples of selected signaling pathways through cell-surface receptors. PK = protein kinase, MAPK = mitogen-activated protein kinase, CREB-cAMP = response element-binding (protein), cAMP = cyclic adenosine monophosphate, PL = phospholipase, MAPK/ERK = complicated signaling pathway coupling growth factor signaling to their specific receptors, ERK = extracellular signal-regulated kinases, PI3K/Akt = signaling pathway required for multiple cellular activities, PI3K = phosphoinositide 3-kinase, Akt = Akt family enzymes are members of the serine/threonine-specific protein kinases, Src = family of proto-oncogenic tyrosine kinases. (A) Nicotinic acetylcholine receptor, nAChR. (From Unwin N. 2005. *J. Mol. Biol.*, 346: 967–989. Reprinted with permission from Elsevier.) (B) G-protein-coupled receptors, GPCRs. (From Holinstat, M., Oldham, W. M., Hamm, H. E. 2006. *EMBO Reports*, 7(9): 866–869. Reprinted with permission from Nature Publishing Group, Macmillan Publishers Ltd.) (C) nerve growth factor receptor, trkA.

In the second example, after ligand binding, TLRs/IL-1Rs dimerize and undergo the conformational change yielding downstream signaling, including recruitment of the adaptor molecule MyD88, IL-1R associated kinases (IRAKs), transforming growth factor-β (TGF-β)-activated kinase (TAK1), TAK1-binding protein (TAB1), TAB2, and TNFR-associated factor 6 (TRAF6). Phosphorylation of IκBs downstream of signaling leads to IκB degradation and, consequently, the release of NF-κB, yielding in activation of respective genes being important in immune response.

In the final example, the activated TGF-βR1 phosphorylates the intracellular effector proteins Smads, a conserved family of signal transducers, containing the domains responsible for DNA binding and protein–protein interactions. Monomeric Smad proteins constantly shuttle in and out of the nucleus; in the nucleus, they bind to DNA and associate with many transcription factors, coactivators, and corepressors, leading to transcriptional induction of repression in a diverse array of genes.

23.2 MIMICRY OF 7TM RECEPTORS WITH SHORT PEPTIDES

Ligand-induced conformational changes of transmembrane helices and intracellular loops promote the activation of associated heteromeric G-proteins. Often, the third intracellular loop (C3 on Figure 23.2A) mediates the coupling between a receptor and G-protein. Synthetic peptides, derived from the intracellular loops of GPCRs, have been recently demonstrated to influence receptor–G-protein interactions with the original receptor in membrane preparations. For example, the peptides mimicking the second and third cytoplasmic loops of rhodopsin and part of its carboxyl-terminus bind to the rod cell G-protein, transducin, and block the receptor coupling.

Synthetic peptides derived from natural proteins have been useful tools in the studies of protein function and protein–peptide interactions. However, this approach has been limited by delivery of peptide to the target location, allowing studies only on membrane preparations. Recently, short peptide sequences within proteins bearing cell-penetrating properties have been defined (cf. Chapter 27), and peptides based on these sequences have shown to be useful in functional studies of the parent protein in living cells and tissues. Moreover, the short mimicking peptides open up novel ways for receptor-targeted drug design by triggering identical physiological effects like agonist activations of respective receptors. These peptides would interact with the same types of G-proteins as original receptors and trigger respective intracellular signaling cascades. This opens up many new possibilities for studying receptor functions, protein–protein interactions, and intracellular signal transduction. Such peptides also may prove useful in medical applications, where up-regulation of a specific physiological response characteristic of the respective GPCRs is required.

FURTHER READING

1. Pangalos, M. N. and Davies, C. H., Eds. 2002. Understanding GPCR and Their Role in the CNS. Oxford: Oxford University Press.
2. Meyers, R. A., Ed. 2007. Proteins. From Analytics to Structural Genomics, Vols. 1 and 2. Weinheim: Wiley-VCH Verlag Gmbh & Co. KGaA.
3. Oldham, W. M. and Hamm, H. E. 2008. *Heterotrimeric G protein activation by G-protein-coupled receptors.* Nat. Rev. Mol. Cell. Biol. 9(1): 60–71.
4. Kobilka B. K. and Deupi X. 2007. *Conformational complexity of G-protein-coupled receptors.* Trends Pharmacol Sci. 28(8): 397–406.
5. Rasmussen, S. G., Choi, H. J., Rosenbaum, D. M. et al. 2007. *Crystal structure of the human beta2 adrenergic G-protein-coupled receptor.* Nature 450(7168): 383–387.

24 Membrane Proteins

Gunnar von Heijne

CONTENTS

24.1 FUNCTIONAL CLASSES OF MEMBRANE PROTEINS

The cell membrane is the essential barrier between the cell and its surroundings, keeping unwanted molecules away and preventing proteins, metabolites, and ions from leaking out. Yet, a completely impenetrable membrane would quickly strangle the cell. The membrane must be permeable to all kinds of molecules, large and small, but the permeability must be precisely regulated. Equally important, the cell must be able to transport molecules against their concentration gradient in order to extract nutrients from the environment or pump out toxic compounds. Transport processes of these kinds are carried out by specific classes of membrane proteins that act as channels (mediating passive diffusion across the membrane) or transporters (mediating active transport against a concentration gradient).

The cell must also be able to respond to signals coming from the outside; signal transduction is often mediated by membrane receptors that couple the binding of an extracellular ligand (e.g., a hormone) to some biochemical reaction inside the cell (e.g., activation of a G-protein). For a discussion of cell-surface receptors, see Chapter 23.

Finally, membrane proteins are central players in bioenergetics. Both photosynthesis and respiration depends on highly evolved membrane protein complexes and machines.

24.2 MEMBRANE PROTEIN STRUCTURE

At the most basic level, membrane proteins can be classified as peripheral and integral. Peripheral membrane proteins only attach to one side of the membrane, either by binding to an integral membrane protein, or by binding directly to the lipid surface. Integral membrane proteins span the lipid bilayer one or more times. Operationally, integral membrane proteins are defined as proteins that cannot be extracted from the membrane fraction by sodium carbonate at pH 11.5.

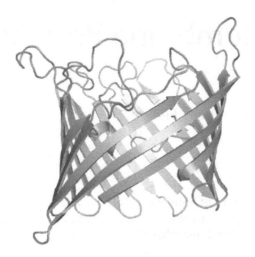

FIGURE 24.1   Porin from Rhodopeudomonas blastica (PDB code 1PRN).

Lipid-binding peripheral membrane proteins are essentially water soluble but present a hydrophobic surface patch, often surrounded by a region with net positive electric charge that interacts favorably with the membrane surface. Aromatic amino acids seem to be particularly suitable for penetrating partly into the membrane.

As far as is known from high-resolution structures of integral membrane proteins, they fold according to two basic architectures: the β-barrel and the helix-bundle. β-Barrel membrane proteins are found in a relatively restricted set of membranes, including the outer membrane of Gram-negative bacteria and the outer membranes of mitochondria and chloroplasts. The basic fold of these proteins is an antiparallel β-sheet rolled up into a barrel where the first and last strands in the β-sheet pair up with each other (Figure 24.1). Depending on the number of strands in the β-sheet, a central channel may form down the middle of the barrel. The smallest known β-barrel protein has only eight strands and is too tightly wound for a sizable channel to form, whereas the largest ones can have more than 20 strands and a central channel that is wide enough to fit a small globular domain formed by one of the loops in the protein.

The lipid-exposed surface of the β-barrel proteins is composed mostly of hydrophobic side chains projecting out from the barrel, with a notable concentration of aromatic side chains in the parts that face the lipid headgroup regions.

β-Barrel proteins have a diverse but still limited repertoire of functions. The smallest ones probably serve as outer-membrane anchors, the medium-sized ones are often substrate-selective diffusion pores for small molecules, and the largest help move bigger substrates such as siderophores or even unfolded protein domains across the membrane.

Helix-bundle membrane proteins are found in essentially all types of membranes, and are estimated to account for 20%–30% of all genes in typical genomes. They contain from one up to more than 20 transmembrane α-helices and often form complexes containing even large numbers of helices (Figure 24.2). The transmembrane α-helices are composed mainly of aliphatic amino acids, but polar and charged

FIGURE 24.2   Aquaporin 1 from bovine red blood cells (PDB code 1J4N).

amino acids are often found scattered along the helices. As in the β-barrel proteins, there is a tendency for aromatic amino acids to be concentrated in the parts facing the lipid headgroups.

The helix-bundle membrane proteins comprise an impressive multitude of functions. Proteins with a single transmembrane helix are often receptors in which the binding of a ligand to the extracellular domain results in the activation of an intracellular signaling domain, triggering a cascade of metabolic reactions and/or gene-activation events. Multispanning helix-bundle proteins can be anything from enzymes to channels and transporters, receptors, photosynthetic proteins, and respiratory proteins.

24.3   MEMBRANE PROTEIN BIOSYNTHESIS

β-Barrel and helix-bundle membrane proteins use very different mechanisms of membrane insertion and folding (see Chapter 7). In Gram-negative bacteria, β-barrel proteins are synthesized with an N-terminal signal peptide that targets them for translocation through the SecYEG translocon across the inner membrane. Periplasmic chaperones and the outer-membrane BamA complex then guides their insertion into the outer membrane in a poorly understood process. Similarly, in mitochondria β-barrel proteins are first imported via the TOM complex into the intermembrane space where they bind to the Tim9/10 chaperone. Finally, the SAM complex (homologous to the bacterial BamA complex) mediates their insertion into the outer membrane. It is unclear how the orientation relative to the membrane of β-barrel proteins is determined during the insertion process, but in nearly all known cases, both the N and C terminus of the protein faces the periplasm (or the intermembrane space in mitochondria).

The biosynthesis of helix-bundle membrane proteins is better understood. Bacterial inner-membrane proteins use the SecYEG translocon for membrane insertion, and

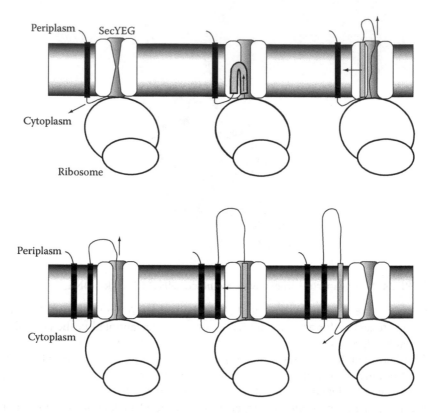

FIGURE 24.3 Simplified model for the SecYEG-mediated insertion of transmembrane helices into the inner membrane of *E. coli*. The top panel shows the insertion of the second transmembrane helix (in gray). Given the orientation of the most N-terminal helix in this particular protein, the second helix triggers the opening of the SecYEG translocon and integrates into the membrane via the lateral gate in the channel. The bottom panel shows the insertion of the third transmembrane helix (in gray). It enters the SecYEG translocon as part of a nascent chain in transit across the membrane, but is diverted into the lipid bilayer via the lateral gate.

all eukaryotic membrane proteins located in the plasma membrane or in organelles along the secretory pathway use the homologous Sec61αβγ complex. In both cases, membrane insertion is co-translational. The nascent chain is fed from the ribosome directly into the translocation channel, and sufficiently hydrophobic segments in the nascent chain are thought to be inserted into the lipid bilayer via a "lateral gate" in SecY (Sec61α).

In the simplest model, transmembrane α-helices are inserted one by one into the membrane (Figure 24.3). The first transmembrane helix (or an N-terminal signal peptide) opens the translocation channel, the second closes the channel, the third opens it again, etc., confining the loops following the helices to either the extra-cytoplasmic or cytoplasmic side of the membrane. This model is probably a good approximation of what happens for proteins with a small number of helices that are far between (or a protein with an N-terminal signal peptide and a single transmembrane helix).

In most multispanning membrane proteins, however, the transmembrane helices are close together in the sequence and the intervening loops are short. In such cases, it is likely that helices interact with each other before leaving the translocon, which means that each individual helix may not in itself be sufficiently hydrophobic to insert into the membrane—"united we stand, divided we fall."

Figure 24.3 may give the impression that the orientation of successive helices is determined by that of the first one. This is a gross oversimplification, however. In fact, there is a strong correlation between the orientation of transmembrane helices and positively charged residues (Lys, Arg) in their immediate flanking regions. The flanking region with the largest number of positively charged residues almost always faces the cytoplasm, and this is true not only for the most N-terminal transmembrane helix but for all transmembrane helices. This so-called positive-inside rule seems to apply to essentially all helix-bundle membrane proteins in all organisms, from the plasma membrane of archea that live at high temperatures in pH 3 to the highly organized photosynthetic membranes in plant chloroplasts. To some extent, the positive-inside rule can override the tendency of hydrophobic segments to insert into the membrane, and thereby affect the whole three dimention structure of the protein. It can also be used as a basis for membrane-protein engineering exercises, and a number of studies where the overall orientation of a protein in the membrane has been flipped by adding or removing positively charged loop residues have been reported.

FURTHER READING

1. Alder, N. N. and Johnson, A. E. (2004). Cotranslational membrane protein biogenesis at the endoplasmic reticulum. J. Biol. Chem. 279, 22787–22790.
2. Bos, M. P., Robert., V., and Tommassen, J. (2007). Annu. Rev. Microbiol. 61, 191–214.
3. Rapoport, T. A., Goder, V., Heinrich, S. U., and Matlack, K. E. (2004). Membrane-protein integration and the role of the translocation channel. Trends Cell Biol. 14, 568–575.
4. von Heijne, G. (2006). Membrane-protein topology. Nature Rev. Mol. Cell. Biol. 7, 909–918.
5. von Heijne, G. (2007). The membrane-protein universe—what's out there and why bother? J. Int. Med. 261, 543–557.
6. White, S.H., and von Heijne, G. (2008). How translocons select transmembrane helices. Ann. Rev. Biophys. Biomolec. Struct., in press.

25 Antibodies

Ülo Langel

CONTENTS

Antibodies or immunoglobulins are proteins in bodily fluids of vertebrates, produced and applied by the immune system in order to identify and neutralize foreign objects, antigens, from bacteria and viruses, and other pathogens. This large class of these important proteins has merited increasing interest in the scientific community due to their extremely high level of diversity in their binding to target antigens. Additionally, antibodies play a major role in many types of diseases such as cancers, infectious diseases, allergy, autoimmune diseases, and inflammation, making them attractive as protein drug therapies. Indeed, more than 20 monoclonal antibody (MAb) drug products are already available on the dynamic and apparently permanently growing market. The popularity of antibody drugs is due to their specific action and their flexibility to conjugate additional therapeutic entities for delivery to target sites or, as in case of conjugated radioisotopes, for specific diagnostic purposes. Production of antibodies is the main function of the humoral immune system, however, this exciting field of immunology is not described here in detail.

25.1 STRUCTURE AND GENERATION OF IMMUNOGLOBULINS

Antibodies can be described as roughly Y-shaped protein molecules consisting of two large heavy chains (H) and two small light chains (L; Figure 25.1). One can differentiate the variable (V) and constant (C) regions; V-region defines antigen-binding properties and C-region is responsible for interactions with effector cells and molecules.

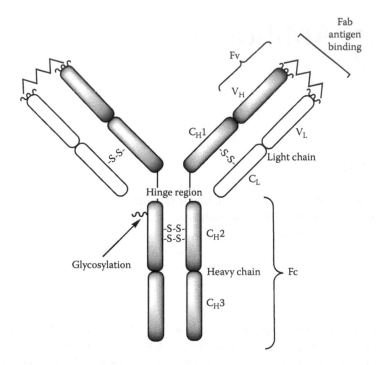

FIGURE 25.1 Immunoglobulins are proteins consisting of two types of polypeptide chains, heavy (H) and light (L) chains (A) exemplified by IgG.

In humans, five classes of immunoglobulins exist: IgA, IgD, IgE, IgM, and IgG (Table 25.1), based on their C-chain structure. Most IgGs are monomers; IgA and IgM are mostly dimers and pentamers, respectively, connected by J-chains. Such diversity aids to perform different roles and diversity of antibodies. IgGs are the most abundant and widely used as therapeutics; therefore, the antibodies will be exemplified here using IgG.

TABLE 25.1
Properties of Immunoglobulin Classes

Class	MW, kDa	Chain Structure
IgG	150	$\kappa_2\gamma_2$ or $\lambda_2\gamma_2$
IgA	180–500	$(\kappa_2\alpha_2)_n$ or $(\lambda_2\alpha_2)_n$, $n = 1$–3
IgM	950	$(\kappa_2\mu_2)_5$ or $(\lambda_2\mu_2)_5$
IgD	175	$\kappa_2\delta_2$ or $\lambda_2\delta_2$
IgE	200	$\kappa_2\varepsilon_2$ or $\lambda_2\varepsilon_2$

Note: All immunoglobulins consist of κ or λ light chains, and α, γ, μ, δ, or ε heavy chains. IgM and oligomers of IgA also contain J chains that connect immunoglobulin molecules.

Antibodies involve, besides heavy and light chains, several additional structural features such as a hinge region, disulfide bonds and glycosylation moieties, Figure 25.1. In IgG, several disulfide bonds connect two heavy chains at the hinge region, heavy and light chains and four intra-chain disulfide bonds, with some exceptions. In IgGs, one oligosaccharide has been found, residing mostly on one of the conserved Asn side-chain in C_H2. Additional carbohydrate modifications have been identified in some cases. Proper disulfide bonds and glycosylation are often critical for functionality of antibodies.

Today, the 3D structures of IgG immunoglobulins have been solved for only a few proteins. In general, they show that anti-parallel β-sheet domains in antibody (their content is around 70%) mainly contribute to its structure. Each domain is around 110 amino acids long and form similar β-barrel folds (Chapter 3) stabilized by disulfide bonds and hydrophobic interactions. These domains, in turn, fold into three spheres linked by a flexible hinge region yielding the Y overall shape of the antibodies. Disulfide bonds and strong noncovalent interactions between the interacting chains stabilize the whole molecule.

IgG subclasses (IgG1-4) contain different heavy chain (γ1-4) and light chain (λ and κ) subtypes (Table 25.1), with different locations and number of disulfide bonds and the length of hinge region. The variable regions (V, divided into three hypervariable sequences) of both chains form the antigen-binding (Fab, "fragment antigen-binding") region, and constant regions form Fc ("fragment crystallizable") region. Structural heterogeneity of purified antibodies has often been registered, caused by gene crossover or posttranslational modifications.

A small region at the Fv (Figure 25.1) of the antibodies, Fab, is extremely variable (the so-called hypervariable region) enabling binding of different targets and antigens, and recognition by the immune system of a wide diversity of these. The unique part of the antigen recognized by an antibody is called an epitope (antigenic determinant). The two V regions in the antibody structure offer two identical antigen-binding sites that recognize antigens specifically by an induced fit mechanism where 5-10 amino acid residues usually contribute to the recognition. Haptens are small molecules that can be recognized by antibody but are only able to stimulate production of antihapten antibodies when linked to a larger protein carrier. An antigen made of two identical hapten molecules joined by a short flexible region can link two or more anti-hapten antibodies, forming dimers, trimers, etc.

The generation of antibodies is a complicated issue due to the complexity of the immune system. The immune system recognizes a huge variety of pathogens dynamically during an infection. To do so, lymphocytes carry out targeted and regulated alterations of genomic structure and sequence. Antibodies are produced by a white blood cell called B cells. Early in the development of B and T cells, antigen receptor V regions are constructed by recombination, followed by Ig expression and secretion by B cells. In the case of the B-cell receptor the C-terminus is a hydrophobic membrane-anchoring sequence, and in the case of antibody it is a hydrophilic sequence that allows secretion. The V regions of heavy and light chains together determine the antigen recognition domain, and the heavy chain C region determines how the antigen is removed from the body.

Hence, the B cells yield the diversity of the Ig molecules and optimize the response to the specific pathogens. The large and diverse population of antibodies is generated by random combinations of a set of gene segments that encode different antigen-binding sites, followed by random mutations in this area of the antibody gene, which creates further diversity. Antibody genes also reorganize in a process called *class switch recombination* that detaches the base of the heavy chain from the C region and joins it to a downstream C region, deleting the DNA between and creating a different isotype of the antibody that retains the antigen-specific variable region. This allows a single antibody to be used by several different parts of the immune system. There are two types of lymphocytes, B and T, and also two similar antibody-binding structures: one is an antigen receptor that stays in the membrane of the cells and another is a free antibody (secreted only by B cells).

25.2   ANTIGEN BINDING OF IMMUNOGLOBULINS

The recognition of an antigen by specific antibody is reversible and mediated by cointeraction of multiple weak noncovalent forces, such as hydrogen bonds, hydrophobic forces, and ionic interactions between the complementary regions of a Fab domain (Figure 25.1) and the corresponding region on the antigen (antigenic determinant). The antigens often consist of more than one different antigenic determinant—that is, they are multivalent.

The adaptive immune system is very flexible and capable of recognizing almost any antigen without having been previously exposed to it. This flexibility is not due to the generation of unique antibody to each antigen, as it was believed at early stage of development of immunology. Instead, as demonstrated more recently, the multispecificity—that is, the ability of a single antibody (or any receptor in general) to engage multiple ligands provides the coverage of huge numbers of antigen–antibody interactions.

Several explanations of mechanisms for multispecific ligand recognition by antibodies (or receptors in general) can be found in literature. In one case, a rigid antibody recognition site interacts with different ligands with different affinities (Figure 25.2A). Alternatively, the antibody can obtain the ligand-induced conformation in an "induced fit" process, enabling diversity in ligand recognition (Figure 25.2B). In antigen–antibody recognition studies, both mechanisms have been demonstrated, depending on the structure of the interaction partners. Additionally, in some cases the reciprocal induction of the antigen structure upon recognition by the antibody has been suggested, especially in case of structurally less defined molecules such as short peptides. The entire range of mechanisms for antibody multispecificity are very likely to contain additional possibilities, and intensive research is ongoing in this field, hopefully leading to better understanding as well as new therapeutic applications.

The antigen–antibody (Ag-Ab) interaction is reversible and can be quantitatively described as any reversible interaction, such as receptor–ligand interaction in Chapter 17. In case of reversible interaction between single antigenic determinant (Ag)

A

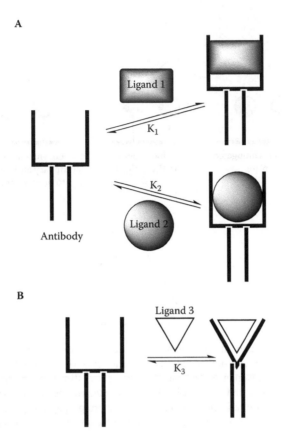

FIGURE 25.2 Schemes for multispecific ligand recognition by antibodies. (A) Rigid adaptation model. (B) Induced fit model.

and single antigen-binding site (Ab) the interaction can be expressed, as in Equation
25.1.

$$Ag + Ab \xrightleftharpoons{K} AgAb \qquad (25.1)$$

The interaction affinity expressed as equilibrium binding constant, K, is, according
to the law of mass action, described as in Equation 25.2.

$$K_a = \frac{[AgAb]}{[Ag][Ab]} \qquad (25.2)$$

Note the similarity with the ligand-receptor binding described in Chapter 17, suggesting similar quantification of the affinity. Several methods exist for quantification of antigen–antibody interactions, the most popular of which are the ELISA and
BIAcore assays.

The enzyme-linked immunosorbent assay (ELISA) is often used to detect and
quantify the antigen–antibody interaction (Figure 25.3A). In general, ELISA method
uses antigen-modified plates (96-well format), which are in complex with a specific

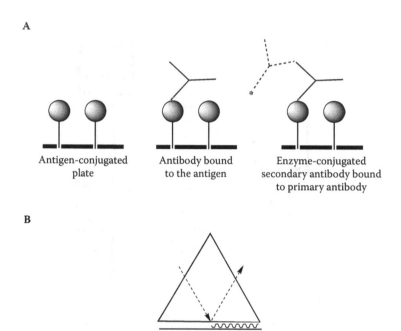

A

| Antigen-conjugated | Antibody bound | Enzyme-conjugated |
plate | to the antigen | secondary antibody bound |
| | | to primary antibody |

B

FIGURE 25.3 Two examples of methods to quantify antibody-antigen interactions. (A) General scheme of enzyme-linked immunosorbent assay (ELISA). (B) Simplified scheme for surface plasmon resonance. The metal (gold, silver) film is evaporated onto the glass block. The light (visible and infrared light, but also electron beam, can be used) penetrates the glass and metal film, and electromagnetic waves are propagated parallel along a metal/vacuum interface. These waves are very sensitive to the adsorption of molecules to the metal surface.

primary antibody. This is followed by the addition of a secondary antibody (with affinity to the primary antibody) conjugated to, for example, an enzyme, such as horseradish peroxidase, or a fluorophore for detection. By adding the antigen or antibody, it is possible to quantify the interaction in a similar way as is done in case of measurements of ligand-receptor interactions, Chapter 17.

The BIAcore surface plasmon resonance (SPR) technology provides another straightforward method of characterizing many aspects of the immune response, on a real-time basis, including antibody concentration, specificity, and binding affinity. SPR is an optical technique that measures the refractive index change at the sensor–fluid interface (Figure 25.3B). A plastic fluidics system plate is pressed into contact with a gold-coated glass chip whose surface is coated with an un-cross-linked polymer matrix, forming micro-reaction flow cells. The gold partial mirror is the optical port as well. Binding of antibodies is quantified by a computer-driven system to facilitate consistency. The injection of analyte (antibody) across a matrix-conjugated ligand (antigen) yields a real-time change in the refractive index associated with the MW change. The affinity quantification is possible by changing the concentration of analyte.

In case of participation of polyvalent antigens or antibodies, the scheme of interaction is naturally more complicated than simply Equation 25.1, and determination of affinity more difficult. For example, the IgM molecule with 10 binding sites will have up to four orders of magnitude or more greater overall affinity (avidity) for a multivalent antigen than an IgG molecule (two binding sites).

The antigen-binding site at Fab is formed by the N-terminal ends of V_H and V_L (Figure 25.1), where the diversity of three small hypervariable regions, each consisting of 5-10 amino acids, provides the structural basis for the diversity of antigen-binding sites. Numerous contacts between both interacting molecules, such as hydrogen bonds, electrostatic interactions, and hydrophobic interactions, combine to yield specific and strong binding. Macromolecules often make more extensive contact while small molecules often bind inside the antigen-binding pocket.

25.3 DIVERSITY OF ANTIBODIES

Natural antibodies are **polyclonal**, derived from different B-cell lines and consisting of antibodies against several epitopes, thus increasing the probability of eliminating the invading pathogen or malignant cell. These antibodies are typically produced by immunization of a suitable mammal, such as a mouse, rabbit, or goat with relevant protein followed by the purification from the mammal's serum to obtain antisera. High-affinity antisera are often used in experimentation or diagnostic tests.

In the 1980s, murine **monoclonal antibodies** (mAb) became available, offering novel possibilities for biomedicine. G. Köhler, C. Milstein, and N. K. Jerne shared the Nobel Prize in physiology and medicine in 1984 for this discovery. Monoclonal antibodies are produced by an immune cell line that are all clones of a single identical parent cell. With this method, it is possible to create specific monoclonal antibodies against almost any substance. Monoclonal antibodies for a given antigen can be used in multiple purposes such as the detection and quantification of the antigen (Western blot, BIAcore, ELISA, immunofluorescence test, immunohistochemistry), and purification by immunoprecipitation and affinity chromatography. One possible way to target cancer is to use mAbs for binding to cancer cell-specific antigens as well as for delivery of a toxins to address tumor treatment or biochemical tools to study cancer mechanisms.

When mABs are produced in animals such as mice, the antibodies need to be "**humanized**" to avoid the recognition of them as foreign in human patients, otherwise they become rapidly removed from circulation or cause systemic inflammatory effects. One approach to achieve humanized antibodies has been to engineer chimeric antibodies where the DNA encoding the binding Fab portion of monoclonal mouse antibodies is merged with human antibody-producing DNA, followed by cell culture expression of these chimeric constructs for monoclonal antibody therapy.

Transition state analogs and catalytic antibodies. All chemical transformations pass through an unstable structure called the transition state, which is poised between the chemical structures of the substrates and products. The transition states (TS) for chemical reactions are proposed to have lifetimes near 0.1 ps, the time for a single bond vibration and, hence, the real transition state structures are impossible to obtain. The transition state of a chemical reaction is a particular configuration defined as the state corresponding to the highest energy along this reaction coordinate—that is, an activated molecule is formed during the reaction at the transition state between

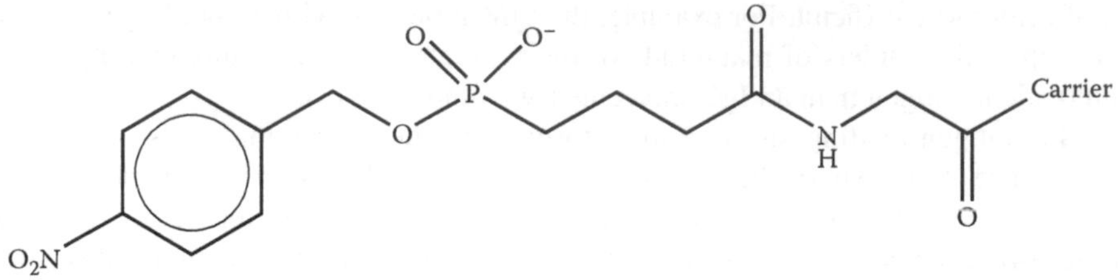

FIGURE 25.4 Hapten-carrier protein conjugate used for immunization to produce catalytic antibody ester hydrolytic properties.

forming products from reactants. In Equation 25.3, a schematic transition state of a hypothetical

$$A + B - C \rightleftharpoons [A...B...C]^{\#} \longrightarrow A - B + C \tag{25.3}$$

$$\text{Transition state}$$

reaction is seen where the transition state molecule $[A...B...C]^{\#}$ and the reactants are in pseudo-equilibrium at the top of the energy barrier.

For the enzymatic reaction, the ES complex converts to products through the transition state ET*, Equation 25.4, which is central to understanding catalysis (cf. Chapter 21).

$$ES \rightleftharpoons ET^* \longrightarrow P \tag{25.4}$$

Enzymes catalyze chemical reactions at rates that are much faster, 10^{10}- to 10^{15}-fold, relative to uncatalyzed chemical reactions at the same conditions. Research on details of enzymatic catalysis, especially understanding the role of TS, has led to the discovery that monoclonal antibodies could be programmed to have enzymatic properties (Equation 25.5).

$$Ab\text{-}S \rightleftharpoons [Ab\text{-}S]^* \longrightarrow Ab\text{-}P \tag{25.5}$$

Catalytic antibodies have become a powerful tool for the efficient and specific catalysis for a wide range of chemical reactions, although the research is ongoing and their design remains a challenging task.

The basic procedure for making catalytic antibody was first proposed in 1969 by W. P. Jencks. The idea was to elicit an immune response to a synthetic transition state analog, TSA, and the resulting antibodies would contain combining sites that would force bound substrate molecules to adopt the geometric and stereoelectronic properties of the transition state. When the TSA is designed and synthesized, it is conjugated to a carrier protein and used for immunization, and the specifically TSA recognizing antibodies are isolated and characterized toward the targeted chemical reaction. Since the initial proof of this concept in 1986 by the groups of R. A. Lerner and P. G. Schultz, more than 100 catalytic antibodies (CAB) or abzymes have been generated and characterized. It has been demonstrated that reasonable TS analogs (TSA) for an organic reaction are likely to yield antibodies with at least modest catalytic efficiency. Hence, the design and synthesis of suitable TSA is a key issue in CAB development.

For example, the mechanism of ester hydrolysis has been extensively studied. The antibody has been developed that catalyses the hydrolysis of p-nitrophenyl alkyl ester and carbonates with rate accelerations above the uncatalyzed reaction of about 10,000. The antibody was elicited to an arylphosphonate TS analog with the intention of generating a binding site that is electronically and geometrically complementary to the anionic tetrahedral TS formed during the hydrolysis reaction. In Figure 25.4, another hapten structure is presented that was used to produce esterolytic CABs with up to 10^5 fold improvement of catalytic activity.

25.4 VACCINES

Vaccines are arguably the most cost-effective public health tools for disease prevention. Even though vaccines are very efficient, they are constantly improved in order to achieve more efficient ones with, for example, better delivery properties. Here, we briefly summarize some modern vaccination strategies.

A vaccine, as known from times of E. Jenner's use of cowpox (*vacca* meaning "cow" in Latin) in the 19th century, is an immunity booster to a certain disease, yielding required antibodies. Vaccines are based on the dead/inactivated organisms or purified products derived from them, used for infection, and the according process of administration of vaccines is referred to as vaccination. Traditional vaccines can contain killed, previously virulent microorganisms (e.g., flu, cholera, hepatitis A); live, cultivated microorganisms to enable the virulent properties (e.g., yellow fever, mumps); toxoids or inactivated toxic compounds causing illness (e.g., tetanus and diphtheria); and subunit vaccines such as fragments of microorganisms (e.g., human papillomavirus). Modern approaches in vaccine development are many, and include, for example, DNA and peptide vaccinations. Modern synthetic vaccines are composed mainly or wholly of synthetic antigens such as peptides, carbohydrates, etc.

In the new and attractive strategy, DNA vaccination (immunization with purified plasmid DNA) is applied to induce immune responses and treat the infections caused by pathogens. The gene encoding the desired antigen is inserted into a (bacterial) plasmid that, upon injection into the host, causes the antigen expression and, consequently, immunity. DNA immunization has been used in case of several viral and parasitic infections, such as influenza, hepatitis B, malaria, and others. Many efforts are currently being made to achieve protective immunity to HIV by DNA immunization. Plasmid immunization offers several advantages over traditional vaccines, such as relative stability of DNA, the specificity of the produced antigen, and cost-effectiveness. The limitation of this technology is that only the encodeable antigens such as proteins and peptides can achieve the immunity induction. Besides the reported success stories in DNA immunization, several risks of this technology have been indicated, such as integration of the plasmid into the host genome, induction of anti-DNA antibodies, activation of cell-mediated immunity, and induction of tolerance—hopefully all avoidable in the future.

DNA-immunizations are usually carried out by mechanical and electrical strategies (e.g., microinjection, DNA-coated particle bombardment by gene guns, or electroporation) or by viral and chemical DNA delivery systems, such as the application of a few micrograms to milligrams of plasmid DNA. Chemically modified, nonviral carrier systems for DNA vaccines have gained much attention recently, among them various molecules with complex DNA that have been used successfully (e.g., positively charged carriers such as cationic lipids and zwitterionic lipids). Since the efficient expression of protein from DNA vaccines depends on the presence of DNA vaccine in the nucleus, the nuclear targeting of plasmid DNA is desirable, and often is achieved by nuclear localization sequences (NLS) forming DNA:NLS complexes. Methods are currently under development for increasing antigen expression by boosting the uptake of the DNA by cells, especially muscle cells, with the hope of improving intramuscular DNA vaccination.

Peptide-based vaccines have shown great potential for the treatment of chronic viral diseases and cancer, although there are no human peptide-based vaccines on the market yet. These vaccines are based on peptide-epitopes of a protein antigen, presenting the minimal immunogenic region of the protein and, hence, enabling precise direction of immune response. The usual problems in peptide therapeutics such as poor immunogenicity, peptide stability, and delivery arise even in development of peptide vaccines and also apply to peptide-based vaccines. It is hoped that these difficulties can be overcome by the available peptide modification strategies such as conjugation of peptides to protein carriers or cell-penetrating peptides, as well as application of noncoded amino acids and PEG-ylation (attachment of one or more chains of polyethylene glycol to a peptide molecule; see Chapter 31).

Recently, a vaccine therapy for Alzheimer's disease, the most common cause of dementia, was developed. A main characteristic of Alzheimer's disease (AD) is a massive cerebral accumulation of amyloid plaque accumulation consisting mainly of fibrillary aggregates of the amyloid β peptide (Aβ 1-42). In this Aβ vaccine strategy, in which a humoral immune response was induced by injection of fibrillar Aβ 1-42 peptide. Even though some patients developed complications after vaccination, the therapy was shown to be effective in reducing clinical and pathological findings. Inspired by these findings, other polypeptide conjugates comprising Aβ-specific epitopes in vaccine structure are under studies for AD treatment.

25.5   ENGINEERED ANTIBODY FRAGMENTS

Genetic engineering techniques permit the construction of many different antibody-related molecules, where different, shorter, engineered IgGs are becoming valuable as research tools as well as therapeutic molecules due to their small size, improved tissue penetrating properties, altered pharmacokinetics, immunogeneicity, specificity, valency, and effector functions. Around 80% of the FDA-approved antibody-based products are the engineered/modified antibody fragments. Some examples of engineered IgG fragments and their modifications are presented in Figure 25.5A.

One important type, scFv, is a truncated Fab comprising only the V_H and V_L domains conjugated by a peptide linker. They can be further modified by being

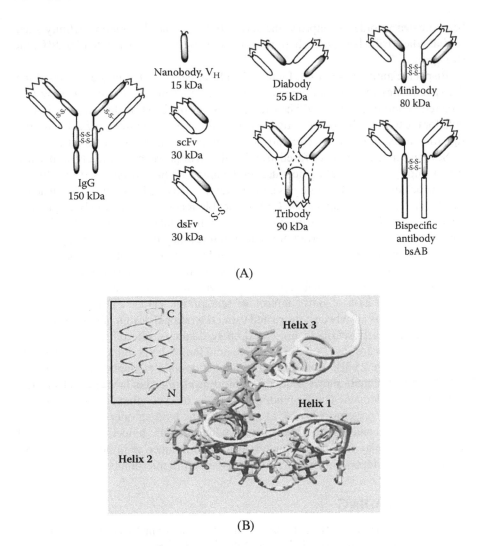

FIGURE 25.5 Engineered antigen recognizing molecules. (A) Schematic structure of IgG and some examples, together with indication of MW, of genetically engineered fragments. (B) Structure of the three-helix bundle Z domain. Top left: side view of backbone ribbon structure of the 3-helix Z domain. Right: Computer-generated image for the 3D NMR structure of the Z domain of SPA. (From Eklund,M., 2002, *Proteins*, 48, 454–462. Reprinted with permission from Wiley-Liss, Inc., subsidiary of John Wiley & Sons, Inc.)

coupled to protein toxins for targeted tumor therapy where the Fv is specific for a tumor antigen. scFv are monovalent in binding with comparable affinity with the parental Fab fragment and their serum half-life is relatively short, making them useful for diagnostic imaging applications. Another analog of scFv is dsFv where a disulphide bond is included in the domain linker. Nanobodies (V_H domains of IgG), the smallest antigen-recognizing domains, have been recently created against various tumor antigens in purpose of tumor targeting. Multimer antibody fragments have been created in order to improve the serum half-life and the binding affinity such as diabodies, tribodies, etc., where several Fv domains are conjugated by different strategies.

Bispecific antibodies (bsAB; Figure 25.5) have been created in order to combine the specificities of two antibodies into a single molecule for immunotherapy and targeted radioimmunotherapy. The strategy designs the bsAB is that one arm is specific for the tumor cell surface antigen, and another arm recognizes and activates the signaling receptor of the effector cells, resulting in the killing of the target cells.

Several peptides have been fused to antibodies to improve their different properties. For example, cell-penetrating peptides (cf. Chapter 31) have been fused to scFvs in order to achieve intracellular targets such as Bcl-2 in mast cells to neutralize the activity of Bcl-2 and induce apoptosis. Cell-permeable recombinant antibody fragments, scFv, have also been named *transbodies*. Such intracellular addressing of antibody–antigen interactions has received much attention recently due to additional possibilities as compared to available technologies of influencing intracellular processes.

High affinity and specificity of antibodies has motivated development of novel artificial binding proteins enabling cost-effective alternative flexibility to structural properties in creation of novel molecular recognition entities. Different antigen-binding fragments have been constructed using the rational design of binding capacities in small protein scaffolds, not based on Ig domains, referred to as affibodies. Affibody binders are based on the 58 amino acid long 3-helix bundle scaffold of the Z domain (Figure 25.5B), originating from the B domain of staphylococcal protein A (SPA). The Z domain shows antibody Fc-binding affinity, but lacks the Fab affinity of the parental domain. The Fc-binding surface is used as template for creation of affibody binding proteins, whereas 13 residues in the binding interface are randomized to create a combinatorial library from which specific binders can be selected using phage display technique. Binders to a large number of protein targets have been reported, with binding affinities in a nanomolar to micromolar range.

FURTHER READING

1. Lendel, C., Dogan, J, Härd, T. 2006. Structural basis for molecular recognition in an affibody:affibody complex. J Mol. Biol., 359(5): 1293–304.
2. Jain, M., Kamal, N., Batra, S. K. 2007. Engineering antibodies for clinical applications. Trends Biotech., 25(7): 307–16.
3. Xu,Y., Yamamoto, N., Janda, K. D. 2004. Catalytic antibodies: hapten design strategies and screening methods. Bioorg. Med. Chem., 12(20): 5247–68.

26 Fibrous Proteins

Ülo Langel

CONTENTS

Fibrous proteins, that is, the proteins consisting of filaments, make up many body fibers and often play structural roles. These proteins are characterized by regular, extended structures that are different from the tertiary structure of other protein classes such as globular and membrane proteins. In fact, one can say that fibrous proteins lack true tertiary structure.

The physicochemical properties of fibrous structures are based on regularities of their amino acid sequences. Multiple structural proteins are known, and they have often medical impact because they play an essential role in motility, elasticity, scaffolding, stabilization, and the protection of cells, tissues, and organisms. Protein fibers are increasingly involved in technical applications. Examples of fibrous proteins include collagens, keratins, actins, fibrinogens, and elastins (see Table 26.1). A distinction can be made between truly fibrous proteins and fibrous structures assembled from essentially globular proteins such as actins and tubulins.

26.1   COMPARATIVE STRUCTURE OF FIBROUS PROTEINS

Fibrous proteins form long rod- or wire-like protein filaments that are usually inert, water insoluble, and found as aggregates due to hydrophobic interactions between amino acid side chains. They often contain α-helix and β-sheet motifs as well as disulfide bonds between chains. Examples of structural organization of fibrous proteins include "coiled coils" of α-helices (keratins), extended antiparallel β sheets (silk fibroin), and triple helical arrangements (collagen family).

The aggregating properties of fibrous proteins are determined by their amino acid compositions. For example, sterically unhindered hydrogen bonding between amino- and carboxyl groups of peptide bonds on adjacent protein chains is possible

TABLE 26.1

Some Fibrous Proteins in Eukaryotes with Reference to Their Fiber Morphology and Function

Protein	Structural Features	Function/Occurrence
Collagen	Long triple-stranded helical rope-like superhelix structure	Involved in connective tissue, skin, and bone
Elastin	Highly cross-linked between Lys residues forming extensive networks; lacks regular secondary structure	Provides elasticity to tissues
Actin	Polar structure with two structurally different ends; tight helix	Component of the contractile apparatus of muscle cells
Fibrinogen	Two-strand protofibrils form after proteolysis of trombin followed by association, forming of thicker fibers and clots	Forms insoluble clot of blood, preventing blood loss and tissue repair after injury
Keratins	Nonpolarized coiled-coil structures	Cell epithelia function, component of nails and hair
Myosin	Thick filaments of myofibrils, 200–400 molecules in filament, coiled-coil structure in tail, motor-domain head	Plays role in muscle contraction, cell division, and generate tension in stress fibers
Tubulin	Form microtubules, which are cylindrical structures, from 13 linear protofilaments to yield polar structure	Determine the location of organelles and cell components; they are used as tracks by particles and motor proteins

due to the frequent occurrence of the "smallest" amino acids, Gly and Ala, as in keratin, collagen, elastin, and silk fibroin, which can contain 75%–80% of Gly and Ala residues. Fibrous proteins, such as keratins, also consist of many Cys residues. These residues often form disulfide bridges that yield additional strength and rigidity by permanent cross-linking and make these proteins insoluble.

In 1953, Francis Crick and Linus Pauling, in their first attempts at modeling the tertiary structure of a protein, proposed the models of supercoiled α helices or "coiled coils." The proposed packing modes of the side chains suggested different periodicities of residues in helical turns and a supercoil in the one or both supercoil senses (see Figure 26.1). Even though Crick's periodicity model ("knobs-into-holes") of seven residues over two helical turns (7/2, heptad repeats) initially prevailed as a description of coiled coils, it was later demonstrated that such protein sequences showed even less common periodicities, such as 4/1, 18/5, 15/4, and 11/3. In Figure 26.2, a model of the dimerization of a tetrameric coiled-coil structure of intermediate filaments is presented.

Extended antiparallel β sheets, such as the silk I fibroin, contain repeated β-turns, type II-like molecules (Figure 26.3) that contain multiple repeats of AGSGAG in the domesticated silkworm Bombyx mori, and alternating Ala- and Gly-rich regions in the wild silkworm Samia Cynthia ricini. The different combinations and proportions of β-sheets are packed into crystalline arrays (30–40 amino acids long, mainly repeating Gly and Ala, or poly-Ala), in loosely packed β-sheets (Ala, Gly, Pro in the

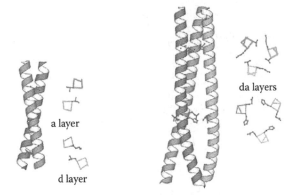

FIGURE 26.1 Periodicities of coiled-coil proteins exemplified with heptad, 7/2 (left, GCN4 leucine zipper), and nonheptad, 18/5 (right, hemagglutinin) periodicities. The cross sections show the canonical knobs-into-holes packing for the regular coiled coil (seven residues per two turns) as well as knobs-to-knobs packing engendered by other periodicities. (From Gruber, M., and Lupas, A. N. 2003. *Trends Biochem. Sci.* 28(12): 679–685. With permission of Elsevier.)

most frequent repeats), α-helices, β-spirals, and spacer regions. Repeating units may occur 50 times, yielding 300–400 kDa silk polypeptides. Interactions of favorably located amino acid side chains give rise to remarkable properties of silk such as strength despite the fine structure. This property of silk can be attributed to a fully extended conformation of β strands; further extension would require the disruption of strong covalent bonds.

The triple helical arrangement of collagen structure was first proposed in 1954 by G. N. Ramachandran and G. Kartha. Today, more than 30 distinct types of collagen have been described showing differences in the amino acid sequences, and with the structure of a triple helix of 3 polypeptide chains (Figure 26.4). Each of these is a left-handed helix (which is very different from α helix), twisted together into a right-handed coiled coil, forming a cooperative quaternary structure stabilized by numerous hydrogen bonds and some covalent crosslinking within and between the triple helices. The collagen structure often involves the repeating of amino acids such as Gly-Pro-Hyp, where Hyp is hydroxyproline. The high content of Pro and Hyp rings, with their geometrically constrained structure, promotes the tendency of the individual polypeptide strands to form left-handed helices spontaneously. A Gly residue is found at every third position, enabling sterically the assembly of the triple helix.

FIGURE 26.2 Model of dimer–dimer association of two antiparallel coiled-coil dimers. (From Strelkov, S. V. 2003. *Bioassays.* 25: 243–251. With permission of Wiley-Liss, Inc., a subsidiary of John Wiley & Sons, Inc.)

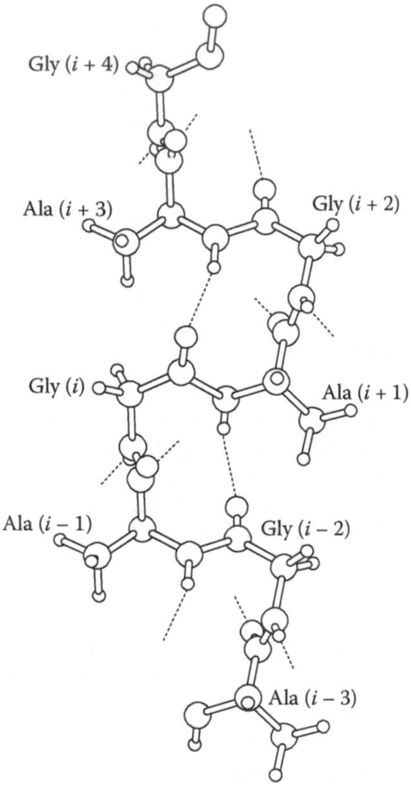

FIGURE 26.3 Conformation of a repeated β-turn type II-like molecule as a model for silk I. There are intramolecular hydrogen bonds between the carbonyl oxygen atom of the ith Gly residue and the amide hydrogen atom of the (i + 3)th Ala residue. (From Yao, J. 2004. *Biomacromolecules* 5(3): 680–688. With permission of the American Chemical Society.)

26.2   EXAMPLES OF FIBROUS PROTEINS AND THEIR FUNCTION

Silk in its natural form is composed of filament core proteins (silk fibroins) and a glue-like coating (sericin proteins) produced by insect pupae and spiders. Silk fibroins are the strongest natural fibers known. They are often classified as keratins with extended β structures (cf. above) that are organized into the multiple adjacent protein chains and which form hard, crystalline regions changing to noncrystalline regions of varying size. Humans have used the silk fiboinprotein produced by the silkworm Bombyx mori to produce high-quality thread and cloth for more than 3000 years. Silk is a lightweight, strong, and elastic biomaterial consisting of protein-based filaments or threads secreted by a specialized abdominal glands connecter to spinnerets, ducts, or spigots, and is used to produce structures for prey capture, reproduction, and locomotion. Silk fibroins are spun at ambient temperatures and pressures using water as the solvent by a complex spinning process where the soluble silk proteins are transformed into solid fibers.

FIGURE 26.4 Collagen structure. Left, polyproline helix; right, triple helical aggregated forms. (From Woolfson, D. N., and Ryadno, M. G. 2006. *Curr. Op. Chem. Biol.* 10: 559–567. With permission of Elsevier.)

Bombyx silk fibroin is synthesized in large quantity by the silk gland of the caterpillar as two polypeptide chains linked by a disulfide bridge. The protein sequences of the 26 kDa light chain and of the 325 kDa heavy chains have been determined, and the latter is Gly-rich, including many GlyAla/Ser repeats. It has received relatively little attention from biochemists.

There are several spider silks with different functions. For instance, the orb web spiders produce at least six different types of silk as well as glue substance having various biological functions such as forming web frame, capture spiral, wrapping silk, and glue coating for capture spiral. Based upon the differential amino acid composition, the silk proteins are assembled to create specific fibers with particular functions. Most research to date is focused on the study of constituents of dragline silk from the spider Nephila clavipes, which is known for its combination of strength and elasticity, and which may inspire the design of novel biomaterials with modified properties. The analysis of spider silks demonstrates small regular peptide motifs such as $(GlyAla)_n/Ala_n$, GlyGlyX, GlyProGlyXX, and spacers (X = various amino acids), where the regions Ala_n and $(GlyAla)n$ form β-sheet crystalline structures. Elasticity is governed by these Gly-rich sequences. Structure-activity studies based on this information can be initiated in order to achieve the engineered silk structures with improved properties. The stability of silk fibers in comparison to globular proteins is due to the extensive hydrogen bonding, the hydrophobic nature of much of the protein, and the significant crystallinity. Silks are insoluble in most solvents, including water.

Keratins are multiple fibrous structural proteins represented by α- and β-keratins. α-Keratins are found in mammalian tissues such as fingernails, hooves, horns, claws, and hair, and consist mainly of α-helically coiled single protein strands (but also β sheets) stabilized by intra-chain hydrogen bonding in "rope" structures. The β-keratins of reptiles (shells, claws) and birds (claws) often consist of twisted β-sheets, stabilized by disulfide bridges. Keratins are additionally the constituents of the baleen plates of whales, the shells and setae in brachiopods, and the gastrointestinal tracts of many animals. Soft epithelial keratins (cytokeratins) and harder hair keratins could be classified. In general, they form in the process of keratinization, where certain skin cells differentiate and become cornified, and prekeratin polypeptides are incorporated into intermediate filaments. This is followed by the disappearance of the nucleus and cytoplasmic organelles, and then programmed cell death. Keratinized epidermal cells are constantly replaced.

Collagen (kolla means glue in Greek) is defined as a fibrous protein of connective tissue and bones that yields gelatine in boiling. Although the importance of collagen was first recognized in prehistory and it became essential in the manufacture of leather and glue, this protein is now regarded as a key component in the mechanisms of the diseases of connective tissues. Collagen is one of the most abundant proteins in the human body, being the major constituent of most connective tissues, giving strength to and maintaining the structural integrity of various tissues and organs. More than 30 types of collagen are known, however, over 90% of the collagen in the body are collagens I (main component of bone), II (cartilage), III (reticular fibers), and IV (basement membranes).

Triple helical, insoluble collagen fibers are metabolically relatively inert, forming structural matrices of these tissues. However, collagens exhibit a variety of bioactivities by interacting with a number of other biomolecules, such as the triple helix of type I human collagen, which has a length of 300 nm and contains binding sites for over 50 other molecules such as membrane receptors and components of extracellular matrix. Collagen can directly function as a signaling ligand through the activation of collagen receptors; signalling by collagen-binding integrins and other receptors (activating intracellular kinase cascades) has also been reported. These multiple important biological functions of the collagen triple helix make collagen an attractive target for matrix engineering and the development of artificial novel biomaterials.

Intermediate filaments (IFs), together with microtubules (25 nm) and actin microfilaments (7–10 nm), and the ~10–12 nm wide intermediate filaments constitute an integrated, dynamic network in the cytoplasm of metazoan cells, which is involved in division, motility, plasticity, and other cellular processes. IF proteins (lamins) are also found in the cell nucleus, forming a meshwork of the filaments on the inside of the nuclear membrane. IFs function in close association with cytoskeletal components such as motor proteins and plakin-type cross-bridging proteins. IFs's unique structural features, including very long elementary units (45 nm) and thin (2–3 nm) rod-like dimmers (Figure 26.5), are exemplified in the structural models of vimentin and lamin dimers. IF proteins are grouped into five types based on amino acid sequence identity, and into three independent groups according to their mode of assembly (keratins, vimentin type, and lamins).

The IF proteins share similar structural features, including a central α-helical rod domain with flanking non-α-helical N- and C-terminal end domains called

Human vimentin

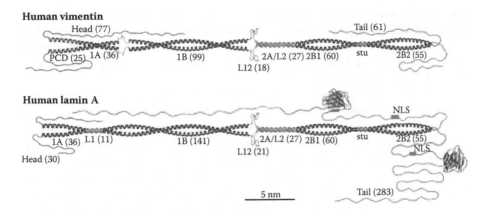

Human lamin A

FIGURE 26.5 Structural model of intermediate filament protein dimmers exemplified by modeling of the human vimentin and the lamin A dimers. The central α-helical rod domain of the individual molecules is subdivided into the coil segments 1A, 1B, 2A, 2B1, and 2B2. Linker segments that connect the individual α-helical segments are indicated: L1, L12, and L2. Regions that are predicted to form nearly parallel α-helical bundles as well as the so-called stutter (stu) region in the heptad repeat pattern are represented. The numbers in brackets refer to the number of amino acids in each respective domain. Scale bar, 5 nm. NLS, nuclear localization signal. (From Herrmann, H., Bär, H., Kreplak, L., Strelkov, S. V., Aebi, U. 2007. Nat. Rev. *Mol. Cell Biol.*, 8: 562–573. With permission of Nature Publishing Group, Macmillan Publishers, Ltd.)

head and tail, respectively (Figure 26.5). Distinct differences exist in the structure of coil 1 and the tail domain of cytoplasmic and nuclear IF proteins such as lamins, which exhibit 42 extra amino acids in coil 1B and a highly conserved immunoglobulin-fold structure of 108 amino acids in the tail. The mechanical properties of IFs are determined, in general, by a structure of coiled coils and, additionally, by cohesive forces between adjacent dimers, yielding the specific nanomechanical behavior of IFs. IFs participate in very important cellular functions and, hence, they also participate in mechanisms of several diseases. For example, mutations in lamin A cause muscular dystrophies, and more than 230 mutations in lamin A have been identified in conjunction with at least 13 different human diseases.

26.3 AMYLOID DISORDERS RELATED TO FIBROUS STRUCTURES

Amyloid fibrils and amyloid-related protein aggregates share specific structural features. The term amyloid was historically coined since the fibrils were erroneously identified as starch (ltn. "amylum"); only later was their protein character with β-sheet structure determined. Abnormal, toxic, accumulation of amyloid in organs such as the brain is associated with several human diseases (amyloidosis, Alzheimer's disease, transmissible spongiform encephalopathy, pheochromocytoma, type 2 diabetes mellitus, and others) and significant understanding of the alternatively folded protein structural motifs is, consequently, of potentially high impact. Additionally, several proteins and peptides are known to make amyloid without any known disease (e.g., calcitonin). More than 10 biologically distinct forms of amyloid are known today. Disorders connected to the misfolding of proteins are exemplified in Chapter 28.

As defined more recently, the amyloid folding motif can be assessed in any polypeptide that adopts a cross-beta sheet quaternary structure, *in vivo* or *in vitro*, with strands from different monomers. Amyloid polymerization is sequence-sensitive, often containing the beta-sheet inducers such as Pro or Gln (e.g., in case of mammalian prions, and Huntington's disease). With other amyloid-forming polypeptides (amylin, amyloid β-protein), the simple consensus sequences have not been identified, and hydrophobic interactions with participation of aromatic-side chains of amino acids seem to be responsible for the generally toxic aggregation. Such aggregation is increasingly the target for drug development for treatment of amyloid disorders.

26.4   NANOSTRUCTURED BIOLOGICAL MATERIALS BASED ON FIBROUS PROTEINS

Artificial polymeric proteins are of great interest in biotechnology. The knowledge of how the structural features of fibrous proteins determine their function often enables the design of novel nanostructured materials. Significant progress has been reported in the design of fibrous proteins that adopt predictable structure and protein folding. With control of amino acid sequences, stereochemical properties, and chain length, the artificial proteins can be designed in order to obtain novel macromolecular materials with novel properties.

The design and commercial synthesis of silk-like artificial proteins is an emerging area in materials science enabling a new generation of biomaterials with broad applications. These novel applications range from the silk substitute synthetic polymer nylon, medical structures, cloth, and ropes to materials for car and airplane construction. To obtain large quantities of silk fibers and improved analogs, the cloning of silk protein genes for overexpression in bacteria, yeast, and mammalian cell cultures has been attempted. The long-term future objective is to produce different silk fibroins that have a wide range of different mechanical properties for commercial applications. Silk fibroin has been increasingly studied for new biomedical applications due to its histocompatibility, slow degradation, and remarkable mechanical properties. The production of synthetic silks, however, has been problematic for a number of different reasons. One is their repetitive sequence (high content of Gly and Ala codons) nature and their length, which can exceed >15 kb, causing the technical barriers when utilizing basic recombinant DNA methodologies. In addition, only partial natural full-length cDNAs encoding silk components have been reported.

Synthetic spider silk sequences of up to 163 kDa that resemble dragline silk genes have been designed and expressed in goats, plants, yeast, and bacteria and insect cell line Sf9. The majority of these studies yielded size heterogeneity of the products

A

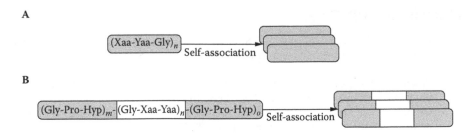

B

FIGURE 26.6 Design of collagen-mimetic peptides by (A) repeating tripeptides, (B) host–guest peptides. (From Koide, T. 2007. *Philos. Trans. R. Soc. Lond. B. Biol. Sci.* 362(1484): 1281–1291. With permission of the Royal Society of London.)

as a result of abortive transcription or translation, so further technical progress is required before large-scale production is possible. Identification of new silk family members is currently carried out using several approaches, including the screening of silk-gland-restricted recombinant plasmid cDNA libraries, or the combining of reverse genetics with MALDI tandem TOF mass spectrometric analysis of tryptic fragments generated from solubilized silk materials. It is the hope that designed silks will have mechanical properties that approach spider fibers found in nature and that this will help the production of synthetic silks in the laboratory.

For functional tissue repair, tissue engineering combines cells and bioactive factors in a defined structure based on biomaterial scaffolds that are maintained in bioreactors under the control of appropriate stimuli. Biomaterial scaffold can be prepared from natural or synthetic polymers, such as native or derivatized collagen, native or regenerated silk fibroin, etc., and used for skin, bone, ligament, or even nerve tissue targeting.

As discussed above, collagens not only maintain the structural integrity of tissues and organs, but also regulate several biological events, such as cell attachment, migration and differentiation, tissue regeneration, and animal development. Thus, synthetic triple-helical peptides that mimic the structure of native collagens have been used to study the individual collagen-protein interactions and matrix engineering. A variety of tailored triple-helical collagen mimetics are available, all based on the knowledge of the native collagen structure (see previous text). In Figure 26.6, two design possibilities of these mimetics are presented. Self-associating open chain peptides (a) and host-guest peptides (b) with the sequences containing $(Xaa-Yaa-Gly)_n$—for example, $(Pro-Hyp-Gly)_n$—or $(Gly-Xaa-Yaa)_n$ are most frequently used due to the ease of synthesis and their self-trimerizing property.

Several types of self-assembling peptides have been systematically studied and applied for scaffolding for tissue repair in regenerative medicine, drug delivery, and biological surface engineering. These include hydrogels formed from natural protein fibers such as collagen, fibrin, and elastin. Peptide nanotubes that allow ions to pass through and to insert themselves into lipid bilayers have been designed from self-assembling peptides. Surfactant-like peptides, which can form vesicles and nanotubes, have also been designed and studied.

FURTHER READING

1. Herrmann, H., Bär, H., Kreplak, L., Strelkov, S. V., and Aebi, U. 2007. Intermediate filaments: from cell architecture to nanomechanics. Nat. Rev. Mol. Cell Biol. 8(7): 562–573.
2. Woolfson, D. N., and Ryadnov, M. G. 2006. Peptide-based fibrous biomaterials: Some things old, new and borrowed. Curr. Op. Chem. Biol., 10(6): 559–567.
3. Gruber, M. and Lupas, A. N. 2003. Historical review: another 50th anniversary—new periodicities in coiled coils. Trends Bioch. Sci. 28(12): 679–685.
4. Wetzel, R., Shivaprasad, S., and Williams, A. D. 2007. Plasticity of amyloid fibrils. Biochemistry, 46(1): 1–10.
5. Hu, X., Vasanthavada, K., Kohler, K., McNary, S., Moore, A. M., and Vierra, C. A. 2006. Molecular mechanisms of spider silk. Cell. Mol. Life Sci., 63: 1986–1999.
6. Nelson, R., and Eisenberg, D. 2006. Recent atomic models of amyloid fibril structure. Curr. Op. Str. Biol. 16(2): 260–265.
7. Zhang, S., Marini, D. M., Hwang, W., and Santoso, S. 2002. Design of nanostructured biological materials through self-assembly of peptides and proteins. Curr. Op. Chem. Biol. 6(6): 865–871.

27 Selected Classes of Bioactive Peptides

Ülo Langel

CONTENTS

In this chapter, several representative classes of bioactive peptides will be presented. The selection has been made among those with possible potential as therapeutics, or those with high impact to the mechanisms of biochemical processes. Some recent examples are included to illustrate the development of peptides into nonpeptide structures, which further improves their applicability *in vivo*.

27.1 PEPTIDE HORMONES AND NEUROPEPTIDES

A hormone is a chemical messenger signaling from one cell (or group of cells) to another in order to carry information to the target cells. Endocrine hormone molecules are secreted (released) directly into the bloodstream, whereas exocrine hormones are secreted directly into a duct, followed by either flow into the bloodstream or flow from cell to cell by diffusion in a process known as *paracrine signaling* (Figure 27.1). Well-known hormone classes are amine-derived hormones (e.g., catecholamines and thyroxine), lipid-derived hormones (e.g., steroid hormones like testosterone and cortisol), and peptide hormones (e.g., vasopressin, growth hormone, and insulin). Peptide hormones are peptides that are secreted into the bloodstream and have endocrine functions in living animals.

All peptide hormones are synthesized as part of a longer pro-hormone that is cleaved by specific proteases to generate the active molecule in a secretory vesicle. Stimulation of signaling cells causes the immediate exocytosis of the stored peptide

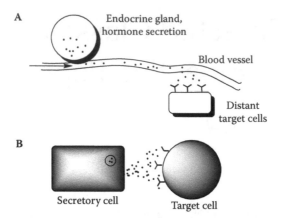

FIGURE 27.1 Schematic illustration of endocrine (A) and exocrine (B) hormone (denoted by dots) action. Receptors that specifically interact with the hormones are indicated on the target cell surface.

hormone into the surrounding medium or the blood. Many peptides hormones also act as neurotransmitters or neuropeptides, a diverse group of more than 100 peptides found in neural tissue, and are released by different populations of neurons in the mammalian brain. Neuropeptides often coexist and are coreleased with other small-molecule neurotransmitters.

Chemical transmission is the major form of neuronal communication, where endogenous substances, neurotransmitters, are released from neurons to act on receptor sites yielding functional change in the properties of the target cell (cf. Chapter 23). Neurotransmitters can be arbitrarily classified as classical (catecholamines, acetylcholine, etc.) and nonclassical (peptides, growth factors, nitric oxide, etc.) transmitters. There are many more peptide transmitters than classical transmitters.

The large number of peptide hormones/neuropeptides have been loosely grouped into several categories, depending on their structural similarity or tissue source: the brain/gut peptides (cholecystokinin, substance P, ghrelin, galanin), opiate peptides (β-endorphin, dynorphin, Leu-enkephalin), pituitary peptides (adrenocorticotropic hormone, growth hormone, β-endorphin, prolactin), hypothalamic-releasing hormones (gonadotropin-releasing hormone, corticotrophin-releasing hormone), and a category containing all other peptides.

Many small peptide hormones/neuropeptides found in nervous tissue have been cloned, and many are known to couple to cell surface receptors (cf. Chapter 23). Moreover, their intracellular signaling pathways are the same as those induced by the classical neurotransmitter systems. Small peptides are usually recognized by GPCRs, whereas other types of cell-surface receptors can recognize longer peptides/proteins. Several neuropeptides act additionally as regulators of nerve cell growth and division. Some examples will follow in order to illustrate the exciting role of some peptide hormones/neuropeptides in biology.

27.1.1 Substance P

The study of neuropeptides can be said to have began in 1931 when U. S. von Euler and J. H. Gaddum discovered substance P, a powerful hypotensive and smooth-muscle contracting agent. Its name refers to *powder* extracts from brain and intestine. This 11-amino-acid peptide (Table 27.1), present in high concentrations in the CNS and also in the gastrointestinal tract, is consequently classified as a brain/gut peptide. Substance P is a sensory neurotransmitter in the spinal cord, where its release can be inhibited by opioid peptides released from spinal cord interneurons, resulting in the suppression of pain. This property makes substance P an attractive target for studies of pain mechanisms as well as for the development of analgesics.

Biological actions of substance P are mediated by tachykinin (neurokinin) receptors belonging to the class of GPCRs, including three subtypes, NK1, NK2, and NK3 receptors. Substance P is the preferred ligand of NK1 receptors, initiating intracellular inositol 1,4,5-trisphosphate (IP3) turnover yielding the increase of intracellular calcium.

A large variety of selective peptide agonists and antagonists for NK1 receptors have been developed. Often these ligands are characterized by nanomolar affinities for specific receptor proteins. Their therapeutic use is, however, compromised by limited oral bioavailability.

As for almost all bioactive peptides, the breakthrough in substance P (NK1) receptor studies was achieved with the discovery of the first non-peptide NK1 receptor antagonist, CP 96 345 in 1991. This agent, in common with a variety of new generation compounds, crosses the blood–brain barrier and so offers new perspectives in our understanding of NK1 receptor mechanisms in disease. For example, their further development as antihyperalgesic, antidepressive, and/or antianxiolytic drugs is ongoing. The multiple bioactivities of substance P enable the targeting and treatment of several disorders through the interaction with NK1 receptors.

The direct activation of G-proteins by substance P, that is, without interaction with specific receptors, has also been reported, and this mechanism has been suggested to cause the activation of mast cells and the release of secretory mediators, including histamine. Such mimicry of GPCRs has also been suggested for several other bioactive peptides introduced below (mastoparan and melittin).

27.1.2 Opioid Peptides

This important category of peptide neurotransmitters is named because they recognize the same postsynaptic receptors as opium, the ancient analgesic with the active component, morphine. Despite its addictive potential, morphine, together with synthetic opiates such as methadone and fentanyl, is still used as an analgesic. The opioid peptides were discovered in the 1970s as mimics of the actions of morphine. More than 20 endogenous ligands (peptides) of the opioid receptors have now been identified and classified as the endorphins, the enkephalins (Table 27.1), and the dynorphins.

TABLE 27.1

Sequences of Some Representative Members of Bioactive Peptide Classes

Peptide	Sequence	Major Effects or Applications
Peptide Hormones and Neuropeptides		
Insulin, human	51 amino acids in B- and A-chains, bound via S-S bonds	Causing liver and muscle cells to take in glucose
Vasopressin, human	CYFQNCPRG amide	Causes water retention in kidney and blood vessel contraction
Oxytocin, human	CYIQNCPLG amide	Causes milk letdown and uterine contraction
Substance P, human	RPKPQQFFGLM amide	Regulates pain, treats mood disorders, anxiety, stress, etc.
Leu-enkephaline	YGGFL*	Endogeneous opiate
Galanin, human	GWTLNSAGYLLGPHAV GNHRSFSDKNGLTS*	Regulates insulin release, pain transmission, treats memory and stress
Peptide Toxins		
ω-Conotoxin	CKGKGAKCSRLMYDCCTGSCRSGKCamide	Inhibits N-type voltage-dependent calcium channels, has analgesic effect
ω-Agatoxin-IVA (ω-Aga-IVA)	48 amino acids	Inhibits current through P-type, but not through N-type or L-type, channels at nanomolar concentrations
α-Neurotoxins	More than 100 identified, 62-74 amino acids, three-loop structure	Many pharmacological activities incl. peripheral and central neurotoxicity, cardiotoxicity, inhibition of enzymes such as acetylcholinesterase and proteinases
α-Bungarotoxin	352 amino acids	Binds irreversibly to the acetylcholine receptor in neuromuscular junction causing paralysis and respiratory failure
Dendrotoxin	65 to 70 amino acids	Block voltage-activated potassium channels leading to increased neurotransmitter release
Antimicrobial Peptides		
Mastoparan, hornet	LKLKSIVSWA KKVL amide	Mast cell degranulating peptide activating phospholipase C coupled G proteins
Mellittin, giant honeybee	GIGAILKVLSTGLPALIS WI KRKRQE amide	Strong hemolytic activity. Integrates into cell membranes. Inhibits Na^+-K^+-ATPase. Increases the permeability of cell membranes to ions, particularly Na^+

(Continued)

TABLE 27.1 (CONTINUED)
Sequences of Some Representative Members of Bioactive Peptide Classes

Peptide	Sequence	Major Effects or Applications
Magainin 1, African clawed frog	GIGKFLHSAGKFGKAFV GEIMKS*	Antibacterial against G⁺ and G⁻ bacteria and fungi
LL37 (CAP18), human	LLGDFFRKSKEKIGKEFK RIVQRIKDFLRNLVPRTES	Antibacterial against G⁺ and G⁻ bacteria, fungi and viruses
Cell-Penetrating Peptides		
Penetratin	RQIKIWFQNRRMKWKK**	Intracellular delivery of wide range of bioactive cargos
Tat(48-60)	GRKKRRQRRRPPQ	Intracellular delivery of fused proteins
Transportan 10	AGYLLGKINLKALAALA KKIL amide	Intracellular delivery of wide range of bioactive cargos
MAP	KLALKLALKALKAALKLA amide	
Arg₉	RRRRRRRRR amide	
pVEC	LLIILRRRIRKQAHAHSK amide	
MPG	GALFLGWLGAAGSTMG APKKKRKV cysteamide	Intracellular delivery of wide range of bioactive cargos

Note: *C-terminal free carboxylic acid; **both, C-terminal acid and amide are active.

The μ, κ, and δ are three genetically and pharmacologically distinct opioid peptide receptors, that differ in their ligand binding and tissue distribution. The δ-receptors have highest affinity for enkephalins, κ-receptors for dynorphins, and μ-receptors for endorphins, including morphine. Fully functional opiate receptors form heterodimers with each other, but also with β-adrenoreceptors, yielding a unique combination of receptors in terms of ligand binding and signaling through different combinations of G-proteins. Opioid peptides and their receptors are widely distributed throughout the brain and are often co-localized with other small-molecule neurotransmitters (GABA, 5-HT) and their receptors.

Opioid peptides exert analgesia (at the spinal level inhibiting substance P release in sensory pain fibers, and in the brain), and are also involved in complex behaviors such as sexual attraction, aggressive/submissive behaviors, and psychiatric disorders such as schizophrenia. The repeated administration of opioids leads to tolerance and addiction, making the studies of the mechanisms of opioid actions very important.

27.1.3 GALANIN

The neuropeptide galanin was first isolated from porcine intestine in 1983 by the group of Viktor Mutt using his chemical method for identification of *C*-terminally amidated bioactive peptides. Similar studies have since identified tens of related peptides. The galanin peptide consists of 29–30 amino acids (Table 27.1), where the *N*-terminal part has been found crucial for its biological activity and is, consequently, conserved between species. Galanin has not been shown to belong to any known family of neuropeptides except for a galanin-related peptide (GALP), which contains the conserved sequence of galanin 1-13 and is recognized by the galanin receptors. The biological effects are mediated by three different G-protein-coupled receptors, GalR1, GalR2, and GalR3, which possess only 40%–50% sequence homology between the receptor subtypes, but show a high degree of homology between different species. Galanin receptors show different functional coupling; GalR1 and GalR3 activate Gi/o, consequently reducing cAMP concentrations, opening G-protein-coupled inward rectifying K^+(GRK) channels, and stimulating mitogen-activated protein kinase (MAPK) activity. The GalR2 receptor predominantly couples to Gq/11, stimulating Ca^{2+} release via inositol phosphate hydrolysis and the opening Ca^{2+}-dependent channels, additionally activating Gi/o. All three galanin receptors demonstrate high affinity for galanin, but are distinguishable by the maintained affinity of GalR2 and GalR3 for N-terminal deletions of the peptide, Figure 27.4A.

Galanin and galanin receptors show a widespread distribution throughout the central and peripheral nervous systems. Because of its extensive distribution and co-localization with several neuromodulators, galanin has shown to be involved in several high-order physiological functions such as cognition, affective behavior, nerve injury and pain, and pancreatic functions. The importance of galaninergic signaling in the neuronal disorders epilepsy, nerve injury, inflammation, depression, and Alzheimer's disease, as well as its role as a neurotrophic factor, points to the possibility of addressing the galaninergic system as a therapeutic target.

27.2 PEPTIDE TOXINS

Poisonous animals, plants, and pathogenic bacteria produce multiple toxins acting by a variety of mechanisms to influence or kill their target organisms; many of these toxins have been identified as peptides and referred to as peptide toxins. Peptide toxins have been identified in animals as diverse as cone snails (conotoxins), sea anemone, snakes, spiders (agatoxins, grammotoxin), insects, worms, scorpions, and frogs. These peptides are directed at multiple pharmacological targets, making them a valuable source of agents to study the biochemical mechanisms of these targets and, additionally, to design novel therapeutics. Most animal venoms comprise a highly complex mixture of tens of peptides, often with diverse pharmacologies, making them a unique source for new therapeutics.

Comparative pharmacodynamic data enable peptide toxins to be classified as neurotoxins, myotoxins, vasoactive peptides, haemolytic, cytolytic, necrotic, hemorrhagic, anti-inflammatory, antitumoral, analgesic, and others. By studying such effects at the molecular level, one can also assign such peptides to other groupings, including ligands of cell surface receptors (agonists and antagonists), ligands for ion channels, and enzyme inhibitors. The following examples will describe some of peptide toxins, their biological actions, and therapeutic developments when available.

The first example is peptide toxins that block voltage-sensitive calcium channels. These have contributed enormously to our understanding of the role of specific calcium channels in normal and pathological conditions. Several members of the ω-conotoxin class of calcium channel blockers are currently in clinical trials for chronic pain.

Spider peptides ω-Aga-IVA and ω-Aga-IVB specifically inhibit the P-type calcium current in the brain by antagonizing activation gating, and represent another class. The structure of ω-Aga-IVB shows that the peptide is structurally stabilized with a network of disulfide bonds. One face of the molecule has a cluster of basic residue that has been suggested to be involved in the binding interaction with the channel.

Another group is the snake toxins of the families Elapidae (e.g., cobras, kraits, coral snakes, mambas) and Hydrophidae (sea snakes) contain a potent neuromuscular toxin α-neurotoxin that produces a nondepolarizing block of the postsynaptic AChR (cf. Chapter 23). Low doses of this toxin in humans can cause weakness and fatigability that resembles acquired myasthenia gravis. Higher doses can lead to complete neuromuscular block, paralysis, respiratory failure, and death. The α-toxins fall into two groups: the long toxins, which have 71 to 74 amino acids and five internal disulfide bonds, and the short toxins, with 60 to 62 amino acids and four internal disulfide bonds. The 74-amino-acid α-bungarotoxin blocks neuromuscular transmission by irreversibly binding to nicotinic ACh receptors, thus preventing ACh from opening postsynaptic ion channels. Paralysis ensues because skeletal muscles can no longer be activated by motor neurons. As a result of its specificity and its high affinity for nicotinic ACh receptors, α-bungarotoxin has contributed greatly to understanding the ACh receptor molecule.

Finally, peptide toxins affecting K^+ channels include dendrotoxin from wasps, apamin from bees, and charybdotoxin from scorpions. All of these toxins block K^+ channels as their primary action.

27.3 ANTIMICROBIAL PEPTIDES (AMP)

Bacterial infections are still a major problem in medicine, and are often complicated by issues surrounding the widespread development of bacterial resistance to common antibiotics. To fight pathogens, an alternative strategy is to use antimicrobial peptides (and proteins). In mammals, the antimicrobial peptides are found in phagocytes and epithelial cells as gene-encoded peptides participating in the immune response, and that demonstrate activity against a wide spectrum of microorganisms. Antimicrobial peptides are produced in many organisms, including bacteria, insects, plants, and vertebrates, where they express defense properties. They are often found in venoms of scorpions (IsCT), wasps (mastoparan), bees (mellittin), spiders, etc. (cf. Table 27.1).

Hundreds of natural cationic AMPs are known today. In humans, three groups of AMP are defined: defensins, cathelicidins, and histatins. By structure, they are divided into three families: α-helical linear peptides, disulfide-bridged cyclic peptides, and peptides with a high content of amino acid residues Pro, Gly, or His. Natural AMPs are often cationic peptides of 13–50 amino acids with a considerable number of hydrophobic residues. Shorter AMPs (<10 residues) have been developed by using natural AMPs as lead compounds or by using combinatorial approaches.

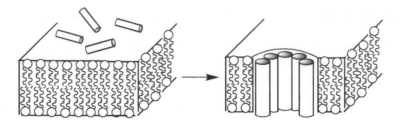

FIGURE 27.2   Pore formation in the plasma membrane by antimicrobial peptides can be one mechanism of toxicity, as exemplified here by the barrel-stave model. Interactions between the positively charged AMP and the negatively charged phospholipids in the cellular membrane (left) leads to membrane disturbance or ordered pore formation (right) and general cellular toxicity. Other suggested models for membrane disruption by AMPs are the wormhole or torroidal model, carpet model, and detergent-like model.

Antimicrobial peptides are usually toxic against Gram-positive and Gram-negative bacteria, fungi, and protozoa. Toxic mechanisms of AMPs can be multiple, and the precise details are still highly debated. In general, it is likely that positively charged peptides interact with the negatively charged cellular membranes of bacterial cells yielding an increase in membrane permeability and cell death. One of the possibilities could be a mechanism where membrane interaction of an AMP leads to pore formation-caused toxicity (Figure 27.2). Additionally, it has been suggested that AMPs can be involved in interactions with intracellular targets such as autolytic enzymes, as well as in the inhibition of transcription and translation.

Antiviral peptides have received much attention due to their high therapeutic potential, and several specific mechanisms have been suggested for their action. These potential mechanisms include the blocking of viral entry by heparin sulfate interaction, or by interaction with specific cellular receptors. Interaction with viral glycoproteins, the membrane, or the viral envelope may also play part in the efficacy of antiviral peptides. Additionally, actions at intracellular targets and host cell stimulation have been suggested as potential mechanisms of AMP antiviral action.

Recently, it was suggested that some AMPs may possess the properties of cellular uptake when incubated with live mammalian cells, suggesting that they could act as cell penetrating peptides (cf. below). However, the potential of AMPs as carriers of bioactive cargos remains to be evaluated, both in terms of efficacy and toxicity.

Together, these multiple mechanisms of AMP action make them attractive targets for the development of therapeutic strategies to treat infectious and inflammatory diseases, and several preclinical and clinical trials are ongoing. Recently, AMP applications as novel cytotoxic agents for cancer treatment were also suggested. Representative examples follow below with the sequences provided in Table 27.1.

Mastoparan is a collective name for a range of membranolytic homologues of small, 14 amino acid amphiphilic peptides isolated from venoms of different species of wasps. Mastoparan was first characterized in 1979 and proposed in the 1990s to mimic receptors by activating G_i-proteins (cf. following text), so inducing histamine release from mast cells. The multiple bioeffects of mastoparan include stimulation of insulin, histamine, serotonin, catecholamine and arachidonic acid release, increase of phosphoinositide metabolism and the mobilization of intracellular Ca^{2+}, and the activation of ion channels. Mastoparans are efficient inducers of membrane lysis.

Melittin, a 26 amino acid peptide, identified in 1950s, is the principal active component (50% of dry weight) of bee venom with powerful antimicrobial and anti-inflammatory properties. Melittin exerts profound inhibitory effects on several pathogenic bacteria and ycasts, thereby suppressing infections. Cellular activities of melittin are multiple, including the action of melittin with membrane proteins (ATP-ases), interaction with ras-oncogene, induction of membrane permeabilization, and inhibition of viral gene expression.

The amphiphilic properties of melittin make it water soluble and still able to associate with cell membranes. This has resulted in melittin being used as a model for lipid-protein interactions as well as for cytolytic peptides. It has been reported in the 1980s that both mastoparan and melittin directly stimulate heterotrimeric G-proteins in a manner similar to that of GPCRs (cf. Chapter 23 and this chapter, Section 27.1, for substance P). Such studies, including the inhibition of adenylyl cyclase in synaptic membranes, suggest that these peptides also signal through $G_{i/o}$ or, additionally, G_s in the case of melittin.

Mammalian cells produce several kinds of AMPs, such as α-defensin in neutrophils, β-defensins in epithelia, histatins in saliva and cathelicidin (CAP18 or LL37) in neutrophils and epithelia, which have been reported to function as antimicrobial agents against Gram-negative and Gram-positive bacteria, fungi, and viruses as well as mediators for inflammation.

Magainins is a class of antiviral α-helical peptides isolated from frog skin that suppress intracellular viral gene targets. Magainin 1 and 2 are 23 amino acid peptides (Table 27.1). Additional amidated forms of these sequences and other related peptides (e.g., PGLa) are also collectively called magainins. Magainins target the lipid matrix, rather than proteins, as demonstrated by studies of peptides composed of D-amino acids that induce membrane permabilization. Besides the broad spectra of antimicrobial activities, magainins also have defined contraceptive potential, inhibiting sperm motility and, hence, conception, as well as embryotoxic properties.

27.4 CELL-PENETRATING PEPTIDES

A relatively new class of peptide transport vectors that have received increasing attention in the past few years is cell-penetrating peptides (CPPs). These cationic and/or amphipathic peptides are usually less than 30 amino acids in length and have the ability to rapidly translocate into most mammalian cells carrying various cargo molecules such as oligonucleotides (ONs), peptides, proteins, plasmids, liposomes, and nanoparticles both *in vitro* and *in vivo*. Thus, CPPs have opened a new avenue in drug delivery, allowing otherwise impermeable therapeutic agents to enter cells and induce biological responses.

The research group of Alain Prochiantz discovered the first CPP in 1994 when a 14 amino acid peptide penetratin (originally called pAntp) was identified and characterized. This peptide corresponds to a third helix of the homeodomain of a Drosophila homeoprotein Antennapedia where its role is to be responsible for the

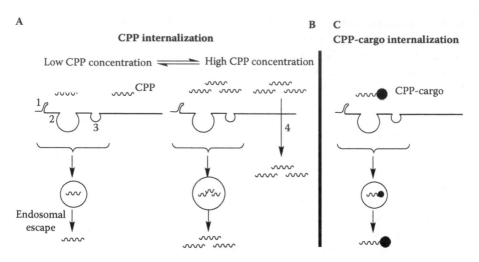

FIGURE 27.3 Possible mechanisms of CPP cellular uptake and summary of cargo delivery by CPP. Note that the uptake mechanism is dependent on the CPP concentration and character, and different uptake mechanisms can be in equilibrium. Endosomal escape seems to be the crucial step in cytosolic delivery of bioactive cargoes. A: At low doses, CPP uptake is preferably endocytotic with a possible combination of macropinocytosis (1), clathrin(2)- and caveolin(3)-dependent endocytosis. B: At higher doses, direct penetration(4) of several CPPs has been reported, in parallel with endocytotic mechanisms. C: Bioactive cargo delivery by conjugation to CPP is usually endocytotic.

DNA recognition and cellular uptake for the whole protein. Later, a short peptide, Tat (a 12 aa peptide derived from HIV-coded Tat regulatory protein), was introduced in 1997, and the era of CPPs had begun. Today, hundreds of CPPs are known, and their wide applications in research and development of therapeutics have been reported. Some "classical" CPP sequences are represented in Table. 27.1. One can see that CPPs are a structurally diverse family of peptides ranging from short highly cationic peptides such as poly-arginine to longer more amphipathic peptides, including transportan. The most studied peptides, Tat (transactivator of transcription) and penetratin, are both derived from naturally occurring shuttling proteins, whereas most other CPPs are either chimeric or totally synthetic peptide sequences.

In respect of cellular uptake mechanisms, CPPs are hard to define (see Figure 27.3). Recent data suggest that endocytosis contributes to the uptake, and caveolae, clathrin-dependent endocytosis, and macropinocytosis have also been suggested to mediate CPP translocation. However, there are conflicting reports claiming that uptake is independent of endocytotic pathways and occurs through transient pore formation.

CPPs, when not attached to a cargo, seem to use different pathways to translocate into the intracellular environment. Endocytotic, as well as nonendocytotic, uptake of CPPs (i.e., "naked" peptides) have been demonstrated. However, according to recent experiments, the endocytotic translocation pathway seems to prevail for cargo-conjugated CPPs. Additionally, by comparison of uptake of different CPPs, it has been suggested that these peptides simultaneously utilize different submechanisms of endocytosis and, at higher concentrations, uptake occurs by an additional rapid translocation process. Novel CPPs with improved endosomolytical properties are suggested for *in vivo* applications in the future. It is also important to study the differential membrane perturbation caused by CPPs attached to different cargoes, and this factor should be considered in the future design and refinement of CPPs as oligonucleotide delivery vectors.

27.5   PROTEIN AND PEPTIDE MIMICS

Protein–protein and peptide–protein interactions (cf. Chapter 17) play the most important role in a variety of biological processes, and the therapeutic strategies targeting these interactions is an attractive area of research. Considerable effort has been spent in order to design proteo- and peptidomimetics capable of disrupting or mimicking these interactions because the controlled interference with these has provided novel research tools and drug targets. Serious challenges have been encountered in this approach due to the very complicated nature of the interactions, for example, the noncontinuous character of the interaction, relatively large area of protein surface responsible for the interaction, etc. Consequently, these efforts are much less mature compared with those of conventional drug discovery, although significant progress has been achieved in the development of peptide-based mimics and inhibitors, scaffold structures, and small-molecule inhibitors of protein–protein interactions.

Natural protein/peptide-derived, single short synthetic peptide sequences with functional motifs have been often used as mimics or inhibitors of parent proteins in studies of protein signaling and interactions. Additionally, the screening of random peptide libraries has been used to target signaling pathways. Among such biologically active peptides and analogs are several peptide families, such as antibiotic peptides, integrin inhibitors, immunoactive, anticancer, neuromodulator, and hormone peptides, peptide-derived semisynthetic vaccines, drug delivery systems, self-assembling peptides, and many others.

One common example of the application of short synthetic functional motifs derived from natural peptides and proteins are the numerous analogs of peptide hormones with improved stability, bioavailability or selectivity at receptor subtypes. Efficient analogs of neuropeptide galanin (cf. previously mentioned) include galanin(1-16) and galanin(2-11), which have been used to produce receptor subtype-selective ligands (Figure 27.4A).

Another possibility is to find short functional sequences by methods such as combinatorial peptide synthesis, screening of phage displayed libraries, or bioinformatics. For example, it has been reported that one can select numerous efficient cell-signaling peptides in this way, such as those selected from GPCRs, G-proteins, platelet modulators, transcription factors (Chapter 22), etc. The finding that short, 10–20 amino acids long peptides from the activation domains of transcription factors autonomously act as protein mimics has fueled interest in the use of small molecules, peptides to low-MW substances to regulate translation. Mimicry or replacement of protein–protein interactions with small organic molecules is of special interest in the search of novel drug candidates as well as in studies of the molecular mechanisms of these interactions.

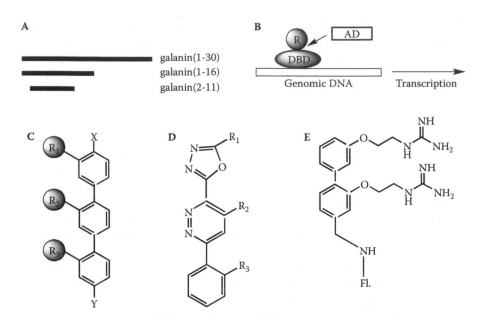

FIGURE 27.4 Examples of mimics of protein–protein interactions. A: Examples of the design of mimics of galanin receptor ligands, mimics of native galanin. B: Regulation of translation with artificial transcription factors, peptide or other small molecule mimics of proteins. C: Terphenyl structure for mimicry of α-helix. D: Oxadiazole-pyridazine-phenyl scaffold, mimic of α-helix. E: Design of the mimic of cell-penetrating peptide, SMoC.

Because the α-helical segments of proteins often participate in several protein–protein interactions, the intensive development of efficient mimetics of α-helices is ongoing. Such design is based on the knowledge that proteins in nature fold into native conformations with a structurally determined combination of functionalities of amino acid side chains; often, only the critical amino acid residues determine the specificity of interactions. Consequently, synthetic molecules mimicking these critical residues in a "right" fold have been demonstrated in several cases to mimic the "parent" proteins. Certainly, the prerequisite for such a design of protein mimics is the knowledge of the native folding, which often can be complicated.

The chemical strategies to design functional and structural mimics of proteins are numerous and involve the combination of computational analysis of structures, and comprehensive methods of chemical design and synthesis. Starting in the 1990s, the group of Andrew D. Hamilton has contributed substantially to the design of small molecule mimics of α-helices by designing and synthesizing complicated chemical structures. To name only a few examples, this group has developed α-helix mimics based on, for example, terphenyl (Figure 27.4C), terpyridine, diphenylindane, and terephthalamide structures. Another important example here is from the group of Julius Rebek, Jr., a synthetic pyridazine-based scaffold (Figure 27.4D) as an α-helix mimic. This group has designed, synthesized, and tested numerous scaffold structures for mimicry of peptidomimetics and succeeded often in demonstrating desired biological activities with them.

The general design principle is often to mimic the backbone of a peptide by a chemical structure that is different from the original peptide structure, and to combine this with functional groups of the peptide.

In 2007, the group of David L. Selwood described the small-molecule mimics of an α-helical cell-penetrating peptide peptide (cf. previous text) penetratin, termed SMoCs (Figure 27.4E), by using molecular modeling techniques. The internalization of the DNA replication licensing repressor geminin was coupled to this novel, exciting cellular transport vector and, *in vitro*, the extracellularly delivered SMoC-geminin showed an antiproliferative effect on human cancer cells. This excellent work paves the way toward the design of a new type of transport vector for research and therapeutics.

In summary, rational design of sufficient and specific mimics of peptide/protein interactions has been started and has proven to be an effective tool in today's design of research tools and drug candidates.

FURTHER READING

1. Duchardt, F., Fotin-Mleczek, M., Schwarz, H., Fischer, R., and Brock, R. 2007. A comprehensive model for the cellular uptake of cationic cell-penetrating peptides. Traffic, 8(7): 848–866.
2. Satake, H. and Kawada, T. 2006. Overview of the primary structure, tissue-distribution, and functions of tachykinins and their receptors. Current Drug Targets, 7(8): 963–974.
3. Lewis, R. J. and Garcia, M. L. 2003. Therapeutic potential of venom peptides. Nature Rev. Drug Discovery, 2(10): 790–802.
4. Yount, N. Y., Bayer, A. S., Xiong, Y. Q., and Yeaman, M. R. 2006. Advances in antimicrobial peptide immunobiology. Biopolymers (Peptide Science). 84(5): 435–458.
5. Davis, J. M., Tsou, L. K., and Hamilton, A. D. 2007. Synthetic non-peptide mimetics of alpha-helices. Chem. Soc. Rev., 36(2): 326–334.
6. Langel, Ü., Ed. 2006. Cell-Penetrating Peptides, 2nd edition. Boca Raton: CRC Press.

Introduction to Part V

Ülo Langel and Astrid Gräslund

At least 30,000 different proteins are produced by human cells, and each has a different role. Each protein is characterized by its own unique sequence and native conformation in order to perform its specific function. Hence, an altered folding, cleavage, or amino acid sequence of proteins may lead to serious disorders. Protein misfolding often leads to protein aggregation and precipitation, with the aggregates adopting either highly ordered (amyloid fibril) or disordered (amorphous) forms leading to disease. For instance, amyloid fibrils are connected to a variety of debilitating diseases, including Alzheimer's, Parkinson's, Huntington's, and Creutzfeldt–Jakob's diseases, as well as type 2 diabetes. There is more and more information available showing the interrelationship between genomic instability and diseases such as cancer initiation and progression, where numerous genomic lesions (point mutations, deletions, and insertions) lead to expression of proteins with altered sequences and properties. In order to be able to fight these devastating disorders, huge research efforts are today devoted to understand the molecular mechanisms behind the disorders. In Part V, Chapters 28–31, the introduction to diseases related to misfolding, miscleavage, and mis-sequences of proteins is presented, together with summary of recent achievements in available applications and future possibilities of using peptides and proteins as drugs.

28 Misfolding-Based Diseases

Astrid Gräslund

CONTENTS

According to Anfinsen's postulate, the amino acid sequence of a protein uniquely defines its native three-dimensional structure and hence its function. This does not contradict the observation that protein folding depends on a variety of environmental factors, such as temperature, pH, solvent, etc. Larger proteins may also require the presence of chaperone proteins to assist in the folding process. In nature, this means that protein misfolding is a relatively common event. The living organism has evolved systems that are, in most cases, able to handle the misfolding events, for example, by proteolytic degradation via the ubiquitin-proteasome pathway of the misfolded protein. But in some circumstances the misfolding leads to serious consequences for the organism, and these are seen in a growing number of human diseases.

28.1 CANCER

One aspect of protein misfolding is particularly important in cancer. In this scenario, inherited or somatic mutations in a tumor repressor protein (the transcription factor p53) are linked to cancer, because the function of this particular mutated protein to bind to its specific DNA sequence is compromised. Malfunction of the p53 protein has been estimated to occur in as much as 50% of human cancers. The role of p53 is to regulate programmed cell death, apoptosis, which is critical for keeping cancer cells from proliferation. This type of protein-linked pathology may be characterized as loss-of-function pathology, in contrast to the amyloid diseases which may be considered as gain-of-function diseases. It has been observed that in many cases there are single mutations in the p53 gene that result in normal levels of expression of full-length protein, which, however, lacks the ability to recognize its proper site on DNA. It has therefore been discussed that one therapeutic approach against cancers associated with a single mutation of p53 could be to use small compounds to try to restore the proper function of the DNA-binding core domain of p53. Another aspect of p53 activity is its connection to the so-called oncogene products. In particular, the mdm2 oncogene product is a protein, which seems to be the main negative regulator of p53 activity by binding to its N-terminus, thereby repressing its ability to act as a transcription factor and by increasing its degradation. The p53-mdm2 protein interaction module seems to be at the center of a signaling network that is crucial for the promotion or destruction of cancer. Therefore, another strategy against cancer could involve modulation of this interaction via intervening small molecules.

28.2   AMYLOID DISEASES

A number of well-known amyloid diseases are listed in Table 28.1. The most important are the neurodegenerative diseases, like Alzheimer's, which affects millions of mostly elderly people over the world. In these diseases, the amyloid material is located in the brain. There are however also systemic amyloid diseases that are linked to accumulation of amyloid material in other parts of the body. In many of these cases, the protein or peptide responsible for the amyloid formation is known (Tables 28.1 and 28.2), although this knowledge is, of course, only one of the biochemical aspects of the diseases. There is a common picture emerging for one part of the biochemistry associated with these diseases, namely, that of protein or peptide aggregation. The word "amyloid" has a Greek origin and means "starch-like," named after the starch-like deposits that were found in the brains of patients after death. These extracellular deposits may look different and have different properties in the different diseases, but they share the common property of having a precipitated protein or peptide of a particular kind as one of the major components. The term *amyloid* remains despite the lack of starch in this context.

TABLE 28.1

Some Neurodegenerative Amyloid Diseases and the Misfolding Protein Involved

Disease	Aggregating Protein or Peptide	Number of Residues
Alzheimer's disease	Amyloid β peptide	40–42
Parkinson's disease	α-synuclein	140
Creutzfeldt–Jacob disease	Prion protein	253
Huntington's disease	Huntingtin with poly Q expansion	3144
Amyotrophic lateral sclerosis (ALS)	Superoxide dismutase	153

TABLE 28.2
Some Systemic Amyloid Diseases and the Misfolding Proteins Involved

Disease	Aggregating Protein or Peptide	Number of Residues
Type II diabetes	Amylin	37
Senile systemic amyloidosis	Transthyretin	127
Hemodialysis-related amyloidosis	β2-microglobulin	99
Cataract	γ-crystallins	variable

28.3   PROTEIN MISFOLDING PATHWAYS

When a protein is produced in the ribosome it exits into the cytoplasm and becomes in benign cases properly folded; it then begins to function. In the event of protein misfolding, disordered aggregates may form and grow in size and finally form insoluble fibrils, characterized by β-structure. The misfolding and aggregation are multistep processes that may occur inside or outside cells, depending on the actual protein or peptide. Often the processes are kinetically described as starting from a small template or seed, which may take a relatively long time to form, but after its formation the aggregates may grow much faster. Figure 28.1 illustrates this classical picture of amyloid formation. On the way to the larger aggregates and fibrils, there are medium-sized soluble entities called protofibrils or oligomers. These aggregation intermediates are nowadays the most suspected species for damaging the neurons and causing neuronal dysfunction in the brain diseases. This is in contrast to an earlier concept when the most easily observable amyloid fibrils were the focus of attention for understanding the diseases. The fibrils could, in principle, even be considered as a "safe" storage place for extracellular garbage, which may, however, function as a reservoir for the soluble species.

28.4   AMYLOID STRUCTURES

The structural understanding of the various species involved in the schemes of amyloid formation is still very incomplete. Fibrils have been characterized from a variety of proteins or peptides, and they seem to share a common structure: Long fibers or threads composed of protein or peptide folded into "cross β-structure" with the peptide chains perpendicular to the axis of the fiber (Figure 28.1). The fibers have nanometer dimensions and can be visualized by, for example, electron microscopy or atomic force microscopy. They have also been studied by other methods such as x-ray fiber diffraction and solid state NMR. The x-ray structure at 1.8Å resolution of a short peptide in the amyloid state has been reported from a study of microcrystals. This structure is a tightly intertwined assembly of peptides in β-structure, parallel and in register. The tightly intertwined side chains ("steric zipper") do not allow any water molecules inside the center of the

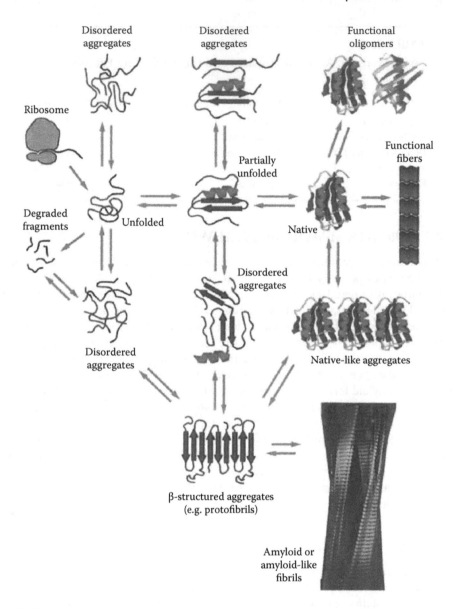

FIGURE 28.1 Schematic description of possible pathways leading to either proper folding or misfolding of a protein after its production in the ribosome. Disordered aggregates, which may be formed at an early stage of the folding process, promote the misfolding processes, which may lead to, for example, β-structured soluble aggregates or insoluble amyloid fibrils. (From Chiti, F. and Dobson, C. M., Protein misfolding, functional amyloid and human disease, Annual Review of Biochemistry, **75** (2006) 333–366. Copyright 2006 by Annual Reviews, www.annureviews.org. With permission.)

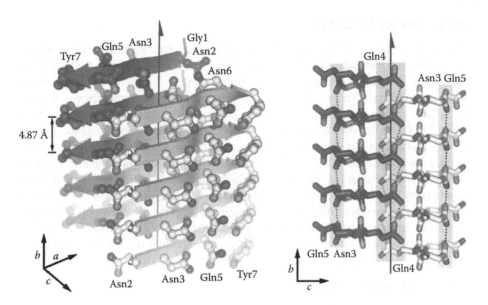

FIGURE 28.2 A ribbon representation of the three-dimensional structure of amyloid fibrils formed by a short peptide sequence GNNQQNY. X-ray crystallography shows that the peptides form a structure that is a parallel, in register β-sheet, with the peptide side chains tightly packed. (Left) The pair-of-sheets structure, showing the backbone of each β-strand as an arrow, with side chains protruding. (Right) The steric zipper viewed down the a axis. (From Rebecca Nelson et al., Structure of the cross-β spine of amyloid-like fibrils, Copyright 2005 Macmillan Publishers Ltd: Nature, **435** (2005) 773–778. With permission.)

assembly. Figure 28.2 shows a ribbon diagram of this structure, which was solved for the peptide GNNQQNY, the sequence taken from the sequence of the yeast prion Sup35.

The oligomeric assemblies are much more elusive than the fibrils and even have many names, such as protofibrils, globulomers, or oligomers. Common features are that they are soluble in aqueous solvent, they vary in size from dimers to 12 mers or more, and they have variable degrees of β-structure. Because of the generic β-structure they will react with certain heterocyclic dye molecules, like Thioflavin T or Congo red. These compounds are therefore considered as markers of amyloid material, fibrils as well as protofibrils. Thioflavin T shows increased fluorescence after binding, whereas Congo red exhibits birefringence. Methods like size exclusion chromatography have been applied to define the sizes of the oligomers. When investigating the properties of oligomers with different histories and therefore different degrees of stability, one must take care not to change their properties during the analysis. The initial monomeric proteins or peptides may be intrinsically disordered, well structured, or partly structured, depending on the protein or peptide in question.

28.5   PRION DISEASES

The prion diseases, or transmissible spongiform encephalopathies, are a special form of neurodegenerative amyloid diseases that are to some extent infectious. They include the Creutzfeld–Jacob disease in humans, which is closely related to similar diseases in animals. The so-called mad cow disease is the most well-known of these, but a large variety of animals have similar diseases. The "scrapie" disease in sheep existed in Great Britain for hundreds of years before the development of the disease in cows some 25 years ago. The bovine form of the prion disease was linked to the variant Creutzfeld–Jacob disease that has afflicted a number of people in England and also elsewhere in Europe. After the initial outbreak of the mad cow disease, severe rules have been applied to cattle feeding, and a large number of animals with suspected disease were killed in Great Britain. This seems to have stopped a threatening epidemic of related human disease. The question of transmission barrier for infectivity has been much debated. It is now believed that the bovine disease was transmitted from sheep, and that the human disease came from cow.

28.5.1   THE PRION PROTEIN

The term *prion* is defined as an infectious protein. Normally, infections are carried by nucleic acids in microorganisms or viruses. In the prion case, the protein-only hypothesis states that only misfolded prion protein in an abnormal isoform is needed for infectivity. The normal, cellular protein PrP^C is a monomeric, protease-sensitive glycoprotein anchored at the cell surface. Its native function is not known. The prion protein is an example of secreted proteins, which after their production are guided into the ER by a signal peptide. The signal peptide is cleaved off and the protein is folded inside the ER, whereupon it is transported via a number of intermediate compartments onto the cell surface. The PrP^C is anchored at the cell surface membrane via a glycosidic anchor and via one transmembrane helix. After conversion to the abnormal "scrapie" isoform, the resulting PrP^{Sc} is rich in β-sheet structure, is aggregated, and is resistant to proteolysis.

Propagation of prions associated with disease are proposed to involve a seeding mechanism, where PrP^{Sc} induces conversion of PrP^C as substrate into more scrapie material. Although this mechanism has been subject to some criticism over the years, it seems to hold, although there may be other protein or membrane partners involved in the conversion processes. Early experiments with transgenic mice showed that they were susceptible to infection by prions if they expressed PrP^C of the same or similar kind as in the prions, but could not be infected if this gene was knocked out. An important new aspect is that infectious mammalian prion protein can be produced *in vitro*. Another new aspect of prions is that the yeast *Saccoromyces cerevisiae* has native prion proteins with functional significance. These proteins have been found to exist in two alternating conformations, and the presence of one can influence protein in the other conformation to change.

28.5.2   PRION PROTEIN STRUCTURE

The three-dimensional structures of several prion proteins have been solved by NMR. The structures of a typical PrP^C is composed of a highly dynamic disordered

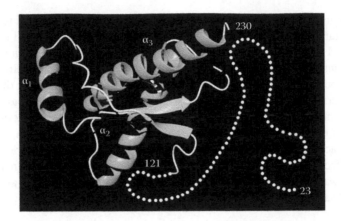

FIGURE 28.3 The solution structure of the prion protein, determined by NMR spectroscopy. Approximately 120 residues in the N-terminus of the protein is unstructured and highly dynamic, whereas residues 121–230 form a globular structure rich in α-helices, and with one small β-sheet. (From Zahn, R. et al., NMR solution structure of the human prion protein. PNAS, **97** (2000) 145–150. Copyright 2000 National Academy of Sciences, U.S.A. With permission.)

N-terminal half, residues 23-124, and a well-structured C-terminal half, residues 125-228 with three α-helices and two short β-strands (Figure 28.3). The structure of PrP^{Sc} is not as well defined but it is clear that the aggregated form has a much higher content of β-structure. The question about the nature of the infectious species of the prion protein has been raised. Available data point to the oligomeric aggregated forms, 300–600 kD, as being the most infectious forms. Less is known about the nature of the neurotoxic form. The question about the physiological role of PrP^C is linked to the question of neurotoxicity, and is also not solved yet.

28.6 AMYOTROPHIC LATERAL SCLEROSIS

Amyotrophic lateral sclerosis (ALS) disease (also known as Lou Gehrig's disease) is a neurodegenerative disease affecting the motor neurons. Although most cases are sporadic, that is, there is no direct link to any mutated protein or infection, about 1%–2% of the cases are linked to mutations and associated misfolding of an enzyme, superoxide dismutase (SOD1). The clinical symptoms of the disease are much the same, in cases with or without mutated SOD1. Therefore, large research efforts have been directed to finding out the role of the enzyme and its mutated forms, in an attempt to understand the pathologic processes.

The homodimeric SOD1 is an abundant cytosolic Cu-Zn-containing metalloenzyme. It is one of the key players in the defense against oxidative stress, and "dismutates" the superoxide radical anion $O_2^{\cdot-}$ to oxygen and peroxide. The superoxide radical anion is formed as an intermediate in the biochemical processes whereby oxygen is reduced to water. However, the current data argue against the hypothesis that it is loss of the antioxidative function of SOD1 that is the major cause of the disease. Since SOD1 is a copper-containing enzyme, and free copper is toxic to cells, other hypothetic mechanisms suggested that the disease is linked to a changed copper metabolism. However, this hypothesis has also been rejected. It is now clear that mutated SOD1 (>100 different mutations linked to ALS have been reported) is generally unstable and more prone to aggregate than the native protein. The more recent hypotheses about ALS is, therefore, that this neurogenerative disease, like several others, is linked to protein misfolding and aggregation, in this particular case of the enzyme SOD1. Conformational changes in the dimers, possibly coupled to dissociation of the dimers into monomers, may facilitate the aggregation, particularly of protein lacking the proper metals bound. There are several cellular processes that may be affected by such protein aggregates. Recently, particular interest has been directed toward damage of the mitochondria caused by the aggregates, and damaged mitochondria may be linked to apoptosis, programmed cell death.

FURTHER READING

1. Murphy, R. 2002. Peptide aggregation in neurodegenerative disease. Annu. Rev. Biomed. Eng. 4, 155–174.
2. Selkoe, D. J. 2003. Folding proteins in fatal ways. Nature 426, 900–904.
3. Cohen, F. E. and Kelly, J. W. Therapeutic approaches to protein-misfolding diseases. Nature 426, 905–909.
4. Temussi, P. A., Masino, L., and Pastore, A. From Alzheimer to Huntingdon: why is a structural understanding so difficult? The EMBO Journal, 22, 355–362.
5. Nelson, R. and Eisenberg, D. 2006. Recent atomic models of amyloid fibril structure. Curr. Opin. Struct. Biol. 16, 260–265.
6. Aguzzi, A., Heikenwalder, M., and Polymenidou, M. 2007. Insights into prion strains and neurotoxicity. Nature Reviews, Molecular Cell Biology 8, 552–561.
7. Wadsworth, J. and Collinge, J. 2007. Update on human prion disease. Biochim. Biophys. Acta 1772, 598–609.

29 Miscleavage-Based Diseases

Astrid Gräslund

CONTENTS

29.1 ALZHEIMER'S DISEASE

Alzheimer's disease is without doubt the most studied among the amyloid diseases because of its great prevalence among the world's growing aging population. Alois Alzheimer, a German physician, had already in 1906 observed brain deposits in an autopsy of a patient suffering from a brain disorder. He described his findings and gave his name to the disease. It took until the 1980s before the molecular nature of the disease could be determined. The deposited extracellular plaques in the brain of deceased patients were analyzed and found to contain large amounts of the amyloid β-peptide (Aβ). Other anomalous deposits in the brains were so-called neurofibrillar tangles found inside cells, and these tangles contained a large amount of a protein called tau. Present knowledge, however, links the disease mainly to the aggregation of the amyloid β-peptide.

29.1.1 AMYLOID β-PEPTIDE

The Aβ peptide is 38–42 residues long. Its sequence is shown in Figure 29.1. The longer peptides are the most aggregation prone. The Aβ peptide is produced by the posttranslational processing of a precursor protein, the Alzheimer amyloid precursor protein APP. APP is a single transmembrane domain protein, coded for by a gene on chromosome 21. It is present in all cells, with an as yet unknown function. The Aβ peptide with a variable and highly hydrophobic C-terminus is formed by cleavage by two enzymes, the β- and γ-secretases. The γ-secretase cleaves the protein inside the cell membrane (Figure 29.1). Another enzyme, α-secretase, also cleaves outside the membrane, in an alternative cleavage to the β-secretase, giving rise to shorter fragments that do not form amyloid. The nature of these three processing enzymes

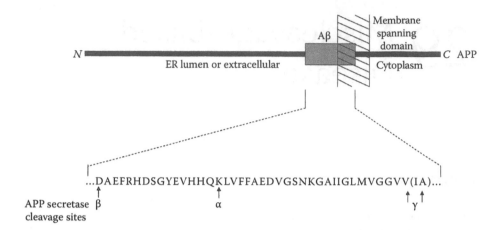

FIGURE 29.1 Overview of the sequence of the amyloid precursor protein (APP) with 770 residues. It has one membrane-spanning helical segment around residue 700, indicated in the figure. The amyloid β-peptide involved in Alzheimer's disease, Aβ, is formed by processing of APP by the enzymes β- and γ-secretase, which create a 40-42-residue-long Aβ peptide. The C-terminus of Aβ is derived from the transmembrane segment of APP. Another APP processing enzyme, the α-secretase, gives rise to other fragments that seem not to be involved in pathological processes.

has been intensely studied. The β- and α-secretases are both membrane-anchored relatively well-defined enzymes. The γ-secretase is in fact a large membrane-bound enzyme complex composed of several different proteins, of which one is the so-called presenilin-1 or -2. The Aβ peptide is produced in a sequence where first the β-secretase cuts off most of the extracellular domain of APP, whereupon γ-secretase cleaves at a not very precise site inside the membrane. Known mutations of APP close to the secretase cleavage sites are known to confer early onset of the disease, presumably through increased preference for the β- and γ-secretase processing pathway and perhaps by producing longer fragments by the γ-secretase cleavage. Other known Aβ mutations causing neuropathological symptoms are centered around the (21–23) segment of the peptide. Table 29.1 gives a list of a few of these mutations. There is no clear-cut overall pattern as to level of Aβ production increase, or increased propensity

TABLE 29.1

Examples of Disease-Associated Aβ Mutations

Mutation	Name
A21G	Flemish
E22G	Arctic
E22Q	Dutch
E22K	Italian
D23N	Iowa

for aggregation or oligomer formation for these mutants. The general tendency seems, however, to be increased aggregation propensity. *In vivo,* several processes may be affected by such mutations.

29.1.2 STRUCTURES OF THE Aβ PEPTIDE

The monomeric form of the peptide is almost without structure in aqueous solution. This is shown by spectroscopic studies, for example, by CD or NMR. Low concentrations are required to prevent spontaneous aggregation of the peptide *in vitro.* Following an aggregation pathway with oligomeric intermediates, the peptide ends up in fibrils that correspond to the brain deposits, the so-called senile plaques. The fibrils are typically linear, with a diameter of 6–10 nm. Their amyloid nature is obvious because they react with thioflavin T and Congo red, and show the typical cross-β conformation (polypeptide chains aligned perpendicular to the fibril axis) in x-ray fiber diffraction studies. A variety of additional techniques, such as solid state NMR and electron microscopy, have been applied to characterize these structures. Figure 29.2 shows a structural model for Aβ(1-40) protofilaments based on such studies.

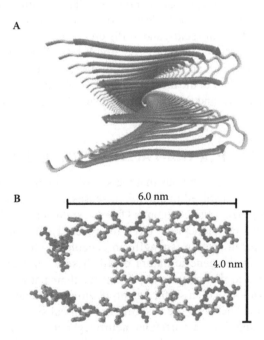

FIGURE 29.2   (A) A ribbon representation of the Aβ fibril structure, based on solid-state and solution NMR studies, and molecular modeling. Parallel peptide hairpins form β-sheets lying perpendicular to the fibril axis. The turn of the hairpin is around residue 25. The N-terminus of the peptide is flexible and not seen in this picture. (B) Atomic representation of residues 1-40. (From Tycko, R., Insights into the amyloid folding problem from solid-state NMR, 2003, Biochemistry 42 (2003): 3151–3159. American Chemical Society. With permission.)

Recent studies, however, strongly point away from the fibrils and toward the oligomeric assemblies as being the neurotoxic species causing synaptic damage associated with disease. A variety of names and descriptions (e.g., oligomers, protofibrils) are probably good evidence of their heterogeneous nature. The common denominator is, however, solubility, β-sheet structure, binding of Congo red and Thioflavin T, and sizes ranging from a few nm to about 100 nm. The number of peptides in one assembly may be anywhere between 2 and 12 or more. The oligomers are resistant to treatment by the detergent SDS. They have been detected in APP transgenic mouse brain as well as the human brain by Western blotting. They are known to affect memory, as studied in live rats. The mechanisms by which they affect the neurons is still under intense study, and several hypothetical possibilities exist, among them interference with specific signaling pathways.

The Arctic mutation AβE22G (Table 29.1), identified from a family in northern Sweden, gives rise to a peptide that is associated with increased formation of oligomeric assemblies compared to common Aβ. The carriers of this mutation exhibit early onset of the disease.

Numerous chemical studies deal with the interactions of the Aβ peptide and external agents, some promoting aggregation, others preventing it. In *in vitro* experiments it is generally seen that phospholipid membranes and cholesterol as well as proteoglycans facilitate Aβ aggregation processes. The presence of metals ions like aluminum, copper, or zinc generally lead to increased peptide aggregation *in vitro*. Apolipoprotein E (299 amino acids) and its different isoforms present a special interaction story. The different isoforms E2, E3, and E4 differ only in two residues, yet the E4 subtype is linked to increased risk of disease, whereas the E2 subtype has a protective effect compared to the most common E3 subtype. The role of apoE seems to be that of a chaperone, and the interaction obviously has a profound effect on the fate of Aβ and its neurotoxic role.

29.2   DISEASES OTHER THAN ALZHEIMER'S DISEASE

It is becoming increasingly obvious that the soluble oligomeric aggregates of several other proteins are involved in amyloid diseases. Antibodies against synthetic Aβ oligomers recognize soluble oligomers from α-synuclein (Parkinson's disease); amylin or islet amyloid polypeptide, IAPP (diabetes II); huntingtin (Huntington's disease); and the prion protein (spongiform encephalopathies). Obviously, there is a commonality in the generic nature of all these assemblies that is manifest in their structures and that may be important for their pathogenic activities. An interesting observation, which so far has no obvious explanation, is the different times of "incubation" between the first hints of clinical symptoms, and the death of a patient suffering from a particular neurodegenerative disease. Parkinson's disease is an example of a neurodegenerative disease that is less severe in its clinical presentation than, for example, amyotrophic lateral sclerosis (ALS), where most afflicted patients die within a couple of years from appearance of the first symptoms. The differences may have to do with what parts of the brain are affected and with the degree of damage caused by the particular disease agents, as well as with the cellular responses toward these agents. One can also reflect on the "infectivity" of the prion diseases. Are all these amyloid diseases to some extent "infectious," the difference being that some misfolded protein aggregates will survive the passage from digestion to brain (prion diseases) and others will not? If one common mechanism involves protein misfolding "seeded" by already misfolded material, this could well be the case. The frequency by which spontaneously formed misfolded "seeds" of a particular disease-associated protein or peptide occur and survive could reflect the average age at which the disease hits the typical patient.

29.3 POSSIBLE THERAPEUTIC INTERVENTIONS

Large efforts, both among scientists and pharmaceutical companies, are devoted to develop therapies to the amyloid diseases. The secretase enzymes were the obvious early targets. It was however found out that the enzymes have more than one function and by inhibiting, for example, the β-secretase, the side effects were not acceptable. Targeting the oligomers by immunotherapy is perhaps the most promising approach today. Both active vaccination by Aβ-peptide and passive treatment with antibodies have been attempted. To date, the side effects on patients were such that the trials had to be aborted, but for at least some patients the early results were positive.

FURTHER READING

1. Sipe, J. P. 2005. Amyloid proteins. The beta sheet conformation and disease. Wiley-VCH Verlag GmbH & Co. KGaA, Weinheim, Germany.
2. Goedert, M. and Spillantini, M. G. 2006. A century of Alzheimer's disease. Science 777–784.
3. Haas, C. and Selkoe, D. J. 2007. Soluble protein oligomers in neurodegeneration: lessons from the Alzheimer's amyloid β-peptide. Nature Reviews, Molecular Cell Biology 8, 101–112.
4. Masters, C. L. and Beyreuther, K. 2006. Alzheminer's centennial legacy: prospects for rational therapeutic intervention targeting the Aβ amyloid pathway. Brain 129, 2823–2839.

30 Missequence-Based Diseases

Astrid Gräslund

CONTENTS

Knowledge of the sequence of the human genome, one must realize that this original DNA sequence was not based on a single individual. Very recently, single individual genomes have been publicized (notably that of Craig Venter, the original entrepreneur behind the first "human genome" description). In the noncoding parts of our genomes, there are distinct differences between all individuals (the SNPs), and the forensic use of DNA evidence is based on these differences. Our coding genes are probably much more similar. Only when sequence variants are associated with mild disease (observable but not really causing a handicap), or even more likely when the variant is associated with susceptibility toward one disease and resistance to another, will we have reason to look for variations in our protein sequences. In addition, evolution will certainly have adapted most of our sequences to the "best" according to our present environment. The examples below will illustrate the phenomenon of known coding gene sequence variations linked to diseases.

30.1 SICKLE-CELL ANEMIA AND OTHER HEMOGLOBIN-RELATED DISEASES

There exist several hemoglobin-related diseases (hemoglobinopathies). As large a proportion as perhaps 7% of the world population could be disease carriers. The first disease that was termed "molecular" is sickle-cell anemia, now known to be caused by a single mutation in the blood protein hemoglobin. As early as the late 1940s, Linus Pauling suggested that the disease was caused by a defective hemoglobin molecule, which was only a few years later was corroborated by experimental observations. Hemoglobin (Hb) has the subunit structure $\alpha_2\beta_2$, and the disease-linked mutation is that of Glu6 in the β chain to Val, giving rise to sickle-cell Hb, HbS. The Glu6 residue is located at the outside of the three-dimensional structure, and it appears to be critical that this residue should be hydrophilic and not

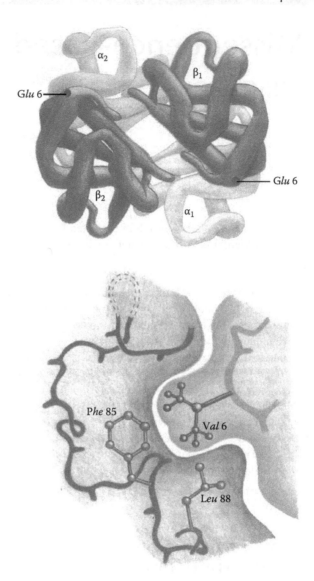

FIGURE 30.1 (Upper) Schematic backbone crystal structure of hemoglobin, showing the site of the mutation (Glu to Val) causing sickle-cell anemia. (Lower) Enlarged structure around the mutation, showing the interface between two Hb molecules that easily form complexes when the negative residue Glu is mutated to the hydrophobic Val. (From Introduction to Protein Structure, *Second edition,* Branden, C. and Tooze, J., 1999 Garland Sciences, Taylor & Francis Group. With permission.)

hydrophobic. In the deoxygenated form, the mutated hemoglobin HbS polymerizes into long fibrillar structures, a process that leads to precipitation of the protein. Figure 30.1 shows a sketch of how the hydrophobic patch created by the Val residue at the surface of one molecule interacts with another hydrophobic patch, Phe85, on another molecule—thus creating the possibility to form "hydrophobic"

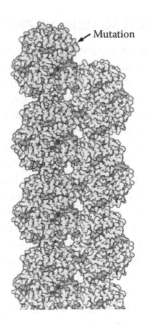

Mutation

FIGURE 30.2 Structure of the polymerized chain of Hb molecules with the sickle-cell mutation. (From Wikipedia "Sickle-cell Disease," source Molecule of the Month by www.pdb.org.)

bonds between successive Hb molecules, which gives rise to the polymerization (Figure 30.2). Because the concentration of Hb is very high in red blood cells, the polymerization process is relatively easy to trigger. The Hb precipitation causes collapse of the red blood cells to a "sickle" shape. The cells loose their native elasticity and become rigid, which reduces their mobility through the blood vessels. The clinical manifestation is that of severe anemia.

An early observation was that the sickle-cell anemia disease is common in areas where malaria is prevalent. This was due to the fact that the malaria parasite does not infect individuals carrying the sickle-cell mutation, thus giving an evolutionary benefit to these individuals. Furthermore, this advantage is also observed for individuals that are heterozygous for the mutated allele (the so-called heterozygote advantage), for which the severe problems with anemia and short life-expectancy are much less prominent than for those individuals who are homozygotes for the mutation. The classical treatment is blood transfusion. The first approved drug is hydroxyurea, a well-known antiproliferative agent. Hydroxyurea seems to activate production of fetal Hb in place of HbS.

Thalassemia is another very frequent type of hemoglobin-related disease. For thalassemia, the molecular explanation is less well defined compared to sickle-cell anemia. Generally, the biochemical problem is low production of one of the types of globin chains, which in turn may lead to assembly of abnormal Hb molecules. The clinical manifestation is anemia. The underlying reason can be mutations in the regulatory pathway for Hb production, but often such defects are coupled to mutations in the Hb gene itself. Again there seems to be a coupling to increased resistance toward malaria for individuals afflicted with thalassemia.

30.2   CYSTIC FIBROSIS, A MEMBRANE PROTEIN-RELATED DISEASE

Cystic fibrosis (CF) is a relatively common and serious hereditary disease. It is associated with imbalance in the mucus production, causing symptoms such as obstruction of lungs and airways, and failure of intestinal and endocrinal functions. It was only in 1988 that the genetic background to CF was found, associated with a membrane-bound protein, namely, a chloride ion channel coded for by a gene called "cystic fibrosis transmembrane conductance regulator" (CFTR). This ion channel is important for the production of some of the body fluids.

The CFTR protein is 1480 residues long and has two domains, each containing six transmembrane helices (Figure 30.3). Energy for its function is provided by ATPase activity. The channel function is regulated by phosphorylation. Single mutations were found to be enough to cause the severe disease symptoms. The most prevalent mutation is that of a deletion of Phe508. This mutation gives rise to an incorrectly folded protein that is easily degraded by the cellular machinery. Other described mutations give rise to too-short proteins or lower production of the protein, the common denominator being impaired function of the CFTR protein as a chloride channel. There are various speculations about the reason for the prevalence of the CF disease, but unlike the situation with sickle-cell anemia and malaria, there is no clear-cut connection between a faulty CFTR gene and resistance to any severe infectious disease.

Like in other diseases associated with single gene mutations, the prospect of gene therapy for CF is potentially promising. Only a relatively modest expression of the normal CFTR gene product would alleviate many of the severe symptoms of the disease. However, the practical problems shared with most other attempts at gene therapy exist also in this case.

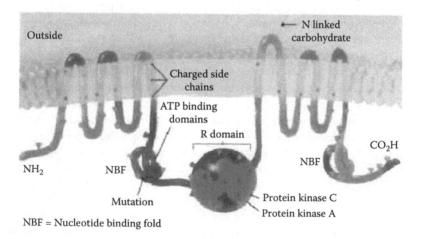

FIGURE 30.3   Cartoon of the crystal structure of the chloride channel membrane protein (CFTR), which has been found with several different types of mutations to be involved in cystic fibrosis. The channel function is impaired in the mutated forms associated with disease. (From Wikipedia "Cystic Fibrosis," CFTR cartoon from 1989 Journal of NIH [National Institutes of Health] Research, defunct since 1997 but previously a NIH [U.S. Government] publication.)

An interesting new aspect of protein variability among individuals is that we have so far only discovered variations that are linked to disease. Recent reports from the huge genomic projects in progress seem to indicate that the variability within a population is larger than hitherto anticipated. Not only do we have so-called SNPs (single nucleotide polymorphism) affecting our noncoding DNA, but also considerable variability among the coding sequences, partly variations without change for which amino acid is coded, but also with amino acid changes that have no obvious functional consequences.

FURTHER READING

1. Vichinsky, E. 2002. New therapies in sickle cell disease. Lancet 360, 629–631.
2. Cianciulli, P. 2008. Treatment of iron overload in thalessemia. Pediatr. Endocrinol. Rev. 6 Suppl 1: 208–213.
3. Gadsby, D. C., Vergani, P. and Csanády, L. 2006. The ABC protein turned chloride channel whose failure causes cystic fibrosis. Nature 440, 477–483.
4. Enquist. K., Fransson, M., Boekel, C., Bengtsson, I., Geiger, K., Lang, L., Pettersson, A., Johansson, S., von Heijne, G. and Nilsson, I. 2009. Membrane-integration characteristics of two ABC transporters, CFTR and P-glycoprotein. J. Mol. Biol. 387, 1153–1164.

31 Peptides and Proteins as Drugs

Ülo Langel

CONTENTS

It is the promise of modern chemistry, biology, and biotechnology to the pharmaceutical industry that basic knowledge can be translated into novel and efficient drugs. However, such transfer of information is very complex and, hence, drug discovery is very expensive and time consuming, and no drugs are yet available for many disorders. Today, only several hundred distinct chemical entities exist as drug molecules (along with several thousand of their analogs) and are available on the market; most of them are low MW (typically of 300–400 Da) substances. Well-known examples from history are morphine (an ingredient of opium, known for 5000 years), aspirin (acetylsalicylic acid first synthesized in 1853; an antianalgesic, which inhibits the enzyme cyclooxygenase), salvarsan (an antimicrobial agent that cures syphilis, introduced by Nobel Prize winner Paul Ehrlich in 1910), β-blockers (introduced in 1960s), HIV protease inhibitors (introduced in 1990s), and other well-known drugs.

In order to be able to cure more diseases more efficiently, the number of novel chemical entities as well as drug targets has to be increased considerably. A new source for both can be peptides and proteins. Both peptides and proteins demonstrate excellent potential as sources for novel candidate drugs, and there is growing research attention paid to this. Indeed, the therapeutic ratio (i.e., ratio of efficacy versus side effects) is arguably higher for naturally occurring (endogenous) protein molecules (MW of 30–180 kDa), also called "biologicals," or "biologics," such as peptide hormones, growth factors, antibodies, enzymes, cytokines, fusion proteins, blood factors, recombinant vaccines, and anticoagulants. Sales of biologics in the United States in 2006 yielded $40 billion, showing an annual growth rate of 20% as compared to an annual growth rate of 6%–8% for the U.S. pharmaceutical market. Almost two-thirds of U.S. drug sales and growth is driven by three types of biologicals: growth factors, monoclonal antibodies, and hormones. Often, these drugs are especially attractive in cases where the traditional small molecule approached has already shown maximal benefit.

Despite the advantages of protein drugs, they can also be characterized by several drawbacks. One such example is that proteins are more difficult to administer due to their digestion by the gastrointestinal tract. This chapter outlines the present situation and the potential future of the field of polypeptides as therapeutics, being a complement to previous chapters where these subjects have been only briefly discussed.

There are many possible molecular targets of peptide and protein drugs since polypeptides are involved in nearly all biochemical processes and, often, their malfunctioning is a part of various pathological processes. Today, the best available drugs target G-protein-coupled receptors, different enzymes, ion channels, nuclear receptors, and protein kinases—drugs with low or high MW that interact with proteins. Novel drug targets, such as multiple specific protein–protein interactions discovered by methods of functional genomics and proteomics, are emerging constantly.

31.1 PROTEIN DRUGS

During the 1980s, the first generation of recombinant biologicals was introduced These were biopolymers with MW of larger than 3–4 kDa with complex two-dimensional structures that are produced and secreted by genetically modified cells. These therapeutic molecules, such as insulin, growth hormone, interferons, erythropoietins (EPOs), and others, have become useful drugs today. It is likely that novel protein drugs will be available soon since multiple clinical trials are ongoing today to develop novel protein therapies. In the following text, we summarize some of the largest classes of protein drugs, adding a discussion about the concerns of researchers.

31.1.1 HURDLES FOR PROTEIN DRUGS

Production of recombinant drugs is complex, including numerous extraction, purification, and concentration steps, eventually leading to protein denaturation and, hence influencing the biological properties. This yields batch-to-batch variations of the recombinant drugs. Production reproducibility is an important issue in this field, and clinical trials are the only way to evaluate and compare the safety and efficacy of biological products.

Many protein-based drugs, approved or in clinical trials, carry some form of posttranslational modification (PTM; cf. Chapter 5). These PTMs can considerably affect the therapeutic properties of protein drugs, and are thus required to be under control or, as it is in many cases, to be understood. Most common PTMs in described protein drugs are glycosylation, carboxylation, hydroxylation, sulfation, disulfide bond formation, amidation, and proteolytic processing. Glycosylation is the most common PTM in biological; around 50% of human proteins are glycosylated, giving rise to multiple possibilities of erroneous gene expression in the glycosylation pathway that can cause disorders. The presence and nature of oligosaccharide modifications of proteins may influence glycoprotein folding, stability, trafficking, and immunogenicity. Several protein drugs, such as erythropoietin, antibodies, blood factors, and some interferons, are potentially sensitive to correct glycosylation (cf. following text).

Immunogenicity is the leading problem in the development of therapeutic proteins, and is considered to be the main difference between biologics and classical, small molecule drugs. Immunoresponses are very individual, and the immunological mechanism of antibody induction depends on the type of the product; foreign proteins usually induce classical immune reaction.

31.1.2 EXAMPLES OF PROTEIN DRUGS

Growth factors are the largest class of therapeutic biologicals on the market, and are represented by the categories EPOs and colony-stimulating factors (CSFs) with such example as epoetin alpha for anemia (Epogen from Amgen, Thousand Oaks, CA), Table 31.1. Growth factor therapeutics are mainly used in oncology. Longer half-life PEGylated or glycosylated versions are developed and available today.

Monoclonal therapeutic antibodies represent the second largest category in the U.S. biotech drug market; more than 20 mAbs are approved in the United States, focused on the treatment of Crohn's disease, non-Hodgkin's lymphoma, colorectal cancer, breast cancer, rheumatoid arthritis, and other disorders in the areas of oncology, and autoimmune and inflammation diseases. All of these approved therapeutic recombinant monoclonal antibodies belong to the IgG class (cf. Chapter 25). Therapeutic antibodies are valuable and differentiate from small molecule drugs by their selectivity and specificity, their unique pharmacokinetics, and their ability to activate the immune system.

Tumor immunotherapy is the most widely advanced area of therapeutic applications of mAbs where the goal is to provide immunity against malignancies by recruiting the immune system to target tumors. This strategy provides patients with less toxic treatments. Several examples are presented in Table 31.1.

The anticancer effects of several antibodies are derived from their blockade of growth factor/receptor interaction and/or down-regulation of oncogenic proteins on the tumor cell surface. Some of these antibodies elicit effector mechanisms of the immune system, such as antibody-dependent cellular cytotoxicity and complement-mediated cytotoxicity. Longer half-life PEGylated or glycosylated (cf. following text) versions of therapeutic mAbs with improved properties are available.

At least 30 different hormones are on the market, such as human growth hormones, insulin, follicle-stimulating hormone, and others, as shown in Table 31.1.

Insulin ("island" in Latin) is a peptide hormone secreted by the pancreatic beta cells of the islets of Langerhans, and its shortage in the body results in diabetes

TABLE 31.1

Some Examples of Protein and Peptide Drugs

Drug or Drug Candidate	Characterization	Targets	Treatment, Applications
		Hematopoietic Cytokines/Growth Factors	
Erythropoietin (EPO)	A secreted, 165-166-amino acid, 18.2 kDa glycoprotein (30–36 kDa as glycosylated), 4 alpha helices	Targets a 66 kDa EPO receptor, activating JAK2, Ras/ MAP kinase, PI3K and STAT transcription factors. Regulates red blood cell production, initiating hemoglobin synthesis; has neuroprotective activity and antiapoptotic functions in several tissue types.	Anemia; also used as blood doping agent
Colony-stimulating factors (CSFs)	Four different secreted glycoproteins		
G-CSF exists as a 174- and 180-amino-acid, 19.6 kDa protein, 4 alpha helices	G-CSF targets G-CSF receptor, activating several signaling transduction pathways including PI3K/Akt, Jak/Stat and MAPK, thereby promoting survival, proliferation, differentiation, and mobilization of hematopoietic stem and progenitor cells	Clinical hematology and oncology (bone marrow transplantation and chemotherapy-associated neutropenia)	
		Monoclonal Therapeutic Antibodies	
Trastuzumab	Chimeric mAb	Against HER2 receptor preventing activation of receptor's intracellular kinase, preventing angiogenesis	Breast cancer
Bevacizumab	mAb	Binds to VEGF preventing its interaction with endothelial cell surface receptors, preventing angiogenesis	Metastatic colorectal cancer
Cetuximab	Chimeric mAb	Against ligand binding domain of EGFR, member of trk	Colorectal cancer, head and neck cancer
		Hormones	
Insulin	51 amino acid long (in 2 chains) peptide/ protein with MW of 5.8 kDa	Activates insulin receptors, initiates signal transduction yielding increased glucose uptake and storage	Diabetes mellitus

Human growth hormone, hGH	191 amino acid single chain protein of 22.1 kDa	Acts by interacting with a specific cell surface receptors. Stimulates growth and cell reproduction in humans and other animals. Synthesized, stored, and secreted by the anterior pituitary gland	Shortness related disorders[a]
Follicle-stimulating hormone, FSH	Glycoprotein of 2 polypeptide units, MW 30 kDa	Interacts with specific FSH receptors activating G-protein signaling. Regulates the development, growth, pubertal maturation, and reproductive processes of the human body	Infertility therapy to stimulate follicular development
Cytokines			
Interferons (IFNs)	Family of 16-26 kDa glycoproteins such as IFN-α, IFN-β and IFN-γ	Activate specific cell-surface receptors and initiate the JAK-STAT signaling pathways	Chronic granulomatous disease (dimeric IFN-γ), chronic hepatitis C and hepatitis B (PEGasys)
Interleukins (ILs)	Family of >30 proteins, 8-30 kDa	Activate specific cell-surface receptors and initiate multiple signaling cascades in immune system, IL-1 is involved in inflammation	IL-1R antagonist for rheumatoid arthritis, IL-2 for renal-cell carcinoma
Therapeutic Enzymes			
PEG-L-asparaginase	Enzyme, 31.7 kDa	Catalyzes the hydrolysis of Asn to Asp	Acute lymphoblastic leukemia
Alpha-galactosidase	Glycoside hydrolase enzyme	Hydrolyzes the terminal alpha-galactosyl moieties from glycolipids and glycoproteins	Fabry disease, agalsidase alpha (Replagal)
Selection of Peptide Drugs			
Angiotensin, AT1	10 amino acid peptide of 1.2 kDa	Inverse agonist at AT1 receptor, interacts with specific AT1 receptors, activating G-protein signaling	Hypertension, heart failure (Losartan)
Oxytocin, OT	9 amino acid peptide of 1.0 kDa	Inverse agonist at OT receptor, interacts with specific OT receptors activating G-protein signaling	Acute postpartum hemorrhage, can induce breastfeeding
Desmopressin (1-desamino-8-d-arginine vasopressin)	9 amino acid peptide of 1.1 kDa	Synthetic replacement for antidiuretic hormone, interacts with specific arginine vasopressin receptor 2 (V2R) activating G-protein signaling	Diabetes insipidus, nocturnal enuresis (DDAVP)

(Continued)

TABLE 31.1 (CONTINUED)
Some Examples of Protein and Peptide Drugs

Drug or Drug Candidate	Characterization	Targets	Treatment, Applications
Glucagon	29 amino acid peptide of 3.5 kDa	Maintain the level of glucose in the blood by interacting with glucagon receptors on hepatocytes, activating G-protein signaling	Severe hypoglycemia
Vancomycin	Branched tricyclic nonribosomal glycopeptide of 1.4 kDa	Antibiotic for infections caused by Gram-positive bacteria	Prophylaxis and treatment of infections
Cyclosporine	Cyclic nonribosomal 11 amino acid peptide 1.2 kDa	Binds to the cytosolic protein cyclophilin of T-lymphocytes inhibiting calcineurin, leading to a reduced function of effector T-cells	Transplant medicine, psoriasis, rheumatoid arthritis
Fuzeon (enfuvirtide)	Synthetic 36 amino acid peptide of 4.5 kDa; gp41(643–678)	Peptide mimic of HIV-1 surface glycoprotein gp41, HIV fusion inhibitor. Acts as decoy of their natural analogs and inhibits the entry of HIV-1 into T-cells	Combination therapy for HIV-1 infection
Integrilin (eptifibatide)	Synthetic cyclic heptapeptide of 0.8 kDa	Antiplatelet drug that selectively blocks the platelet glycoprotein IIb/IIIa receptor	Thrombocytopenia, renal insufficiency, severe, uncontrolled hypertension.
Calcitonin	Amidated 32-amino acid peptide of 3.4 kDa	Interacts with specific calcitonin receptors activating G-protein signaling. Acts to reduce blood calcium (Ca^{2+}), opposing the effects of parathyroid hormone (PTH)	Osteoporosis, Paget's disease, hypercalcemia

Note: EGFR—epidermal growth factor receptor; GCSF—granylocyte colony-stimulating factor; GLP-1—glucagon-like peptide-1; gp41—a glycoprotein providing the mechanism by which HIV enters the cell; HER2—a member of the ErbB protein family, more commonly known as the epidermal growth factor receptor family; JAK2, Ras/MAP kinase, PI3K—different protein kinases; STAT—transcription factor type; VEGF—vascular endothelial growth factor

[a] Turner syndrome, chronic renal failure, intrauterine growth retardation, and others.

mellitus (*diabetes* meaning "siphon" in Greek; *mellitus* meaning "sweet" or "honey" in Latin) or the sugar disease. Insulin was discovered in animal pancreatic extracts in 1921 by F. G. Banting and C. H. Best in Toronto and successfully tested in diabetic dogs. First successful human tests of insulin started already in 1922, giving immediate hope to diabetic patients who had had no cure previously. Isolation of animal insulin in early days and its human application was only possible because human insulin is highly homologous to used animal (cattle, pigs) insulin. The production of insulin today uses modern technology in order to improve the efficacy and delivery of insulin. Several types of insulin analogs are available today: long- (taken once daily), rapid- (can act in less than 15 min), and intermediate-acting; an inhalable form of insulin has recently become available. Several insulin level-controlling polypeptides are available today such as extendin-1 and a long-acting mimetic of glucagon-like peptide-1, GLP1.

Cytokines is another large category of therapeutic biologicals and could be classified into four major types of molecules: interferon (IFN) alpha, interferon beta, interferon gamma, and interleukins (ILs). IFN-alpha is prescribed for hepatitis C and some cancers, IFN-beta for treatment of multiple sclerosis, IL-1 receptor antagonist for rheumatoid arthritis, IL-2 for renal-cell carcinoma (Table 31.1). The cytokine family consists mainly of smaller, water-soluble proteins and glycoproteins with MW of 8–30 kDa.

Therapeutic enzymes mainly include recombinant enzymes for treatment of rare genetic disorders such as Fabry disease, Gaucher disease, Pompe disease, Hunter syndrome, and others (Table 31.1).

31.2 PEPTIDE DRUGS

It is perhaps surprising that only few therapeutic peptides are available today, several decades after the discovery of solid phase synthesis by B. Merrifield (cf. Chapter 15). The hurdles facing peptide therapeutics have been many, such as low stability, solubility, delivery and administration problems, quick body clearance, production costs, etc. However, screening and optimizing peptide leads have yielded a huge number of lead peptides, and this gives promise that many more peptide drugs will be available soon. The shortcomings of peptides can be manipulated by applications of novel synthesis approaches. Obvious advantages of peptide drugs are their high potency, activity and specificity, low accumulation in tissues, low toxicity, and high biological and chemical diversity.

Currently, there are several types of novel peptide therapies in development for antibiotic diseases, viral infections, immune system disorders, cardiovascular diseases, neurological disorders, and cancer. Three years ago, in March 2005,[8] more than 40 marketed peptide drugs were available worldwide; around 270 peptides were in clinical phase testing and about 400 in advanced preclinical phases. Many researchers and pharmaceutical companies believe that peptide drugs will represent a huge market in the near future.

Table 31.1 presents several examples of different peptide drugs that are currently used in the therapy of multiple disorders. Since semantically the border between peptides and proteins is rather diffuse, usually longer than 50 (or 100) amino acid

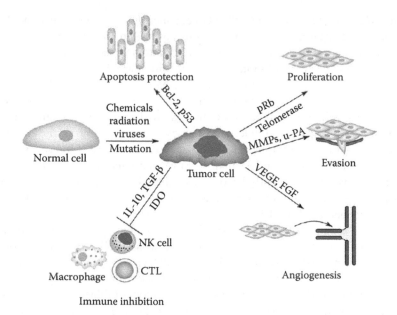

FIGURE 31.1 Therapeutic targets for tumor regression. Abbreviations: CTL, cytotoxic T lymphocyte; FGF, fibroblast growth factor; IDO, indolamine 2,3 dioxygenase; IL-10, interleukin-10; MMP, matrix metalloprotease; NK cell, natural killer cell; pRb, retinoblastoma protein; TGF-β, transforming growth factor-β; u-PA, urokinase-type plasminogen activator; VEGF, vascular endothelial growth factor. (From Bhutia, S. K., 2008, Trends Biotech., 26(4): 210–217. Elsevier. With permission.)

long polypeptides are called proteins. Also, several peptide drugs are the representatives of hormones. Hence, the division in the Table 31.1 may sometimes seem controversial, such as with insulin. Several peptide drugs are recognized by different cell-surface receptors such as G-protein-coupled receptors (in case of peptide hormones) or integrins (in case of integrilin).

Many anticancer peptides are in development, that is, in ongoing preclinical or clinical trials. Because of the extremely complicated nature of cancers and the multiple possibilities of interacting with its mechanisms, the number of possible drugs, including the peptides, is huge. Tumor cells are different from "normal" cells in the sense that they have a nonlimited, replicative potential, self-sufficiency to growth signals, low sensitivity to growth-inhibitory signals, avoidance of cell death, sustained angiogenesis, and create changes in the immune system. Although these factors contribute to their lethality, these characteristics of tumors can also be exploited for development of cancer therapies (cf. Figure 31.1). In this simplified scheme of the tumor mechanisms, peptides for interrupting tumor-related mechanisms are available, such as the peptide inhibitors of MMP-2 and MMP-9 suppressing the migration and invasion of tumors, and muramyl dipeptide (peptidoglycan immunoadjuvant originally isolated from bacterial cell wall fragments) stimulating the release of proinflammatory cytokines in tumor-suppressed hosts. Based on possible mechanisms of action of the anticancer peptides presented in Figure 31.1, they could be divided, from the point of view of cell death regulation, into peptides with necrotic activity (cell membrane lytic, positively charged peptides such as cecropins, melittin, and defensins) and peptides with apoptotic activity (lactofericin, magainin, angiotensin, peptides derived from p53 protein, and others).

Other cancer-related, function-blocking peptides are available, participating in mechanisms of receptor interactions, call adhesion, and metastasis. Tumor cell receptor targeting is a rapidly developing area utilizing hormone (somatostatin, bombesin/gastrin-releasing hormone, luteinizing hormone-releasing hormone, bradykinin antagonists) receptors, and several hormone receptor analogs exhibit strong anticancer activity. Tumor targeting peptides have also been identified from random phage-display libraries such as tumor homing peptides, peptides inhibiting VEGF and VEGFR interaction, and aminopeptidase N interacting peptides. Peptides inhibiting cell adhesion properties are known to alter metastasis and angiogenesis, such as tripeptidic RGD integrin-recognition motif blocking integrin-mediated cell adhesion, and tumor invasion.

Peptide inhibitors of protein kinases and proteases (Figure 31.1) are attractive targets for anticancer therapy. Short peptide inhibitors of matrix metalloproteases (MMPs), such as cyclic HWGF peptide, have been shown to suppress tumor migration and invasion.

31.3   IMPROVEMENT OF STABILITY AND AVAILABILITY

Modifications of biologics with PEGylation, glycosylation, or the creation of novel analogs are currently the strategies to further improve drug development and will be briefly explored in the following text.

31.3.1   PEGYLATION

Polyethylene glycol (PEG) conjugation to proteins is a common form of engineering to achieve increased bioavailability, optimized pharmacokinetics, shielding of potential immunogenic epitopes, protection from proteolytic cleavage, and the slowing of body clearance or increase water solubility of proteins. Many PEGylated protein drugs are currently in use, such as alpha interferons (PEGasys, PEG-Intron), G-CSF (PEG-Neupogen), human growth hormone (PEG-hGH), and GC-CSF (Neulasta), all with improved plasma half-life and, consequently, with reduced frequency of administration.

PEGylation is the covalent conjugation of the polyethylene glycol (PEG) polymer to another molecule by enzymatic or chemical approaches (cf. Figure 31.2). PEG is a linear or branched polyether with terminal hydroxyl groups with MW of up to tens of thousands. Monomethoxy PEG (mPEG) is most often used for polypeptide modification. Required functional groups can be introduced as end groups of PEG.

PEG is soluble in both aqueous solutions and organic solvents, and this feature makes PEG very flexible for end-group chemical modification as well as conjugation to biomolecules under mild conditions. Additionally, PEG structure is highly flexible and binds 2–3 water molecules per ethylene oxide unit, making it act more

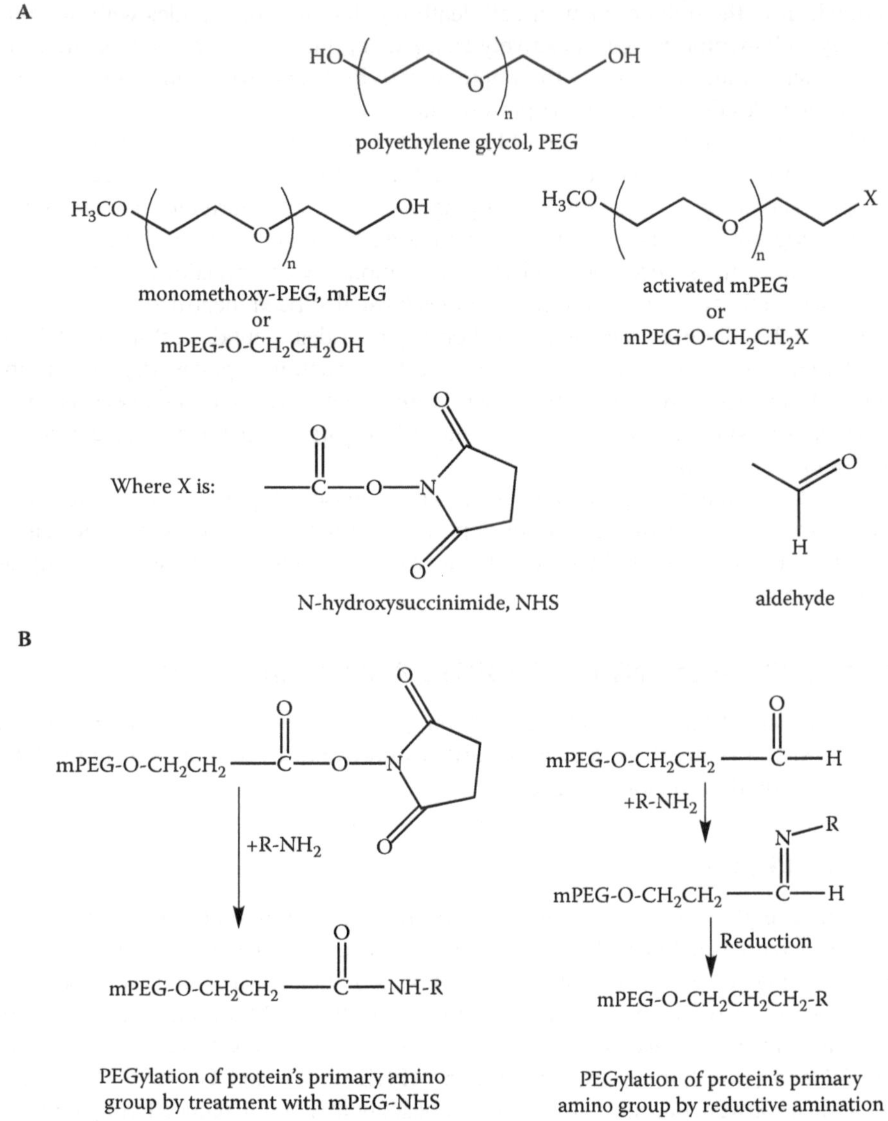
A
HO
OH
n
polyethylene glycol, PEG
H3CO
OH
n
monomethoxy-PEG, mPEG
or
mPEG-O-CH2CH2OH
H3CO
X
n
activated mPEG
or
mPEG-O-CH2CH2X
Where X is:
N-hydroxysuccinimide, NHS
aldehyde
B
mPEG-O-CH2CH2
+R-NH2
mPEG-O-CH2CH2 —C—NH-R
mPEG-O-CH2CH2 —C—H
Reduction
mPEG-O-CH2CH2CH2-R
PEGylation of protein's primary amino group by treatment with mPEG-NHS
PEGylation of protein's primary amino group by reductive amination

FIGURE 31.2 Examples of chemistry for peptide and protein PEGylation. A: Structures of polyethylene glycol, PEG, and modified PEG. B: Examples of PEGylation of peptide/protein modifications.

effectively as compared to a soluble protein with the same MW, and yielding unique properties. PEGs with MW above 1000 Da are not toxic, and PEG is rapidly cleared in vivo without degradation. PEG is only weakly immunogenic, making its use for protein modifications convenient.

PEG (or mPEG) is activated prior to protein/peptide modification (Figure 31.2A) by modification with a suitable functional group X, chosen by the type of the functional group in the polypeptide to be modified. PEGylation of primary amino groups on proteins is exemplified in Figure 31.2B by activating mPEG with N-hydroxysuccinimide or aldehyde moieties. However, multiple activating groups are available to modify side chains of Lys, Cys, His, Arg, Asp, Glu, Ser, Thr, and Tyr; all common strategies to label bioactive polypeptides are used (cf. Chapter 17) in PEGylation.

31.3.2 NANOCARRIERS

Pharmaceutical nanocarriers such as micelles, liposomes, nanoemulsions, solid lipid nanoparticles, polymeric nanoparticles, functionalized nanoparticles, nanocrystals, cyclodextrins, dendrimers, nanotubes, etc., refer to nanoscaled drug delivery devices. These nanocarriers are currently applied to deliver almost all types of pharmaceuticals; however, their impact is very high in delivery of peptides, recombinant proteins, vaccines, and nucleotides. Nanocarriers have shown to be beneficial in drug delivery due to changes in body distribution and other physicochemical properties leading to selective delivery at specific sites and reduced side effects yielding improvement in therapeutic efficiency.

Surface modifications of hydrophobic nanocarriers is usually a way to control their required biological properties and make them to perform various important functions such as increased stability, required biodistribution, targeting to pathological zone, responsiveness to local stimuli, efficient intracellular drug delivery, etc. (cf. Figure 31.3). In order to prepare such multifunctional nanocarriers, different

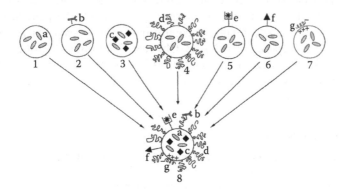

FIGURE 31.3 Schematic structures of the multifunctional pharmaceutical nanocarrier. a—drug loaded into the carrier; b—specific targeting ligand, usually a monoclonal antibody, attached to the carrier surface; c—magnetic particles loaded into the carrier together with the drug and allowing for the carrier sensitivity toward the external magnetic field and its use as a contrast agent for magnetic resonance imaging; d—surface-attached protecting polymer (usually PEG) allowing for prolonged circulation of the nanocarrier in the blood; e—heavy metal atom—[111]In, ^{99m}Tc, Gd, Mn—loaded onto the nanocarrier via the carrier-incorporated chelating moiety for gamma- or MR-imaging application; f—cell-penetrating peptide, CPP, attached to the carrier surface and allowing for the carrier enhanced uptake by the cells; g—DNA complexed by the carrier via the carrier surface positive charge. (From Torchilin, V. P., 2006, *Adv. Drug Del. Rev.*, 58, 1532–1555. Elsevier.)

chemical moieties providing individual required properties are assembled on the surface of a nanoparticle.

PEGylation is a popular method to modify nanocarriers in order to improve the half-life of a drug clearance due to the protection of drug molecules from biofluids such as blood components, providing solubility and low toxicity etc. (cf. previous mention). Targeted nanocarriers are obtained by attaching the targeting moiety on the surface of nanocarrier, such as ligands specific to targeted cells, targeting antibodies, sugars, peptides, etc. Polymeric components with pH-cleavable bonds are applied to produce stimuli-responsive drug-delivery nanocarriers, which are stable in the circulation or normal tissues, but degrade and release the drug in areas with lower pH (tumors, infarcts, inflammation zones, etc.). Intracellular delivery by nanocarriers has been addressed recently by applications of cell-penetrating peptides (see Chapter 27).

31.3.3 IMPROVING PEPTIDE EFFICIENCY

Inhibitory peptides are applied for inhibition of protein networks in cancer and other diseases and often there is a need to engineer desired properties of compounds using the native peptide as the starting point. Because of flexibility of peptide chemistry, several possibilities exist to improve the therapeutic potential of the native peptides, in addition to PEGylations and applications of nanocarriers.

Structure-activity studies followed by shortened, substituted, or cyclic peptides is a common way to modify potential peptide drugs. Another way to achieve better oral-active peptides is incorporation of *D*-amino acids. Peptide engineering yielding chimeric peptides is another modern way to achieve the peptide drugs with combined properties. For example, by combining the peptides with anticancer properties with cell-penetrating peptides (cf. Chapter 27) or tumor-homing peptides has yielded potential anticancer drug candidates with improved properties. Certainly, it is also a possibility to design tumor cell-selective drugs, which is a future goal of tumor drug development.

FURTHER READING

1. Bartfai, T. and Lees, G. V., 2006, Drug Discovery from Bedside to Wall Street. San Diego: Elsevier Academic Press.
2. Gutte, B., Ed. 1995. Peptides. Synthesis, Structures, and Applications. San Diego: Academic Press.
3. Pagé, M., Ed. 2002. Tumor Targeting in Cancer Therapy. Totowa, New Jersey: Humana Press, Inc.
4. Aggarwal, S. 2007. *What's fueling the biotech engine?* Nature Biotech., 25(10): 1097–1104.
5. Roberts, M. J., Bentley, M. D., and Harris, J. M. 2002. *Chemistry for peptide and protein PEGylation.* Adv. Drug Del. Rev., 54(4): 459–476.
6. King, J., Waxman, J., and Stauss, H. 2008. *Advances in tumour immunotherapy.* Q. J. Med. advance access: QJM, doi:10.1093/qjmed/hcn050, pp. 1–9.
7. Sharkey, R. M. and Goldenberg, D. M. 2006. *Targeted therapy of cancer: new prospects for antibodies and immunoconjugates.* CA Cancer J. Clin., 56(4): 226–243.
8. Marx, V. 2005. *Watching peptide drugs grow up.* Chem. Eng. News., 83(11): 17–24.

9. Bhutia, S. K. and Maiti, T. K. 2008. *Targeting tumors with peptides from natural sources.* Trends Biotech., 26(4): 210–217.
10. Amiji, M. M., Ed. 2007. Nanotechnology for Cancer Therapy. Boca Raton: CRC Press.
11. Nie, S., Xing, Y., Kim, G. J., and Simons, J. W. 2007. *Nanotechnology applications in cancer.* Annu. Rev. Biomed. Eng., 9: 257–288.
12. Torchilin, V. P. 2006. *Multifunctional nanocarriers.* Ad. Drug. Del. Rev., 58(14): 1532–1555.

Index

9 780367 384876